Evolving Designs, Applications, Technological Advances, and the Future of Qualitative Research

Liliane Cristina Segura
Universidade Presbiteriana Mackenzie, Brazil

Ana Lucia Fontes de Souza Vasconcelos
Universidade Federal de Pernambuco, Brazil

Vilma Geni Slomski
Fundação Escola de Comércio Álvares Penteado University Center, Brazil

IGI Global
Scientific Publishing
Publishing Tomorrow's Research Today

Vice President of Editorial	Melissa Wagner
Director of Acquisitions	Mikaela Felty
Director of Book Development	Jocelynn Hessler
Production Manager	Mike Brehm
Cover Design	Jose Rosado

Published in the United States of America by
IGI Global Scientific Publishing
701 East Chocolate Avenue
Hershey, PA, 17033, USA
Tel: 717-533-8845
Fax: 717-533-7115
Website: https://www.igi-global.com E-mail: cust@igi-global.com

Library of Congress Cataloging-in-Publication Data

LCCN: 2025021281 (CIP Data Pending)
ISBN13: 9798337344324
Isbn13Softcover: 9798337344331
EISBN13: 9798337344348

British Cataloguing in Publication Data
A Cataloguing in Publication record for this book is available from the British Library.

Table of Contents

Detailed Table of Contents

This chapter provides a comparative analysis of NVivo and ATLAS.ti, two prominent software tools for qualitative data analysis. It evaluates the core functionalities, user interfaces, and analytical capabilities of both platforms, focusing on their strengths and limitations. The chapter details how each tool supports data management, coding, and analysis, and considers their impact on research efficiency and effectiveness. Key factors such as usability, cost, integration with other research tools, and user support are discussed to assist researchers in selecting the most suitable software for their qualitative research needs. The analysis aims to offer practical insights and guidance for optimizing the qualitative research process

This chapter aims to discuss the methodological design for the adoption of discourse analysis. The methodological approach that may decide for discourse analysis is an important part of the structure of a study and; it is hinged on various methodological decisions but most importantly five, namely: the choice of the type of research whether qualitative or quantitative; describing its philosophical underpinnings that nurture the study; the approach and strategy to be adopted say discourse analysis in conducting the investigation; means used for data collection and in the language of qualitative research data building, say content analysis, interviews, imagery reviews, social media posts analysis to mention just a few; distinguishing the form of analysis; and evaluating ethical considerations. This chapter is structured as follows: Introduction, Position of a Speech in discourse analysis, Appropriateness of usage of Discourse Analysis, Act of data analysis within the contexts of Discourse Analysis and Conclusion.

The chapter unpacks the evolution of discourse analysis and its use in qualitative research. The chapter explored the evolution of discourse from modernity to

This paper explores the Cartography of Controversies as a methodological extension of Actor-Network Theory (ANT) to investigate complex socio-technical systems empirically. Through a case study on regulating clinical trials in Brazil, focusing on legislative debates over PLS 200/2015 and PL 7082/2017, the study demonstrates how controversy mapping can visualize the interactions of heterogeneous networks that shape and are shaped by regulatory processes. The findings show that regulation functions as both an actant and an outcome of interwoven relationships, highlighting socio-technical assemblages' dynamic and performative nature. This approach underscores the utility of cartography for operationalizing ANT, enabling a richer understanding of fluid and evolving phenomena in qualitative research.

Qualitative research has been focused on accumulating data for providing a realistic view of phenomena under study and form understanding from the perspectives of respondents. Uncertain situations such as COVID-19 pandemic has seen extraordinary social distancing measures thereby becoming a barrier in the data collection process. However, the progression of a IT showed favorable repercussions. Virtual methods have rapidly gained momentum. Limited research comprehensively explores the techniques for qualitative data collection and analysis through digital mode, opportunities and challenges faced. The PRISMA framework was used for presentation. The review was initiated with a scoping search to identify the existing articles. Articles were reviewed deeply to determine if the inclusion conditions were met. Conducting qualitative research through the usage of digital technologies could be perplexing for both parties and the debate on ethics and reliability have found mixed results calling paving the ground for future scope of the research.

The research is explanatory, using the "Engaged Scholarship" framework (Van de Ven, 2007). Data collection was divided into evidence from primary sources, obtained through semi-structured interviews and focus group sessions with experts from Brazil's National Financial System (NFS), and secondary sources, based on documentary analysis. The data was processed using the Atlas.ti computer tool (Friese, 2014), applying a protocol based on the proposals of Saldaña (2016) and Bardin (2016), triangulating the results (Yin, 2014). The analysis of the results, aided by automated Atlas.ti reports, was carried out based on the content analysis of each data source in relation to the assumptions of the study's theoretical approach. The methodological platform applied in the thesis proved to be innovative due to the use of the Engaged Scholarship approach, supported by the suggestion of a protocol for analyzing and triangulating evidence, with the aid of a computer platform.

This study investigates the impact of neighborhood influences on farmers' pesticide use behavior in rural Chilli cultivation in Andhra Pradesh, India. Recognizing that local farming practices, climate conditions, and community dynamics play significant roles in shaping individual farming decisions, we focus on developing advanced machine learning models to mitigate these influences. Using a combination of Bi-Stacked LSTM (Long Short-Term Memory) and Bi-Stacked GRU (Gated Recurrent Unit) networks, we analyze a dataset collected from Chilli farmers, incorporating variables related to pesticide use, economic factors, farming practices, and health impacts. The models capture both temporal and contextual dependencies, isolating individual farmers' behavior while accounting for broader neighborhood influences. The results of this study aim to enhance understanding of sustainable farming practices and the economic implications of pesticide use, with a particular focus on improving health and safety standards in rural agriculture.

COVID-19 pandemic is an important disease of recent decades, affecting the health of the population, as well as the global economies, social interaction patterns, and human psychology. To cope with this uncertain and stressful situation people devised different strategies. This study aims to reveal the coping strategies adopted by the consumers of an emerging country, namely Türkiye. It is a qualitative study aiming to understand the way Turkish consumers coped with the COVID-19 pandemic. Data was collected through conducting 60 in depth interviews and content analysis was applied using QDA Miner 6 software package. The findings reveal that mostly preferred coping strategies are avoidance, active coping, emotional support and positive thinking whereas rational thinking, emotional venting, and instrumental support are the least preferred coping strategies of the Turkish sample. The findings contribute to the literature by adapting the multidimensional model of consumer coping developed by Duhachek (2005) and proposing coping strategies that are specific to a pandemic.

Field interviews and qualitative studies play a crucial role in tourism research, offering in-depth insights into the experiences and perspectives of key stakeholders. This study explores the tourism landscape of Fez, Morocco, using a multi-method qualitative approach. In-depth interviews were conducted with tourism professionals, including hoteliers, tour guides, and restaurant owners, as well as government officials. These interviews provided valuable perspectives on local tourism challenges, opportunities, and the role of technology in smart tourism management. Additionally, stakeholder matrices were employed to systematically analyze the roles, relationships, and influences of different actors in the tourism sector. By integrating these qualitative methodologies, this research offers a comprehensive understanding of tourism dynamics in Fez, contributing to strategic planning and sustainable tourism development.

Preface

As editors of *Evolving Designs, Applications, Technological Advances, and the Future of Qualitative Research*, we are pleased to present a work that reflects not only the current landscape but also the growing complexities and future directions of qualitative research in the field of business. This volume is the result of a collective effort—an academic dialogue involving national and international scholars committed to the continuous development of the qualitative research paradigm.

The motivation for organizing this book stems from the increasing recognition that qualitative research extends far beyond methodological application. It represents a way of seeing and engaging with the world as a dynamic process that involves critical reflection, rigorous theoretical and methodological choices, and a deep engagement with the complexities of human and organizational behavior. As such, we sought to curate a volume that does not merely introduce qualitative research techniques, but one that deeply explores the epistemological, philosophical, and applied dimensions that define and enrich this form of inquiry.

Throughout this book, contributors from diverse academic and professional backgrounds share their experiences in navigating qualitative research in business contexts. The chapters highlight the intricate decisions researchers make in designing studies, selecting, and applying methods, analyzing data, and constructing and communicating meaning. This reflects the reality that qualitative research is not linear but iterative, adaptive, and often shaped by the contexts and relationships embedded in the research process.

CHAPTER OVERVIEWS

Chapter 1: Philosophical Underpinnings of Qualitative Research: Transcending Modernity

This foundational chapter explores the philosophical roots of qualitative research, positioning it as a response to the dominance of positivism and the assumptions of modernity. Drawing from constructivist and interpretivist traditions, it examines qualitative genres such as ethnography, phenomenology, and narrative inquiry, while addressing issues of axiology, sampling, and data analysis. The chapter offers a critical reflection on both the strengths and limitations of qualitative research, grounding the volume in philosophical depth.

Chapter 2: Development of Research Design

This chapter provides a comprehensive guide to the development of research design in qualitative studies, emphasizing its role in producing meaningful and ethical scholarship. It covers the formulation of research questions, participant selection, data collection strategies, and budgeting. Combining theoretical grounding with practical instruction, it serves as a valuable resource for researchers seeking methodological clarity.

Chapter 3: Selection of Participants in Qualitative Research

Focusing on participant selection, this chapter underscores the importance of purposive and snowball sampling in qualitative inquiry. It addresses ethical considerations and the influence of digital technologies on recruitment strategies, encouraging flexibility and inclusivity in response to evolving research contexts.

Chapter 4: The Role of Technology in Modern Qualitative Research

This chapter explores how emerging technologies are reshaping qualitative research practices. It introduces tools and software that support data collection and analysis—such as transcription tools, artificial intelligence, and virtual reality—while highlighting opportunities and ethical concerns in the digital research environment.

Chapter 5: Evaluating Qualitative Data Analysis Tools: A Comparative Study of NVivo and ATLAS.ti

Offering a comparative evaluation of NVivo and ATLAS.ti, this chapter presents practical insights into software usability, functionality, and cost. It helps guide researchers in selecting the most appropriate tools to enhance their qualitative analysis and overall research quality.

Chapter 6: Discourse Analysis

This chapter outlines a structured approach to discourse analysis, discussing key design elements including philosophical underpinnings, data collection, and ethical considerations. It offers practical guidance for applying discourse analysis to examine language, meaning, and social context within qualitative frameworks.

Chapter 7: Constraints and Challenges of Conducting Discourse Analysis in Qualitative Research

Building on previous discussions of discourse analysis, this chapter highlights theoretical and practical challenges encountered in its application. It addresses issues such as definitional ambiguity, epistemological tensions, and methodological complexity, offering strategies for researchers to navigate and adapt within this evolving approach.

Chapter 8: Research Protocol in Studies with Semi-Structured Interviews

This chapter presents a detailed roadmap for constructing research protocols involving semi-structured interviews. It includes steps for developing interview scripts, ensuring validity, and maintaining ethical standards, making it an essential manual for applied qualitative research.

Chapter 9: Phenomenological Research in Education Studies: Merits and Pitfalls

This chapter investigates the potential of phenomenological research within the field of education. It examines philosophical foundations and practical applications, emphasizing how constructivist and existentialist perspectives can illuminate teaching and learning challenges. The discussion highlights both the benefits and constraints of phenomenology in educational inquiry.

Chapter 10: Mapping Complexity as a Qualitative Approach to Operationalizing Actor-Network Theory

Using Cartography of Controversies, this chapter applies Actor-Network Theory to examine the regulation of clinical trials. It showcases how socio-technical systems can be empirically mapped and understood through qualitative research, offering a novel approach to studying complexity and organizational dynamics.

Chapter 11: Qualitative Methodology Techniques and Approaches of Business Research in the Digital Era

This chapter explores the challenges and opportunities of conducting qualitative business research during digital transformations and global disruptions such as the COVID-19 pandemic. It utilizes the PRISMA framework to examine virtual data collection and emphasizes ethical concerns and methodological gaps that require future research.

Chapter 12: Content Analysis and Triangulation of Results through Coding Protocols

Presenting a detailed analytical framework, this chapter illustrates how Atlas.ti and the Engaged Scholarship model can be used to triangulate qualitative data. Using data from the financial sector, it demonstrates how interviews, focus groups, and document analysis can be integrated to improve reliability and generate actionable insights.

Chapter 13: Improving Rural Business Practices through Advanced Modeling

This chapter investigates the use of machine learning to analyze pesticide usage patterns among rural farmers in India. While grounded in quantitative methods, the research is framed within a qualitative context, offering insights into sustainability, decision-making, and behavioral change in rural economies.

Chapter 14: A Qualitative Study on Consumer Coping Strategies During a Pandemic

This chapter examines consumer coping strategies during the COVID-19 pandemic using in-depth interviews. Drawing on a well-established coping model, it uncovers culturally specific behaviors and emotional responses to crisis, contributing valuable insights into consumer psychology and resilience.

Chapter 15: Exploring Tourism Studies through Strategic Stakeholder Matrices

Focusing on tourism development in Fez, Morocco, this chapter utilizes stakeholder matrices and qualitative interviews to assess the role of local actors in shaping tourism strategies. It highlights the intersections of smart technology, sustainable planning, and community engagement in tourism management.

We selected the title *Evolving Designs, Applications, Technological Advances, and the Future of Qualitative Research* to reflect the multifaceted and evolving nature of this field. It encapsulates our intention to guide readers through both foundational elements and advanced considerations—ranging from philosophical underpinnings to practical applications—while addressing emerging challenges and opportunities in research. Importantly, this book emphasizes the value of methodological rigor and reflexivity in producing robust, meaningful, and ethically sound research outcomes.

This work is particularly designed with the academic community in mind—graduate students, faculty members, researchers, and professionals seeking to refine their qualitative research skills. By presenting detailed examples, case studies, and conceptual discussions, the book supports readers in understanding how theoretical choices translate into research practice and how these choices affect the quality and relevance of their findings.

Moreover, we hope this book contributes to the demystification of qualitative research by illustrating its relevance, reliability, and potential to generate profound insights in organizational studies. We aim to inspire confidence and foster curiosity among those who may be new to the paradigm, while also offering seasoned researchers a space for reflection and continued learning.

Ultimately, our goal is to strengthen the collective understanding of qualitative methodologies and encourage their adoption in diverse business contexts, thus advancing not only academic knowledge but also practical impact in organizational settings. We are confident that the chapters that follow will serve as valuable resources for those navigating the rich, complex, and ever-evolving terrain of qualitative research in business.

Chapter 1
Philosophical Unerprinnings of Qualitative Research:
Trranscending Modernity

William Chakabwata
https://orcid.org/0000-0002-4224-5239
University of South Africa, South Africa

ABSTRACT

The chapter explored the development of qualitative research which emerged as a response to the claims about positivism regarding the nature of reality and epistemology. The chapter associated modernity with positivism, especially the scientific inventions that accompanied industrial revolution. In the chapter the axiology of qualitative research are explored. The epistemology of qualitative research genres such as ethnography, phenomenology and narrative inquiry are discussed. The chapter also discusses sampling and data analysis methods in qualitative research. The strength and weaknesses of qualitative research are also highlighted. The theoretical underpinning of the chapter is constructivism and interpretivism.

INTRODUCTION

Qualitative research emerged as a reaction to the assumptions of modernity and positivism regarding the ontology, epistemology and axiology of knowledge. Positivism conceptualised reality as showcasing a single version and is univocal, whereas qualitative research embraces the multiplicity of reality and is also multifocal. In this chapter the multiple strands of qualitative research are explored, the link between

DOI: 10.4018/979-8-3373-4432-4.ch001

modernity and post positivism is spotlighted and the ontology, epistemology and axiology of qualitative research are discussed. The methodologies that are subsumed under qualitative research which includes Grounded Theory (GT), ethnography, narrative inquiry, virtual ethnography among many others are probed. The chapter also explores sampling methods used in qualitative research and data analysis. The chapter then ends with a discussion, recommendations and conclusions.

BACKGROUND

Cohen et al.,(2018) explicated qualitative research refers to a cornocupia of research approaches which covers a wide range of research methods. Lincoln and Guba (1985, pp. 226-47) enumerated the attributes of qualitative research design as constituting:

- Deciding on an area of research
- Deciding if the paradigm selected is appropriate for the research
- Deciding if the paradigm selected is appropriately alinged to theory
- Deciding on the participants for the research
- Deciding on the stages that the research will take
- Making a decision on instrumentation
- Deciding on how data will be collected and recorded
- Deciding on data analysis procedures
- Making logistical plans in diverse areas which includes plans for the project, for going into the field and closure and ending of the project
- Planning for trustworthiness

Qualitative research may proceed with research questions that are clear, but which are subject to modification during the research process. In addition, it is vital that caution is taken to ensure that the research questions employed are not severely constricted that they yield very little data and that they are not too expansive that they lack focus (Cohen et al., 2018). Moreover qualitative research does not depend on the use of hypothesis as in quantitative research approach, as it widely known that the researcher has an invidious effect on the research process. Qualitative research especially Grounded Theory, holds that theory must emerge from the data and is the endpoint for research and not the starting point. Furthermore, qualitative research maintains an open mind to data, an attitude of willingness to modify research questions and willingness to display reflexivity.

MODERNITY

At a philosophical level modernity was associated with positivism. Meridian University (2023) explicated positivism as a philosophical movement that held that knowledge of reality is situated in phenomenon that we can measure and observe as in traditional research in agriculture. The proponents of positivism are identified as Auguste Comte, and Jeremy Bentham. In addition positivism held the premise that knowledge is attained through sensory impressions and scientific investigations. The 19th century was notable for revolutions which encompass both industrial and political. There was a marked shift from knowledge as perceived from a scientific view to knowledge as established through scientific methods.

Petty et al., (2012) noted that positivism or post –postivism which is also known as empirical science evolved during the period of enlightenment when rationalism and reason gained ascendency and torpedoed the religious hegemonic claims on epistemology. In addition, post postivism was based on the assumptions that knowledge could be created through a rigourous process that can lead to the creation of facts, which extrude a single objective reality. Post-positivism believed in the creation of knowledge that can be replicated and which can be generalised to all situations.

Qualitative research emerged as a reaction to the epistemological and ontological claims of modernity on knowledge. Creswell and Creswell (2018) indicated that the ontology and epistemology of postpostivism was queried by social constructivists who averred that people endevour to comprehend the world in which they live in and function. Social constructionism is perceived to be linked to intepretivism, and is a part of qualitative research approaches. In addition is explicated that people seek to explain reality from their subjective perspective. Social constructivism received its impetus from the work of Berger and Luckmann's (1967) "The social construction of reality" and Lincoln and Guba (1985) " Naturalistic inquiry". The central tenet of social constructivism is that people's view of the world are diverse and multiple culminating in researchers seek to comprehend reality from this multiplicity of perspectives as opposed to a single narrow lens. The subjected intepretations that people emerge with in research are negotiated and situated in social and historical context. This marked a shift from postpostivism and modernity to postmodernism.

Leavy (2017) expounded that constructivism means that people engage in the creation of experiences in mundane interactions. Focus is placed on ways in which meanings is attributed to occurances and daily engagements. Phenomenology owes its development to Edmund Husserl (1913/1963), Martin Heidegger (1927/1982) and Maurice Mealeu-Ponty among many others. Phenomenologists focalise on how conscious is experienced. Ethnomethodology holds the assumptions that social life is fluid and is created and then recreated. Dramaturgy was developed by Erving Goffman, who averred that people utilise numerous strategies to create meaning for their lives. Life is perceived as a stage where people project a favourable impression before people and then a reality of their experiences when they are on their own. All these approaches charted a path that took epistemology from being conceptualised as a single version of reality to a level where reality is contested, multiple and diverse. Miles et al., (2014) averred that there are multiple genres of qualitative researches which can add up to twenty, and some of them have established traditions which includes poetic inquiry,ethnodrama, autoethnography, duoethnography.

PHILOSOPHICAL UNDERPINNINGS OF QUALITATIVE RESEARCH

Sol and Heng (2022) explicated epistemology as branch of science that focalise on the theory of knowledge. It is perceived as central to philosophy as it spotlights knowledge construction. Potter (2017, p1) averred that epistemology owes its first use to James Frederick Ferrier, a Scottish philosopher (1804-1864), in the mid-19th century derived from two Greek terms episteme and logos. Episteme was translated to knowledge, while logos connotes a study of. At a general level epistemology is a component of philosophy that is preoccupied with a study of forms of knowledge. Epistemology seeks to address the question of how knowledge is constructed. Knowledge transcends belief and must be justified. Epistemologists tend to view knowledge from numerous angles, but in terms of research, we can spotlight propositional knowledge and procedural knowledge. Epistemology in research may be categorised as positivism, pragmatism and interpretivism.

Asmawi and Alam (2024) approached epistemology of qualitative research from the framework of relativism, pragmatism, interpretivism, constructivism, and empiricism. Ramandhan et al., (2021) indicated that a pragmatic approach may be used with qualitative research which means that Implementation Science (IS) utilises a pragmatic approach to selecting an appropriate analytical method for using qualitative data (Ramanadhan et al.,2021). This may involve the implementation in sciences of multiple methods such as grounded theory, framework analysis and interpretive phenomenological analysis in a practical way. Goldsmith (2021) asserted that pragmatic analysis involves analysis of diverse forms of data and is applied for a variety of reasons. Data that can be probed under framework analysis includes in-depth interviews, focus groups, data from observations, policy documentation, visual data, and online discussion data, among many others. Framework analysis is a growing form of pragmatic research in health sciences. Framework analysis constitute generating an analytical framework and utilising that framework.

Asmawi and Alam, (2024) defines a pragmatic as retrieving elements from qualitative research that are strategic for completing a study in Computer science (IS). Wagenaan et al., (2022) indicated that pragmatism acknowledges that the meaning of a research phenomenon is unveiled in action and not on some a priori before research. The following section explores ontology in qualitative research.

Research philosophy is also termed research paradigm. The word paradigm was first used by Kuhn (1962) in a book titled " The structure of scientific revolutions". Kuhn averred that scientific work transpires in a framework of agreement on basic principles or paradigms (Kuhn, 1962). Lincoln and Guba (1985) appropriated the term paradigm to characterise a worldview that was radically different from the traditional approach to sciences. In addition they averred a radically different approach to research that a rebuttal of the fundamental principles that postivism and postpostivism were founded on. Furthermore the authors postulated a new paradigm that was termed naturalist, which was premised on social science research principles.

Saunders et al., (2019,p130) averred that a paradigm refers to "a system of beliefs and assumptions about the development of knowledge". A researcher functions by making assumptions which are either explicit or implicit about the nature of reality and axiology and epistemology. A research philosophy that is clearly articulated helps to provide credility in research. Creswell (2014) acknowledged four research paradigms which included postpositivism, constructivism, transformative,interpretivism and pragmatism. The postpositivism worldview is related to what is conceptualised as traditional scientific research and is more inclined to quantitative research. Interpretive research is perceived to be subjective by design. This section of the chapter will foreground intepretivism and social constructivism as paradigms for qualitative research.

ONTOLOGY

Spencer et al.,(2014,p82) expounded that "ontology is the study of reality". Qualitative research holds assumptions that reality is subjective and is situated and a universal apprehension of reality cannot be realised because phenomenon is situated in diverse context. Ontology in qualitative research is encapsulated in terms such as emic and etic. Ugwu and Eze (2023) asserted that qualitative research is examination of phenomenon, which encompass their quality, varied forms, the place where they are located and tends to utilise data in the context of words as opposed to numbers. Qualitative data utilises data such as audio, words, videos, pictures and documents.

Asmawi and Alam (2024) indicated that ontology belong a branch of science that seeks to address the nature of things. Research premises are constituted by epistemology, ontology and axiology. Ontology seeks to furnish legitimation of human knowledge. This branch of philosophy also explores the degree of limitations of human knowledge. The nature of ontology in qualitative research is central as the author positionality plays a central role in the research process. A researcher ontological position will frame the way they frame research questions, the design of data gathering tools and the way data is analysed and reported.

AXIOLOGY

Guraya et al., (2023) axiology explores ways in which a researcher value may impact on the research process.A researcher worldview can impact on a researcher in terms what is right or wrong. Interpretivism showcase the presence of values in research. In a nutshell axiology deals with the ethical conduct of the researcher during the data collection process and data analysis and intepretation. The values in research are informed by the Mill's utilitarian disposition which adduced that underscored that human beings have a right to choose and that must be upheld (Stanford Encylopedia of Philosophy, 2009).

Axiology foregrounds ways in which a researcher values and assumptions impinge on the research process. Deane (2018) include an examination of values and or theorising on the nature of values.

PHENOMENOLOGY

Phenomenology is a part of the qualitative research approach and it owes its origins to the work of Hursell as a method to comprehend one lived experience (Alyssa & Anderman). Phenomenology has its roots in the work of Kant (1781) who opined that objects we encounter in the world (noumena) are separate from the phenomena, which means object as experienced by human consciousness. Kant maintained that it is not possible to the objects directly as we human beings experience reality through mental filters and we can therefore experience phenomenon (Kant, 1999). Hegel (1807) built on the presupposition of Kant and actually conceptualised phenomenology as a methodological manner of outlining what one is perceving and utilising these observation to edge towards absolute knowledge, which in reality can never be attained. Husserl enunciated an approach based on the field of philosophy that arduously examined conscious in a way that is not imbued by subjective experiences or theory. In order to suspend subjectivity the researcher

Multiple theorists from diverse fields of learning have expanded the concepts to make them applicable to numerous disciplines. Van Manen (1990) expanded on phenomenology by theorising on hemerneutical phenomenology, which spotlighted on the reality of the participants and how they interprete it. Moustakas (1994) delved into what is termed trascendental phenomenology that downplayed the interpretation of the researcher lived reality foregrounding the lived experience. In addition, he mooted the term that is commonly termed bracketing. Dorfler and Stierand (2020,p2.) defined bracketing as " researcher's attempt to hold in abeyance their preunderstandings and assumptions to attain experiences before making sense of them". The concept of bracketing is both a philosophical and theoretical abstraction, it is indispensable in phenomenological studies. Bracketing is central tenet of phenomenology, albeit exothermic. The main reason for this contest is that phenomenology has furnished a theoretical justification for interpretivism and also for positivism, making the meaning of the term relative on the philosophical underpinning at play. In interpretivism which is the core focus of this chapter,function closely with the two concepts of reflection and reflexivity to are key to interpretive research and are conceived as a way of ensuring rigour in qualitative research.

Alase (2017) expounded that interpretive phenomenological inquiry (IPA) allows participants to articulate their feelings and the way that experience their reality. Rajasinghea et al (2024) postulated that in IPA people are considered as a repository of knowledge and the researcher and the participants actively take part in the generation of knowledge. Participants are given the latitude to articulate their lived expressions without threat of censure or persecution. IPA had its genesis with the work of Jonathan Smith, when he urged for the formation of an approach to psychology that would be able to create an intersection between the experience and qualitative research and psychology. It is for that reason that IPA evolved as a component of qualitative approach in psychology. However, from its inception it has spread its tentacles to clinical and counselling psychology and educational psychology. A grounded theory a genre under qualitative research is discussed in the next section.

GROUNDED THEORY

Grounded theory emanated from sociology and had extensive applications in multiple disciplines which includes nursing, physiotherapy, education, anthropology, physiotherapy, software engineering, education and information systems among many others (Mohajan & Mohajan, 2022). In Grounded theory (GT), the research data are methodically, repeatedly, arduously collected and are then probed to construct theory. Cullen and Brennan (2021,p1) averred that Grounded Theory (GT) is a "a qualitative research methodology used to develop theory". GT had its origins to the work of Glasser and Strauss 1967, although they latter parted ways with Strauss modifying his conceptualisation of the intial concepts. Different perceptions of

GT have evolved over the years. A person intending to conduct research using GT has to be cognisant of each of the strands of GT and ways in which they may be employed in research. Singh and Estefan (2018) indicated that GT has evolved into four strands with distinct features which includes articulations by Barney Glasser, pragmatism of Anselm Strauss and Juliet Corbin, symbolic interactionism and Karthy Charmaz constructivism. These four approaches use the same semantics, although there are subtle nuances among them. Glaser's philosophical assumptions are inclined towards positivism and holds that there is a unitary reality, Strauss and Corbin advocate a post-positivistic angle that holds that dispute claims of perceiving reality as it actually is and Charmaz who articulated multiple perspectives of reality using a constructivist worldview (Mohajan & Mohajan, 2022; McCall & Edwards, 2021; Babchuk & Boswell, 2023).

Glasser' s GT holds that personal worldview may taint data and holds that a researcher must spotlight on a single view of reality. Corbin and Strauss averred that personal views may pollute reality and that they it is unavoidable, but can be curtailed. In the constructivist worldview espoused by Charmaz, it is expounded that personal experience and knowledge can help a person to learn. GT supersedes the process of creating codes and generating themes in qualitative research and postulates a theory or model from the interlinkages of concepts (Mohajan & Mohajan, 2022). A completed GT must evince attributes which includes an intimate alignment with data, instrumentality, conceptually compact, must stand the test of time, malleable over time and analytically informative. Enthnography is discussed in the next section.

ETHNOGRAPHY

Ethnography had its genesis in the 19th century, when athropologists began to study cultures that were different from the western culture (Adhikari, 2023). Ethnography uses mostly observation, interviews and document analysis (Cohen,et al.,2018). The term ethnography is derived from Greek words "ethnos" which denote people and "graphia" which denote writing (Adhikari, 2023, p. 1). Moreover, in contemporary times, the approach has been used to study both western and non western cultures. Jerolmack and Khan (2018) defined ethnography as "a method of social science inquiry in which the researcher embeds herself in the ongoing interactions of a particular social setting or set of relationships in order to understand and explain the members' lived experience". In autoethnography the spotlight is on an individual, who is studied in a social setting. Although the data collection is primarily qualitative, it is possible to use quantitative data. Irshad & Chaudhry (2023) postulated that enthnography is essentially a qualitative research in which a researcher engages with the participants of the research in their natural setting.

. The key attributes of ethnographic study are enumerated as:

- It's a study that can transpire over a long period such as month or even years;
- It involves participants and non participants observation, The nature of the research questions and the context of study,will determine the type of method that will be employed;
- Research is effected in the natural setting of the participant. In this way a researcher is able to secure trust from the participants.
- Use of emic and etic perspective. The emic perspective assist the researcher to comprehend the way a group perceive the world from using an insider perspective and the etic perspective involves the outsider perspective where a researcher seeks to comprehend what is universal in a group using an objective angle.
- Data is also collected by utilising indepth interviews and document analysis. In the context of disciplines such as education document analysis can supplement information that is not accessible through through observation and interviews.

In the next segment, narrative inquiry, which is a form of qualitative research is probed.

The next section examines narrative inquiry.

NARRATIVE INQUIRY

Lima (2023) indicated that there is a burgeoning of narrative inquiry and narrative interviewing among qualitative researchers in disciplines which includes nursing health sciences and education. This approach focalise on the strength of stories in influencing human understanding, creating meanings and in the formulation of identity. Moreover, narrative inquiry uses story to scrutinise human experiences including those of marginalised groups. Ghanbar et al., (2024) has gained universality in the field of nursing, education and health sciences.

Narrative inquiry is captures the experiences of a group,or an individual chiefly through the use of an interview, which then be structured chronologically. In addition narrative inquiry many capture biographical information, life history or oral tradition. Narrative inquiry has benefits such as furnishes details of people's lived experience, and it spotlight the lived experiences of a marginalised group (Deakin University Library, 2024). The next section examines critical ethnography.

CRITICAL ETHNOGRAPHY

Cohen et al., (2018) explicate critical ethnography as drawing from critical theory. Critical ethnography is perceived as being subversive due to the way it employees research in order to realise societal transformation. In addition, the method embraces multiple approaches such as critical race theory, queer theory, participatory action research, feminist lenses, postcolonial theories among many others (Rashid & Goldszmidt, 2024; Madison, 2020;). Critical ethnography helps to empower research participants with skills to identify structures of inequality in society and then empower people to challenge them.

Semigina (2022) averred that critical ethnography critiques the hidden values in a society and power dynamics that consolidate inequality of a given social group and social inequality. A researcher is allowed to articulate a personal ontological and epistemological view to justify their action and to theorise on an issue and immerse themselves in a research context. The agenda for research is to seek transformation or social change (Ryder, 2021). The next section examine virtual enthnography which is also a major form of qualitative research designs.

VIRTUAL ETHNOGRAPHY

Cohen et al., (2018) averred that the internet as a repository of data has also become a source of information for ethnographic research. Multiple names have been used to describe online ethnographic studies which includes The internet is constituted of people with real and virtual cultures, with its own norms and values, ways of initiation into rights of passage and roles that are negotiated. The internet presents itself as a virtual space that is socially constituted. Naping (2020) the social media also presents virtual platforms for ethnographic research,by the way it enhances interaction among people from different communities across the globe.

The research methods used includes the study of visual discussion themes, comments made on social media, use of emoticons and links and images. It also encompass engagements that occur online and off-line (De Seta, 2020; Pritzker, 2020; Barendregt, 2021). The following section looks at case studies.

CASE STUDIES

Chowdhury and Shil (2021) expounded that the use of a case study in qualitative research enables the conduction of a holistic research. The use of case study can be traced from the disciplines such as legal, medical and psychology. A case study may be implemented in a single site or in multiple sites. A single site presents an opportunity to test a well developed theory, while multiple sites seeks to comprehend the complexity of a research issue. In a case study a researcher may be able to extract data from diverse sources such as written documents, in-depth open -ended interviews and observation (Takahashi & Araujo, 2020). Lune & Berg (2017) asserted that case studies may be approached research designs that are exploratory, explanatory and descriptive. In any exploratory case study data gathering may commence with no previously defined research questions. This may be perceived as a prelude to a subsequent scientific research. Explanatory case studies are employed when it is required to use inferential statistics to determine a plurality of factors as in multivariate studies. A descriptive case study a researcher establish an appropriate theory before setting out on a research agenda. The subsequent section looks at sampling in qualitative research.

SAMPLING IN QUALITATIVE RESEARCH

Bhardwaj (2019,p.158) defined sampling as " a group of objects, people, or items that are taken from the larger population for measurement." Sampling is very important in research when working with a large population. Sampling in research is divided into probability and non- probability sampling based on the research epistemology, ontology and axiology. Kumar et al.,(2020) qualitative research in tandem with its epistemological and ontological pressuposition on the nature of reality relies on human judgements to decide on the sample size for research. In qualitative research the decision regarding sample size and the participants in a research depends on the judgments made by the researcher (Creswell & Creswell,2018). Non –probability sampling means choosing of a sample from a population without relying on random sampling. Non- probability sampling that are used in qualitative research are convenience or haphazard sampling, volunteer sampling, judgement sampling, convenience sampling, quota sampling, snowball or network sampling, web panels and Quota sampling (Statistics Canada, 2024).

Lune and Berg (2017) indicated that covenience sampling or available or accidental sampling involves the selection of participants who are readily available. Purposive sampling or judgemental sampling involves the selection of participants based on their expertise on the topic being researched on. Snowball sampling in qualitative research approach can help to identify participants who have the required features. This sampling strategy may be used to select participants on highly discreet research issues (Cohen et al.,2018; Creswell, 2014;Bhardwaj, 2019). The fundamental approach is to identify a few participants who have similar attributes and then utilise them as referrals to locate other participants. The process is repeated until an adequate number of participants is attained. In quota sampling participants have to meet a certain number of attributes and requirements. The use of Quota sampling ensures that people with the required condition are made a part of the study and marginalised groups may be included.

It is submitted that the sample size in qualitative research is a function of the research questions and the methodology selected. Qualitative research does not seek to apply its findings to universal context in the form of generalisations and consequently it is not burdened by constriction in quantitaive research approach (Cohen et al.,2018; Creswell & Creswell, 2018). Mocăna u(2020) noted that accessibility to the population may impact on the size of the sample.

The data collection procedures in qualitative research are examined in the next section.

DATA COLLECTION METHODS

Ugwu and Eze (2023) enumerated some of the ways in which data can be collected when using qualitative research approaches which includes:

- Interviews
- Focus group interviews
- Surveys
- Secondary research
- Participant observation
- Indirect observation
- Diaries

Rajasinghea et al.,(2024) opined that in IPA data is collected from a miniscule sample that epitomise the issue of interests and which is homogeneous.

DATA ANALYSIS

Cohen et al., (2018) expounded that there is no homogeneous way of analysizing qualitative data. Different genres may employee different approaches to data analysis. The purpose of data analysis will impact on the approach used to do data analysis. For instance an analysis of a biographical study or case study may assume the form of a desciptive nature which is chronological. In the case of a ethnographical study, it may take the guise of stories or narratives that are not necessarily chronological foregrounding explication, analysis and explanation of common themes in the research reflecting a culture of a people. In the case of grounded theory a study may proceed in the form of methodical analysis that includes coding and identifying categories until a theory begins to emerge from the data. The amount of the data set has an effect on the way the data will also intepreted.

CODING

Tsindos (2023,p185) explicated coding as " a process whereby a researcher applies a label to describe the meaning of a text used in qualitative research". This chapter embraces the definition postulatd by Erlingsson and Brysiewicz (2017,p93) on key terms such as condensation, category and theme. The same authors opined that condensation entails the process of providing the essence of a text without comprising the integrity of its meaning. A code is defined as a title or name that can be expressed in one ot two words. A theme is perceived a display an overarching meaning among categories. A theme is normally reflected as a statement rather than a word.

Coding may be viewed in two perspectives either as deductive or inductive (Saldina, 2015). In qualitative research both inductive and deductive coding are utilised. Qualitative analysis approaches which includes grounded theory and thematic analysis depends largely on inductive analysis, where as content analysis and programe evaluation depend on deductive analysis (Delve, 2024). Deductive coding connotes that the researcher commences the research process with predermined codes, and then extract data that fits the codes, whereas in inductive coding analysis commences with no codes and you generate the codes from the dataset. Inductive coding as been perceived as a bottom up approach due to the way it commences with no codes, whereas deductive coding as been conceptualised as a top to down approach because of the way it commences with explict codes.

ANALYSING PRAGMATIC DATA IN QUALITATIVE STUDIES

Goldsmith (2021) declared that Framework Analysis is also termed framework approach, framework technique and framework method connotes the use of themes obtained deductively and inductively to perform a cross –sectional analysis an amalgam of data description and rarefied analysis. These two key phases takes place through five stages which are:

i) Getting familiar with the data- The provides a researcher with an opportunity to get acquainted with the data. The major themes from the data may be located at this stage, but without any coding at this stage.
ii) Deciding on a thematic framework- This stage involves identification of themes and also more rarefied concepts. The process is iterative and construct themes from apriori as well from the data. The major themes are noted as well as related sub components.
iii) Indexing all the data against the framework- involves the creation of a framework and using administering it on the data.
iv) Charting to create a sumary of the recorded data- involves organising the indexed data and treating the data by subjecting it high level analysis of concepts in order to methodicaly analyse the data. The key objctive of charting is create a concise version of the data.
v) Mapping and interpreting patterns within the framework (Goldsmith, 2021, p. 2061)- Data is analysed in a way that transverses data sets and within frameworks. Differences within groups and sub-groups of data are compared in order to identify if possible new patterns.

The last three stages of framework analysis are the most decisive. Framework analysis is rooted in pragmatism and can be used to analyse diverse data types.

INTERPRETIVE PHENOMENOLOGICAL ANALYSIS

Alase (2017) expounded that interpretative phenomenological analysis (IPA) involves the presentation of a detailed account of an individual's experience and the interpretation of the researcher. Pietkiewicz and Smith (2012) adduced that the central focus of (IPA) is to apprehend ways in which people interpret their reality. This implies that people are continously engaging in intepreting objects, events or experiences in their lives. IPA derives its tenets from phenomenology, hermeneutics, and idiograph.

Phenomenology as enunciated by Hurssel utilise an eidetic method which focalise on identifying the vital components of a phenomena, in place of describing objects or events from a predetermined criteria.When a researcher engages in this process they use bracketing, which is a practice of inhibiting one worldviews and preconceptions to impinge on the research (Smith et al., (2020). Heidegger and other writers such as Merleau Ponty submitted that attaining absolute bracketing as espoused by Hurssel may not be possible.

CONTENT ANALYSIS

Dawadi (2020) involves the conducting methodical analysis of complex data sets. It involves a process of identifying themes that can be retrieved from the data sets. Themes are identified through reading and transcription of data sets. Lune and Berg (2017,p.182) explicate content analysis as a methodological analysis of artefacts in order to isolate "themes, meanings and assumptions". A variety of artefacts which includes written documents, audiotapes, photographs or films among. The purpose of content analysis is to code data in a way that it can be utilised to respond to the research questions. Coding helps to remould the content from focus group interviews into data.

Content analysis may assume the guise of conventional, directed and summative. In conventional content analysis there is no assumptions, categories and coding (Vespestad & Clancy, 2021). It is implemented in a context where the research question is to expansive and the appoach permits flexibility in exploratory research. Theory is anticipanted to emerge from the data. Directed analysis is also termed the deductive methods and is effected by using a theoretical framework or predetermined categories (Mayring, 2019; Berkovic, 2024). The approach facilitates the testing of hypothesis. Summative analysis is employed in quantitative studies. The following section will explore thematic analysis in qualitative research.

THEMATIC ANALYSIS

Naeem et al., (2023) explicated thematic analysis as encapusulating the choosing of concepts and quotations and themes which will require intense scrutiny and the generation of a conceptual model. Thematic analysis is able to contribute to new insights. Naeem et al., (2023,p4) adduced that thematic analysis may assume the following steps:

- The process of transcription, getting acquainted with the data and selection of central quotes;
- Identification of key words- Data that comes from focus group interviews, visual content among other sources is then isolated to create key words or recurring patterns in the case of pictures;
- Coding-involves the creation of key words or phrases is effected and these are aligned to the research questions;
- The codes that are identified are then used to generate more abstract meanings by creating themes, which are then linked to research questions and data.

- Conceptualisation –involves the concepts that are situated in the data are explained.Relations between concepts is demonstrated diagramatically.
- The development of a conceptual model- This reflects the final step in thematic data analysis.

Lochmiller (2021) explicated the way themes can be generated from data that is derived from focus group interviews.

JUDGING THE QUALITY OF QUALITATIVE RESEARCH

There are a plethola of submissions on criteria of judging qualitative research. Some the criteria included criteria such as nature of topic, degree of rigour, sincerity, credibility of the research findings, resonance, degree of resonance, nature of contribution, adherence to ethics and coherence of the work. Devakurubai (2020) indicated that the issue regarding quality in qualitative research is exorthemic. This has culminated in the burgeoning of multiple ways of judging quality and terminologies. Dingwall et al.,(1998) postulated that quality may be determined by taking an oppositional view to three items:

- Rigour and validity constrasted against trustworthiness
- Generic contrasted with specific traditional norms
- Quality during the research process versus quality at the end of the research process

The arguments on rigour has roots in postpostivism. Validity in qualitative research supplanted by trustworthiness which means that the degree at which the researcher is definite that the data and the findings " credible, transferable and dependable" (Devakurubai, 2020, p. 77). If the rigidity is emphasized it can impact creativity and innovation of the researcher. Denzin and Lincoln (2014) noted that triangulation in qualitative research also enhances validity. Triangulation is vital in qualitative research in order to reduce control bias.

Stahl and King (2020) adduced that qualitative research is not committed to replicability of research. Lincoln and Guba (1985) postulated a framework of four criteria in order to ensure quality in qualitative research namely credility and trustworthiness, dependability as a component of trustworthiness, transferability and confirmability. The concept of credibility seeks to determine the extent to which a research findings correspond to reality on the ground (Noble & Smith, 2015;). Methodological, investigator, theoretical triangulation, environmental triangulation and data triangulation helps to enhance credility between the researcher and the reader (Asogwa et al, 2023; Vivek etal.,2023; Campbell et al.,2020; Moon, 2020; Donkoh & Mensah, 2023). Methodological triangulation alludes to the use of both qualitative and quantitative research approaches in what is termed mixed studies. Data triangulation involves the collection of data from people at different time, space and people perspective. This collection of data from the same population at different time intervals for example over six month. Theoretical tringulation refers to the entails the utilisation of numerous hypthesis / and or theories in a research (Campbell, et al.,2020). Theoretical tringulation allows numerous theories to be tested using the same data. Methodological tringulation can occur at various levels such as sampling,data collection instruments and data analysis. Researcher triangulation refers to the engagement of multiple researchers in a single study. The researcher can engage in data gathering process such as interviewing the participants or data analysis (Moon, 2020).Environmental tringulation alludes to the utilisation of numerous methods of collecting data and sources for the purpose of ascertaining credibility. Environmental triangulation may assume the guises of many data sources, many researchers or numerous researchers to secure a holistic view of a phenomenon. Credibility can also be enhnced by self- reflexive practice where a researcher maintains an ongoing attitude of reflecting on their work and also recording events occuring in the field.

Lincoln and Guba (1985) also foregrounds trustworthiness as a criteria for quality in qualitative research. It is opined that recurring themes and descriptions noted in one scenario may be applied to another environment that has similar features. Transfer may be realised when an extensive descriptions that can be applied to other contexts. Dependability as a component of trustworthiness implies that activities such as peer interrogation and peer probing of the work establish a level of trust in the work. Trust can be built by allowing peer review of the field notes and also the positionality of the researcher can help embed trust in the work. Peer review of the complete work can also facilitate bracketing, which connotes the distingushing between the observation and intepretation.

Researchers can also engage in reflexive auditing in order to enhance dependability. Transferability as a component of trustworthiness, which connotes the researcher becoming as intimate as possible to objective reality. Transferability alludes to the researcher endevouring to be as precise as possible and also receiving input from peers. This chapter concurs with the perspective of Yadav (2022, p. 685) who averred that "one size-fits-all guidelines for evaluating quality of qualitative research doesn't exist doesnot exist". This means that there are no rigid normative standards for evaluating quality in qualitative research.

PRACTICAL APPLICATION OF QUALITATIVE RESEARCH IN BUSINESS WORLD

In this section, an explanation is offered on the application of qualitative research in business studies. Walle (2015) noted that a report published by Carnegie Corporation in 1959 asserted that business education generated more of technicians as opposed to strategic thinkers. This led to more more research that was basedon quantitative research as business sought to like other fields. Toda there is a growing recognition that using qualitative research paradigms has trade offs and there is no perfect way of doing research. Qualitative research that uses naturalistic observations are growing in the business world, as it allows employees or managers to be observed in a natural setting. The business sector has also employed the use of business anthropology to conduct research and enthnographic studies which explore how people use a particular product. Narrative inquiry has been used to study the growth and challenges of Sm all and Medium Enterprises (SMEs) (Kaimathiri, 2020; Mertova & Webster, 2020).

Consumer research has been dominated by the use of qualitative research methods (Walle, 2015). (Turner et al.,2021). Case studies that use indepth interviews have been used in qualitative research to address research questions. Qualitative research is not about numbers, but focalise on the people and the way they engage with various services and products from the business world. There are multiple opportunities for consumers to express perceptions they have towards company logo, product or services using qualitative studies. Grounded theories can be applied to business research in order to understand management behaviour.

EMERGING TRENDS AND CHALLENGES OF USING QUALITATIVE RESEARCH

There are a number of emerging challenges and emerging issues in qualitative studies in the business sector. One of the emerging issue is the use of software to analyse data such as Quality Data Analysis Software (CAQDAS). There multiple software packages that can be used in business world to analyse data in an inerative way which includes ATLAS.ti, MAXQDA, Nvivo and Quirkos which can help analyse huge datasets. Managers, markerters and executives can present data in a format that can be easy to follow using softwares for data analysis. One of the emerging challenge in qualitative research in business studies is the use of mixed methods, which combine qualitative and quantitative data in the same study. Mixed research methods has the potential to strengthen validity, enhance generalisability and address diverse forms of research questions.

WEAKNESS AND STRENGTH OF QUALITATIVE RESEARCH

A number of authors have cited the weaknesses of qualitative research (Priya, 2016; Mwita, 2022; Jeri, 2023; Madrigal & McClain, 2024). Some of the limitations that have been adduced in literature regarding the use of qualitative research includes that the research process is demand ample time to complete it, experiential knowledge has an effect on data intepretation and analysis, the research is more open ended and this grants the participates more control over the research outcome. It is also averred that qualitative research is demanding in terms of labour and the coding process. Finally, it is declared that different researchers may extricate different conclusions from the same data.

Tenny et al.,(2022) asserted that the validity and reliability of qualitative studies has also been questioned. However, reliability and validity of qualitative research may be strenghthened by triangulation, peer examination, use of thick description and the creation of an audit trail.

DISCUSSION

In the chapter, it was showcased that there are qualitative research subsumes numerous methodologies which includes ethnography, GT, narrative inquiry and virtual ethnography among many others. It was also noted that qualitative research emerged as a reaction to positivism particularly the way it foregrounded the neutrality of the researcher and also the importance of generating reality that can be applied universally. Qualitative As already spotlighted in this work qualitative research has no pretense to neutrality and it acknowledges the role of the researcher positionality and the way it can impinge on data collection, interpretation and reporting. It is for that reason that qualitative research relies on reflection and reflexivity to manage a researcher biases.

Qualitative research is informed is informed by a constructivist philosophy which averred that there is no external objective reality and that reality is socially constructed. This view is extruded as a realist perspective that holds that people's perspectives are fashioned by their unique experiences which are also dynamic. Qualitative research is therefore conceptualized as being either inductive or deductive.

The chapter also highlighted the way qualitative studies are burgeoning in a variety of fields such as health, education, psychology and anthropology by virtue of the manner, they illuminate the lived conditions of the participants. Concerns have also been raised over the quality of qualitative research, particularly in terms of being devoid of scientific rigour and also lacking clear sampling guidelines and generated research that is biased. As we also noted there are diverse genres of qualitative research, and it is imperative for a researcher to become familiar with the each genres and its guiding principles before using it in a study.

Data collection in qualitative research involves use of in-depth interviews, observation, review of documents, focus group interviews and use of open-ended questionnaire. The sample tends to be small in qualitative research depending each genre. Data collection continues until saturation is attained.

It was also noticeable that qualitative research is preoccupied with the description of phenomenon, which may encapsulate social experiences, attitudes needs as the research seeks to find an appropriate design to address human challenges. Contrary to the practices in quantitative research where the data is expected to be standardized, in qualitative research, demands flexibility. While genres such as GT and phenomenology share certain assumptions they also tend to reflect differences that has a potential to confuse a novice researcher.

RECOMMENDATIONS

Given the growing popularity of qualitative researches it is important to take into accounts the concerns that people have raised regarding its weaknesses. In order to offset these weakness it is vital that researchers take the following recommendations into account:

i) The important of indicating researcher's positionality in a way that helps the reader to appreciate the influence that the researcher had on the research process;
ii) Researchers must also display reflection and reflexivity as they engage in data analysis, data interpretation and presentation of reports;
iii) Given the complexity and diversity among the genres of qualitative research methods, it is vital that a researcher is able to show holistic understanding of each methods before utilizing it in research;
iv) A qualitative researcher must also be familiar with diverse ways of ensuring rigour in qualitative research which includes transferability, dependability, credibility and confirmability.
v) Researchers in qualitative research must also be familiar with the diverse data coding and analysis and sample selection procedures for each genre of qualitative research;
vi) Novice researcher must take time to distinguish the underlying assumptions of GT, ethnography and phenomenology in order to avoid mixing up the methods.

CONCLUSIONS

There are numerous genres in qualitative studies and it is vital that a researcher be familiar with each of them in order to inform their practices. Because of its ability to study human experiences qualitative research methods are burgeoning. The researchers also need to get acquainted with the way validity and reliability, sampling are approached in qualitative research. A researcher needs to be familiar with the various nuances of the research methods in qualitative research work.

REFERENCES

Adhikari, E. (2023). *Ethnography as a research methods: A critique*. Kathmandu University.

Alase, A. (2017). The interpretative phenomenological analysis (IPA): A guide to a good qualitative research approach. *International Journal of Education and Literacy Studies*, *5*(2), 9–19. DOI: 10.7575/aiac.ijels.v.5n.2p.9

Alyssa, E., & Anderman, L. (n.d.). Using interpretive phenomenological analysis to advance theory and research in educational psychology,. Retrieved 09 21, 2024, from DOI: 10.1080/00461520.2020.1787170

Asmawi, A., & Alam, S. (2024). Qualitative research: Understanding its underlying philosophies. *Forum for Education Studies*, *2*(2), 1–14. DOI: 10.59400/fes.v2i2.1320

Asogwa, V., Hamisu, M., & Ede, M. (2023). Methodological triangulation in educational research: Pros and cons. *Propellers Journal of Education*, *2*(1), 79–87.

Ayton, D. (2023). Thematic analysis. In Ayton, D., Tsindos, T., & Berkovic, D. (Eds.), *Qualitative Research – a practical guide for health and social care researchers and practitioners* (pp. 199–209). Monash University.

Babchuk, W., & Boswell, E. (2023). Grounded theory. *International Encyclopedia of Education*, 107-122.

Barendregt, B. (2021). Digital ethnography, or 'deep hanging out' in the age of big data. In Grasseni, C., & Barendregt, E. (Eds.), *F. de Maaker, D. Musso, A, M. Littlejohn, . . . M. Westmoreland, Audiovisual and digital ethnography. A practical guide* (pp. 168–190). Routledge. DOI: 10.4324/9781003132417-8

Berkovic, D. (2024, 10 6). *Qualitative Research: A practical guide to for health and Social Care Researchers and Practitioners*. Retrieved from Open Educational Resources Collective: https://oercollective.caul.edu.au/qualitative-research/chapter/__unknown__-21/

Bhardwaj, P. (2019). Types of sampling in research . *Journal of the Practice of Cardiovascular Sciences ¦ Volume 5 ¦ Issue 3 ¦*, 157-163.

Braun, V., & Clarke, V. (2006). Using thematic analysis in psychology. *Qualitative Research in Psychology*, *3*(2), 77–101. DOI: 10.1191/1478088706qp063oa

Braun, V., & Clarke, V. (2022). Toward good practice in thematic analysis: Voiding common problems and be(com)ing a knowing researcher. *International Journal of Transgender Health, 24*(1), 1–6. DOI: 10.1080/26895269.2022.2129597 PMID: 36713144

Bridges, D. (2017). *Philosophy in education al research: Epistemology, ethics and quality*. Springer. DOI: 10.1007/978-3-319-49212-4

Brinkmann, S. Jacobsen., & Kristiansen, S. (2014). Historical overview of qualitative research in the social sciences. In P. Leavy, *The Oxford Handbook of qualitative research* (pp. 17-42). Oxford: Oxford University Press.

Campbell, R., Goodman-Williams, R., Fehler-Cabral, G., & Feeney, H. (2020). Assessing Triangulation Across Methodologies, Methods, and Stakeholder Groups: The Joys, Woes, and Politics of interpreting convergent and divergent data. *The American Journal of Evaluation, 41*(1), 125–144. DOI: 10.1177/1098214018804195

Chowdhury, A., & Shil, N. (2021). Thinking 'Qualitative' through a case study: Homework for a researcher. *American Journal of Qualitative Research Vol. 5 No. 2, p,* 190-210.

Cohen, L., Manion, L., & Morrison, K. (2018). *Research methods in education.* Routledge.

Cooper, K., & White, R. (2012). *Qualitative research in the post-modern era.* Springer. DOI: 10.1007/978-94-007-2339-9

Creswell, J. (2014). *Research design: Qualitative, quantitative and mixed approaches.* Sage.

Creswell, J., & Creswell, J. (2018). *Qualitative,quantitative, and mixed methods approaches.* Sage.

Cullen, M., & Brennan, N. (2021). Grounded Theory: Description, divergences and application. *Accounting. Finance & Governance Review, 27,* 1–19. DOI: 10.52399/001c.22173

Dawadi, S. (2020). Thematic analysis approach: A step by stepm guide for ELT research practitioners. *Journal of NELTA, 25*(1-2), 62–71. DOI: 10.3126/nelta.v25i1-2.49731

De Seta, G. (2020). Three lies of digital ethnography vol 2 No 1. *Journal of Digital Social Research, 2*(1), 77–97. DOI: 10.33621/jdsr.v2i1.24

Deakin University Library. (2024, 10 5). *Qualitative study design*. Retrieved from Deakin University Library: https://deakin.libguides.com/qualitative-study-designs/narrative-inquiry#:~:text=Definition,ordered%20into%20a%20chronological%20 narrative

Deane, P. (2018, 05 22). *A guide for interdisciplinary researchers: Adding axiology alongside ontology and epistemology*. Retrieved from A communiIntegration and Implementation Insights blog and repository of resources for improving research impact on complex real-world problems: https://i2insights.org/2018/05/22/axiology-and-interdisciplinarity/

Delve. (2024, 09 17). *Deductive and Inductive Coding*. Retrieved from Delve: https://delvetool.com/blog/deductiveinductive

Denzin, N., & Lincoln, Y. (2014). *The SAGE handbook of qualitative research*. Sage.

Devakurubai, E. (2020). Quality in qualitative research. *Indian Journal of Continuing Nursing Education*, *21*(1), 76–80. DOI: 10.4103/IJCN.IJCN_48_20

Dingwall, R. M., Watson, P., Greatbatch, D., & Parker, S. (1998). Cathing goldfish: Quality in quality research. *Journal of Health Services Research & Policy*, *3*(3), 167–171. DOI: 10.1177/135581969800300308 PMID: 10185376

Donkoh, S., & Mensah, J. (2023). Application of triangulation in qualitative research. *Journal of Applied Biotechnology and Bioengineering*, *10*(1), 6–10. DOI: 10.15406/jabb.2023.10.00319

Dorfler, V., & Stierand, M. (2020). Bracketing: A phenomenological theory applied through transpersonal reflexvity. *Journal of Organisational Management*, 1-22.

Erlingsson, C., & Brysiewicz, P. (2017). A hands-on guide to doing content analysis. *African Journal of Emergency Medicine : Revue Africaine de la Médecine d'Urgence*, *7*(3), 93–99. DOI: 10.1016/j.afjem.2017.08.001 PMID: 30456117

Gary, A., & Holmes, D. (2020). Researcher Positionality - A Consideration of Its influence and place in qualitative research - A New researcher guide. *Shanlax International Journal of Education*, *8*(4), 1–10. DOI: 10.34293/education.v8i4.3232

Ghanbar, H., Cinaglia, C., Randez, R., & De Costa, P. I. (2024). A methodological synthesis of narrative inquiry research in applied linguistics: What's the story? *International Journal of Applied Linguistics*, *34*(4), 1–27. DOI: 10.1111/ijal.12591

Glaser, B., & Strauss, A. (1967). (*The discovery of Grounded Theory strategies for qualitative research*). CA Sociology Press.

Goldsmith, L. (2021). Using Framework Analysis in Applied Qualitative Research. *The Qualitative Report*, *26*, 2061–2076. DOI: 10.46743/2160-3715/2021.5011

Guraya, S., Harkin, D., Yusoff, M., & Guraya, S. (2023). Paradigms unfolded – developing, validating, and evaluating the Medical Education e-Professionalism framework from a philosophical perspective. *Frontiers in Medicine*, *10*, 1–13. DOI: 10.3389/fmed.2023.1230620

Irshad, M., & Chaudhry, A. (2023). Principle nature of ethnography: A systematic literature review of qualitative research methods in anthropology. *Pakistan Journal of Social Research*, *5*(4), 1–9. DOI: 10.52567/pjsr.v5i04.1326.

Jamieson, M., Govaart, G., & Pownall, M. (2023). Reflexivity in quantitative research: A rationale and beginner's guide. *Social and Personality Psychology Compass*, *17*(4), 1–15. DOI: 10.1111/spc3.12735

Jayathilaka, A. (2021). Ethnography and organisational ethnography: Research methodology. *Open Journal of Business and Management*, *9*(1), 91–102. DOI: 10.4236/ojbm.2021.91005

Jeri, L. (2023). Strengthening the weakness of a generic qualitative research design. *The Qualitative Report*, *28*(10), 1759–1768.

Jerolmack, C., & Khan, S. (2018). An introduction: An analytical approach ethnography. In Jerolmack, C., & Khan, S. (Eds.), *Approaches to ethnography analysis and representation in participant observation* (pp. 8–21). Oxford university Press.

Kaimathiri, P. (2020). *A Narrative Inquir e Inquiry into the Daily Challenges t o the Daily Challenges to the Business o the Business Growth of W owth of Women's Microenterprises in Rur oenterprises in Rural Kenya.* A doctoral thesis submitted to the College of Management and Technology: Walden University.

Kant, I. (1999). *A critique of pure reason*. Big City Publishers. (Original work published 1781)

Karcher, K., McCuaig, J., & King-Hill, S. (2024). (Self-) Reflection / Reflexivity in Sensitive Qualitative Research: A Scoping Review. *International Journal of Qualitative Methods*, *23*, 1–15. DOI: 10.1177/16094069241261860

Kuhn, T. (1962). *The structure of scientific revolutions*. International Encyclopedia of Unified Science.

Kumar, S., Kumar, S., Govindaraj, M., & Prabhu, N. (2020). Sampling framework for personal interviews in research . *Journal Of Archaeology Of Egypt/Egyptology 17(7)*, 7102-7114.

Leavy, P. (2017). *Research design Qualitative, Mixed Methods,Arts-Based, and Community-Based Participatory Research Approaches.* The Guildford Press.

Lima, A. (2023). Understanding narrative inquiry through life story interviews with former prisoners. *Irish Educational Studies*, *42*(4), 775–786. DOI: 10.1080/03323315.2023.2257673

Lincoln, Y., & Guba, E. (1985). The naturalistic inquiry. Tousand Oaks: Sage. DOI: 10.1016/0147-1767(85)90062-8

Lochmiller, C. (2021). Conducting thematic analysis with qualitative data. *The Qualitative Report*, *26*(6), 2029–2044.

Lune, H., & Berg, B. (2017). *Qualitative research methods for the social sciences.* Pearson.

Madison, D. S. (2020). *Critical ethnography: Method, ethics, and performance.* Sage. DOI: 10.4135/9781071878965

Madrigal, D., & McClain, B. (2024, 10 8). *Strength and weaknesses of qualitative Research.* Retrieved from UX Matters: https://www.uxmatters.com/mt/archives/2012/09/strengths-and-weaknesses-of-quantitative-and-qualitative-research.php

Mayring, P. (2019). Qualitative content analysis: Demarcation, varieties, developments. *Forum Qualitative Social Research*, *20*(3), 1–14.

McCall, C., & Edwards, C. (2021). Theory new perspectives for implementing Grounded Theory. *Studies in Engineering Education*, *1*(2), 93–107. DOI: 10.21061/see.49

Meridian University. (2023, 10 2). *How positivism shaped our understanding of reality.* Retrieved from Meridian University: https://meridianuniversity.edu/content/how-positivism-shaped-our-understanding-of-reality

Mertova, P., & Webster, L. (2020). *Using narrative inquiry as a research method: An introduction to critical event narrative analysis in research teaching and professional practice.* Routledge.

Miles, M., Huberman, A., & Saladana, J. (2014). *Qualitative data analysis.* Sage Publications.

Mocăna u, D. (2020). Determining sample size qualitative research. *International Multidisciplinary Scientific Conference on the Dialogue between Sciences & Arts, Religion & Education*, 181-187.

Mohajan, D., & Mohajan, H. (2022). Development of Grounded Theory in social sciences: A Qualitative Approach. *Munich Personal RePEc Archive (MPRA)*, 1-28.

Moon, M. D. (2020). Triangulation: A Method to Increase Validity, Reliability, and Legitimation in Clinical Research. *Journal of Emergency Nursing: JEN*, *45*(1), 103–105. DOI: 10.1016/j.jen.2018.11.004 PMID: 30616761

Mwita, K. (2022). Strengths and weaknesses of qualitative research in social science studies. *Research in Business & Social Science*, *11*(6), 618–625. DOI: 10.20525/ijrbs.v11i6.1920

Naeem, M., Ozuem, W., Howell, K., & Ranfagni, S. (2023). A Step-by-Step Process of Thematic Analysis to Develop a Conceptual Model in Qualitative Research. *International Journal of Qualitative Methods*, *22*, 1–18. DOI: 10.1177/16094069231205789

Naping, M. (2020). Cleaning up the local environment: A study of virtual ethnography in the social media community of Makassah,Indonesia. *European Journal of Research in Social Sciences*, *8*(2), 56–61.

Noble, H., & Smith, J. (2015). Issues of validity and reliability in qualitative research. *Evidence-Based Nursing*, *18*(2), 1–7. DOI: 10.1136/eb-2015-102054 PMID: 25653237

Olmos-Vega, F., Stalmeijer, R., Varpio, L., & Kahlke, R. (2023). A practical guide to reflexivity in qualitative research: AMEE Guide No. 149. *AMEE Guide No. 149. Medical Teacher*, *45*(3), 241–251. DOI: 10.1080/0142159X.2022.2057287 PMID: 35389310

Petty, N., Thomson, O., & Stew, G. (2012). Ready for a paradigm shift? Part 1: Introducing the philosophy of qualitative research. *National Institute of Health*, 1-8.

Pietkiewicz, I., & Smith, J. (2012). Praktyczny przewodnik interpretacyjnej analizy fenomenologicznej w badaniach jakościowych w psychologii. *Czasopismo Psychologiczne*, *18*(2), 361–369.

Pillow, W. (2003). Confession, catharsis, or cure? Rethinking the uses of reflexivity as methodological power in qualitative research. *International Journal of Qualitative Studies in Education : QSE*, *16*(2), 175–196. DOI: 10.1080/0951839032000060635

Potter, W. J. (2017). Epistemology. In Matthes, S., & Potter, R. (Eds.), *The international encyclopedia of comunication research methods* (pp. 1–14). John Wiley.

Pritzker, S. (2020). *Doing an online ethnography: An overview of central issues and methodological strategies*. University of Alabama.

Priya, C. (2016, 09 11). *Limitations of qualitative research methods*. Retrieved from PG Project: https://www.projectguru.in/limitations-qualitative-research/

Rajasinghea, D., Garvey, B., Burt, S., Barosa-Pereira, A., & Clutterbuck, D. (2024). Innovative interpretative phenomenological analysis (IPA) approach in a coaching research project: Implications for future qualitative coaching research and beyond. *An International Journal of Theory. Coaching (Abingdon, UK)*, *17*(2), 301–318. DOI: 10.1080/17521882.2024.2358771

Ramanadhan, S., Revette, A., Lee, R., & Aveling, E. (2021). Pragmatic approaches to analyzing qualitative data for implementation science: An introduction. *mplementation. Science Communication*, *2*(70), 1–10.

Ramandhan, S., Revette, A., Lee, R. M., & Aveling, E. (2021). Pragmatic approaches to analyzing qualitative data for implementation science: An introduction. *Implementation Science Communications volume 2. Article*, (70), 1–10.

Rashid, M., & Goldszmidt, M. (2024). Critical ethnography: Implications for medical education research and scholarship. *Medical Education*, *58*(10), 1185–1191. DOI: 10.1111/medu.15401 PMID: 38600755

Rotem, N. (2024). Historical ethnography: Key characteristics and the Journey before, during, and after the archival field. *Forum Qualitative Social Research*, *25*(2), 1–10.

Ryder, A. (2021). Critical ethnography and research relationships: Some ethical dilemmas. *Anthropology and Humanism*, *46*(2), 300–314. DOI: 10.1111/anhu.12337

Saldina, J. (2015). *The coding manual for qualitative researchers*. Sage.

Saunders, M., Lewis, P., & Thornhill, A. (2019). *Research methods for business students*. Pearson.

Semigina, T. (2022). Critical ethnography: Opportunities for social work research. *Social Work Education*, *9*(3), 405–421. DOI: 10.25128/2520-6230.22.3.7

Singh, S., & Estefan, A. (2018). Selecting a Grounded Theory approach for nursing research. *Global Qualitative Nursing Research*, *5*, 1–9. DOI: 10.1177/2333393618799571 PMID: 30306099

Smith, S. M., Cotterill, S., & Brown, H. (2020). An interpretative phenomenological analysis of performance influencing factors within the practice environment. *Journal of Physical Education and Sport ® (JPES), Vol.20 (4)*, 1646 - 1657.

Sol, K., & Heng, K. (2022). Understanding epistemology and its key approaches in research. *Cambodian Journal of Educational Research*, *2*(2), 80–99. DOI: 10.62037/cjer.2022.02.02.05

Spencer, R., Pryce, J., & Walsh, J. (2014). Philosophical approach to qualitative research. In Leavy, P. (Ed.), *The Oxford of qualitative research* (pp. 81–97). OUP.

Stahl, N., & King, J. (2020). Expanding approaches to research: Understanding and using trustworthiness in qualitative research. *Journal of Developmental Education*, *44*(1), 26–28.

Stanford Encylopedia of Philosophy. (2009, 03 27). *The history of utilitarianism*. Retrieved from Stanford Encylopedia of Philosophy: https://plato.stanford.edu/entries/utilitarianism-history/

Statistics Canada. (2024, 10 6). *Non-probability sampling*. Retrieved from Statistics Canada: https://www150.statcan.gc.ca/n1/edu/power-pouvoir/ch13/nonprob/5214898-eng.htm

Takahashi, A., & Araujo, L. (2020). Case study research: Opening up research opportunities. *RAUSP Management Journal*, *55*(1), 100–111. DOI: 10.1108/RAUSP-05-2019-0109

Tenny, S., Brannan, J., & Brannan, G. (2022). Qualitative study. *National Library of Medicine*, 1-4.

Tsindos, T. (2023). Coding approaches. In Ayton, D., Tsindos, T., & Berkovic, D. (Eds.), *Qualitative Research – a practical guide for health and social care researchers and practitioners* (pp. 185–192). Monash University.

Turner, D., Ting, H., Wong, M., Lim, T.-Y., & Tan, K.-L. (2021). Asian Journal of Business Research Volume 11, Issue 3. *Applying Qualitative Approach in Business Research*, 1-13.

Ugwu, C., & Eze, V. (2023). Qualitative research . *International Digital Organisation for Scientific Research vol 8 Issue (1)*, 20-35.

van Manen, M. (1997). From meaning to method. *Qualitative Health Research*, *7*(3), 345–369. DOI: 10.1177/104973239700700303

Vespestad, M., & Clancy, A. (2021). Exploring the use of content analysis methodology in consumer research. *Journal of Retailing and Consumer Services*, *59*, 1–9. DOI: 10.1016/j.jretconser.2020.102427

Vivek, R., Nanthagopan, Y., & Piriyatharshan, S. (2023). Beyond methodology: Theoretical foundations of triangulation in qualitative and multi-method research: A literature review. *Scientific Studios on Social and Political Psychology*, *29*(2), 53–62. DOI: 10.61727/sssppj/2.2023.53

Vivek, R., Nanthagopan, Y., & Piriyatharshan, S. (2023). Beyond methods: Theoretical underpinnings of triangulation in qualitative and multi-method studies. *SEEU Review, 18*(2), 105–122. DOI: 10.2478/seeur-2023-0088

Wagenaan, H., Kiesclich, K., & Hangel, N., & Zimmerman. (2022). Collaborative comparisons: A pragmatist approach towards designing large-scale, comparative qualitative research. *SSM. Qualitative Research in Health*, •••, 1–11.

Walle, A. (2015). *Qualitative research in business: A practical overview*. Cambridge: Cambridge scholars .

Yadav, D. (2022). Criteria for Good Qualitative Research: A comprehensive review. *The Asia-Pacific Education Researcher, 31*(6), 679–689. DOI: 10.1007/s40299-021-00619-0

KEY TERMS AND DEFINITIONS

Positionality: a person location in the structure of power in society

Existentialism: It's a philosophy that strengthens the freedom of individuals and their right to decide their own path

Ontology: The nature of reality

Epistemology: The way we know what we know and how we know

Axiology: is a theory about values that researcher bring to the research process which also impact on their work

Confirmability: the extent to which the results of a qualitative work can be confirmed by others

Dependability: the extent to which people can rely on the findings of a research as reflecting reality

Phenomenology: the study of consciousness and experience

Chapter 2
Development of Research Design

Satya Subrahmanyam
https://orcid.org/0000-0003-0441-2742
Holy Spirit University of Kaslik, Lebanon

ABSTRACT

This chapter explores the development of research design in qualitative research, emphasizing its crucial role in shaping effective studies. It begins with an overview of key components and progresses through the formulation of research questions, including their significance and refinement process. The selection of participants and sampling techniques is examined, highlighting criteria, methods, and sample size considerations. Planning and structuring research projects are detailed, covering objectives, research frameworks, data collection methods, timelines, and budgeting. The chapter also outlines the development of a qualitative research plan, including writing proposals and ethical considerations. It concludes by reinforcing the importance of meticulous and ethical research design in achieving credible and valuable outcomes.

INTRODUCTION

Research design is a critical component in the realm of qualitative research, serving as the blueprint that guides the entire research process. It determines the methods and procedures for collecting and analyzing data, ensuring that the research questions are addressed effectively. The importance of research design cannot be overstated, as it provides a structured approach to inquiry, enabling researchers to draw meaningful and valid conclusions. This chapter delves into the significance

DOI: 10.4018/979-8-3373-4432-4.ch002

of research design in qualitative research and offers a brief overview of its key components.

Importance of Research Design in Qualitative Research

Qualitative research aims to explore and understand complex phenomena by capturing the richness and depth of human experiences, behaviors, and interactions. Unlike quantitative research, which relies on numerical data and statistical analysis, qualitative research employs methods such as interviews, focus groups, and participant observation to gather non-numeric data. The strength of qualitative research lies in its ability to provide detailed, contextualized insights into specific contexts and populations.

A well-constructed research design is essential in qualitative research for several reasons. Firstly, it ensures the coherence and consistency of the study, linking the research questions to the methods and procedures used for data collection and analysis (Maxwell, 2013). Secondly, it enhances the credibility and trustworthiness of the findings by outlining clear steps and justifications for the chosen methods (Lincoln & Guba, 1985). Thirdly, it provides a roadmap for the researcher, helping to navigate the complexities and challenges that may arise during the research process (Merriam & Tisdell, 2016).

Key Components of a Research Design

A comprehensive research design encompasses several key components that collectively ensure the study's rigor and validity. These components include the research problem and questions, theoretical framework, methodology, data collection methods, data analysis procedures, and ethical considerations.

Research Problem and Questions

The foundation of any research design is the formulation of a clear and concise research problem and corresponding research questions. The research problem identifies the gap in existing knowledge or the issue that the study aims to address. The research questions, derived from the research problem, guide the inquiry and shape the direction of the study (Creswell & Poth, 2018). In qualitative research, these questions are typically open-ended, allowing for an in-depth exploration of the phenomenon under investigation.

Theoretical Framework

The theoretical framework provides the underlying lens through which the research is conducted and interpreted. It consists of theories, concepts, and models that inform the study's design and analysis. The theoretical framework helps to situate the research within a broader context, linking it to existing literature and guiding the interpretation of findings. It also aids in identifying key variables and relationships that are pertinent to the study (Ravitch & Riggan, 2017).

METHODOLOGY

The methodology section outlines the overall approach and strategy for conducting the research. In qualitative research, common methodologies include ethnography, phenomenology, grounded theory, and case study (Merriam & Tisdell, 2016). Each methodology has distinct features and procedures, and the choice of methodology depends on the nature of the research questions and the goals of the study. The methodology section should provide a rationale for the chosen approach and describe how it will be implemented.

Data Collection Methods

Data collection methods are the tools and techniques used to gather information from participants. In qualitative research, common data collection methods include interviews, focus groups, participant observation, and document analysis (Creswell & Poth, 2018). The choice of data collection methods should align with the research questions and the overall research design. It is important to provide detailed descriptions of the procedures for data collection, including the sampling strategy, recruitment of participants, and the steps taken to ensure the reliability and validity of the data.

Data Analysis Procedures

Data analysis in qualitative research involves organizing, coding, and interpreting the collected data to identify patterns, themes, and insights. The data analysis procedures should be clearly outlined in the research design, including the specific techniques and software used. Common approaches to qualitative data analysis include thematic analysis, content analysis, and narrative analysis. The research design should also address how the findings will be validated, such as through member checking, triangulation, or peer debriefing (Lincoln & Guba, 1985).

Ethical Considerations

Ethical considerations are a crucial component of research design, ensuring that the study is conducted in a responsible and ethical manner. This includes obtaining informed consent from participants, ensuring confidentiality and anonymity, and addressing any potential risks or harm (Creswell & Poth, 2018). The research design should outline the steps taken to protect participants' rights and well-being and adhere to ethical guidelines and standards set by relevant institutions and organizations.

Research design is a fundamental aspect of qualitative research, providing the structure and direction needed to conduct a rigorous and credible study. By carefully considering and addressing the key components of research design, researchers can ensure that their studies are coherent, trustworthy, and ethically sound. This chapter has highlighted the importance of research design in qualitative research and provided a brief overview of its essential components, laying the foundation for a deeper exploration of each element in subsequent sections.

FORMULATION OF RESEARCH QUESTIONS

Importance of Research Questions

Definition and Significance of Well-Formulated Research Questions

Research questions are the cornerstone of any scholarly investigation, serving as the foundation upon which the entire study is built. They define the scope and aim of the research, guiding the methodological approach and determining the focus of the data collection and analysis processes. According to Creswell and Poth (2018), well-formulated research questions provide clarity and direction, ensuring that the study remains coherent and systematically addresses the problem at hand. These questions encapsulate the essence of the inquiry, distilling broad research interests into specific, manageable, and researchable components.

The significance of well-formulated research questions cannot be overstated. They help to avoid ambiguities and ensure that the study is feasible and aligned with the available resources and time constraints (Maxwell, 2013). Furthermore, clear and focused research questions enhance the study's validity and reliability by providing a clear framework for data collection and analysis, thereby facilitating the drawing of meaningful and defensible conclusions (Yin, 2016).

Impact on the Direction and Focus of the Study

Research questions play a pivotal role in shaping the direction and focus of a study. They act as a navigational tool, steering the research process toward relevant and significant areas of inquiry. Well-crafted questions help to narrow down the scope of the research, ensuring that it remains concentrated on specific issues or phenomena, thus preventing the study from becoming overly broad or diffuse (Merriam & Tisdell, 2016). This focused approach not only makes the research process more manageable but also increases the depth and rigor of the investigation.

Moreover, research questions influence the selection of research methods and the design of data collection instruments. They determine whether the study will employ qualitative, quantitative, or mixed methods, and guide the choice of specific techniques such as interviews, surveys, or experiments (Creswell & Poth, 2018). By providing a clear focus, research questions also facilitate the interpretation of findings, enabling researchers to draw connections between the data and the overarching aims of the study.

Steps in Formulating Research Questions

Identifying the Research Problem

The first step in formulating research questions is to identify a clear and concise research problem. The research problem is a specific issue, challenge, or gap in existing knowledge that the study aims to address. This involves a critical examination of current literature and practical experiences to pinpoint areas where further investigation is needed. Identifying the research problem is crucial as it lays the groundwork for developing focused and relevant research questions.

Conducting a Literature Review

Once the research problem has been identified, the next step is to conduct a comprehensive literature review. This process involves a systematic examination of existing studies and theoretical frameworks related to the research problem. The literature review helps to contextualize the research problem, highlighting what is already known and where gaps or inconsistencies exist. It also provides insights into how previous researchers have approached similar issues, which can inform the formulation of new research questions.

Refining and Narrowing Down the Scope

After conducting the literature review, the research problem is often too broad to be addressed by a single study. Therefore, it is necessary to refine and narrow down the scope of the research. This involves breaking down the broader problem into specific aspects or dimensions that can be investigated more thoroughly (Creswell & Poth, 2018). Narrowing the scope ensures that the research questions are manageable and that the study remains focused on critical areas of interest.

Ensuring Questions are Clear, Focused, and Researchable

The final step in formulating research questions is to ensure that they are clear, focused, and researchable. Clarity is essential to avoid misunderstandings and ensure that the questions are interpreted consistently by all stakeholders. The questions should be specific and focused, addressing particular aspects of the research problem rather than broad or vague issues. Additionally, the questions must be researchable, meaning that they can be answered through empirical investigation using available resources and methodologies (Yin, 2016). Ensuring that research questions meet these criteria increases the feasibility and effectiveness of the study.

Examples of Effective Research Questions

Case Studies or Examples from Existing Research

Examining examples of effective research questions from existing studies can provide valuable insights into the characteristics of well-formulated questions. One such example is a study on the impact of social media on adolescent self-esteem. The research questions in this study were:

1. How does social media usage affect the self-esteem of adolescents aged 13-18?
2. What are the specific aspects of social media that contribute to changes in self-esteem among adolescents?

These questions are clear, focused, and directly address specific aspects of the broader research problem.

Another example is a study on workplace diversity and organizational performance. The research questions included:

1. How does the level of diversity in an organization influence its overall performance?

2. What are the mediating factors that affect the relationship between diversity and performance?

These questions are well-defined, targeting specific elements of the broader issue of workplace diversity.

Analysis of What Makes These Questions Effective

The effectiveness of these research questions can be attributed to several key factors. Firstly, they are specific and focused, addressing particular dimensions of the broader research problem. This specificity ensures that the study remains manageable and that the data collection and analysis processes are directed toward relevant and significant areas of inquiry (Creswell & Poth, 2018). Secondly, the questions are clear and unambiguous, which helps to avoid misunderstandings and ensures consistency in interpretation (Maxwell, 2013). Thirdly, the questions are researchable, meaning that they can be answered through empirical investigation using appropriate methodologies and available resources (Yin, 2016).

Additionally, these research questions are grounded in existing literature and theoretical frameworks, which enhances their relevance and significance. By building on previous studies and addressing gaps or inconsistencies in current knowledge, these questions contribute to the advancement of the field and provide a solid foundation for drawing meaningful and valid conclusions.

The formulation of research questions is a critical step in the research process, shaping the direction and focus of the study. Well-formulated research questions provide clarity, coherence, and direction, ensuring that the study addresses specific and significant issues. By following a systematic process of identifying the research problem, conducting a literature review, refining the scope, and ensuring clarity and focus, researchers can develop effective research questions that enhance the rigor and validity of their studies.

SELECTION OF PARTICIPANTS AND SAMPLING

Criteria for Selecting Participants

Relevance to the Research Questions

Selecting participants who are relevant to the research questions is paramount in qualitative research. The aim is to gather in-depth insights from individuals who have direct experience or knowledge pertinent to the study's focus (Creswell

& Poth, 2018). The relevance criterion ensures that the data collected will be rich and informative, contributing significantly to addressing the research questions. This involves identifying specific characteristics or criteria that participants must meet to be included in the study. For example, if the research aims to explore the experiences of first-generation college students, participants should be those who are currently enrolled as first-generation students.

Diversity and Representation

Diversity and representation are critical to capturing a broad range of perspectives and experiences. Ensuring that the participant pool reflects various backgrounds, demographics, and viewpoints enhances the comprehensiveness of the findings (Patton, 2015). In qualitative research, this often involves seeking participants from different age groups, genders, ethnicities, socio-economic statuses, and other relevant categories. This diversity helps to illuminate the multiple dimensions of the research problem and contributes to a more nuanced understanding of the phenomenon under study.

Accessibility and Willingness to Participate

Accessibility and willingness to participate are practical considerations that influence participant selection. Participants must be accessible to the researcher, meaning they can be contacted and are available for interviews or other data collection methods. Additionally, participants must be willing to engage openly and honestly with the researcher, which requires building trust and ensuring that they feel comfortable sharing their experiences (Merriam & Tisdell, 2016). Researchers must also consider ethical issues, ensuring that participation is voluntary and informed consent is obtained.

Sampling Techniques

Overview of Sampling Methods in Qualitative Research

In qualitative research, sampling methods are employed to select participants who can provide rich, detailed, and relevant data. Common sampling techniques include purposive sampling, snowball sampling, and theoretical sampling.

Purposive Sampling

Purposive sampling involves deliberately selecting participants who possess specific characteristics or have experienced the phenomenon of interest. This method is widely used in qualitative research due to its focus on obtaining information-rich cases. Researchers use their judgment to identify participants who are most likely to provide valuable insights, ensuring that the sample aligns closely with the research objectives.

Snowball Sampling

Snowball sampling, also known as chain referral sampling, involves identifying a few initial participants who meet the study criteria and then asking them to refer other potential participants. This method is particularly useful for accessing hard-to-reach or hidden populations. It relies on the social networks of the initial participants to expand the sample, gradually increasing the number of participants through referrals.

Theoretical Sampling

Theoretical sampling is primarily used in grounded theory research, where the aim is to develop a theoretical framework based on the data collected. This method involves selecting participants based on emerging themes and concepts during the data collection process. The sample evolves as the study progresses, with participants chosen to elaborate on or test emerging theoretical constructs.

Strengths and Limitations of Each Method

Each sampling method has its strengths and limitations, impacting its suitability for different types of qualitative research.

Purposive Sampling

Strengths:

- Ensures the inclusion of participants who are directly relevant to the research questions.
- Can provide rich, in-depth data from information-rich cases.

Limitations:

- Potential for researcher bias in selecting participants.
- May not be representative of the broader population.

Snowball Sampling
Strengths:

- Effective for reaching hidden or hard-to-access populations.
- Can build trust and rapport through referrals from known contacts.

Limitations:

- May lead to a homogenous sample if initial participants have similar characteristics.
- Limited control over the composition of the sample.

Theoretical Sampling

Strengths:

- Allows for the development and refinement of theories based on emerging data.
- Provides flexibility to adjust the sample as new insights arise.

Limitations:

- Requires ongoing data analysis, which can be time-consuming.
- May be challenging to implement without experience in grounded theory methodology.

Sample Size Considerations

Determining the Appropriate Sample Size

Determining the appropriate sample size in qualitative research is influenced by the study's aims, the nature of the research questions, and the methodological approach. Unlike quantitative research, which often relies on statistical power calculations, qualitative research focuses on achieving depth and saturation (Mason, 2010). Saturation occurs when additional data no longer contribute new insights or themes to the study, indicating that the sample size is sufficient.

Factors Influencing Sample Size in Qualitative Research

Several factors influence the determination of sample size in qualitative research:

1. **Research Objectives:** Studies aiming for in-depth exploration of specific phenomena may require fewer participants than those seeking to cover a broader range of experiences.
2. **Methodological Approach:** Different qualitative methods have varying sample size requirements. For example, phenomenological studies typically involve smaller samples (e.g., 5-10 participants), while ethnographic studies may require larger samples to capture cultural variations.
3. **Data Saturation:** The point of saturation is a key determinant of sample size. Researchers often conduct data collection and analysis simultaneously to identify when saturation is reached.
4. **Available Resources:** Practical considerations, such as time, budget, and access to participants, also impact sample size decisions. Researchers must balance the need for comprehensive data with the constraints of their study context.

Ensuring Credibility and Transferability

Strategies to Enhance the Credibility and Transferability of the Findings

Ensuring the credibility and transferability of findings is essential for the rigor and validity of qualitative research. Credibility refers to the trustworthiness and authenticity of the findings, while transferability pertains to the applicability of the results to other contexts.

Credibility

To enhance credibility, researchers can employ several strategies:

1. **Triangulation:** Using multiple data sources, methods, or investigators to cross-verify findings. For example, combining interviews, observations, and document analysis can provide a more comprehensive understanding of the research problem.
2. **Member Checking:** Involving participants in the validation of findings by sharing preliminary results and seeking their feedback. This process helps to ensure that the interpretations accurately reflect participants' perspectives.

3. **Prolonged Engagement:** Spending sufficient time in the field to build rapport with participants and gain a deeper understanding of the context. This immersion enhances the richness and authenticity of the data.
4. **Peer Debriefing:** Engaging with colleagues or mentors to review and critique the research process and findings. Peer debriefing provides an external perspective, helping to identify potential biases and strengthen the study's credibility.

Transferability

To enhance transferability, researchers should provide detailed descriptions of the research context and participants, enabling others to determine the applicability of the findings to their own settings (Lincoln & Guba, 1985). This involves:

1. **Thick Description:** Providing rich, detailed accounts of the research context, participants, and findings. Thick description allows readers to understand the nuances of the study and assess its relevance to their own contexts.
2. **Contextualization:** Situating the findings within the broader theoretical and empirical literature. By linking the study to existing knowledge, researchers can highlight the potential transferability of their findings to similar settings.
3. **Explicit Sampling Criteria:** Clearly outlining the criteria for participant selection and the sampling process. This transparency enables other researchers to assess the relevance and applicability of the sample to different contexts.

The selection of participants and sampling techniques are critical aspects of qualitative research that significantly impact the study's validity and relevance. By carefully considering criteria for participant selection, employing appropriate sampling methods, and addressing sample size considerations, researchers can ensure that their studies are robust and meaningful. Additionally, strategies to enhance the credibility and transferability of findings further strengthen the rigor of qualitative research, contributing to the advancement of knowledge in the field.

PLANNING AND STRUCTURING QUALITATIVE RESEARCH PROJECTS

Setting Research Objectives

Defining Clear and Achievable Objectives

The first step in planning a qualitative research project is to define clear and achievable research objectives. These objectives serve as the guiding principles for the study, outlining what the researcher aims to accomplish. Clear objectives help to maintain the focus and direction of the research, ensuring that all efforts are aligned toward achieving specific outcomes (Creswell & Poth, 2018). To formulate effective objectives, researchers should ensure that they are specific, measurable, achievable, relevant, and time-bound (SMART). Specificity involves detailing exactly what the research intends to explore. Measurability ensures that the objectives can be assessed and evaluated. Achievability considers the feasibility of attaining the objectives within the available resources and timeframe. Relevance ensures that the objectives are pertinent to the research questions and the broader field of study. Finally, being time-bound means setting a clear timeline for achieving the objectives.

Aligning Objectives with Research Questions

Research objectives must align closely with the research questions. This alignment ensures coherence and consistency throughout the research process, linking the aims of the study with the methods and data collection strategies. Research questions guide the inquiry by focusing on specific issues, while objectives outline the steps needed to address these questions (Maxwell, 2013). For example, if a research question aims to explore how teachers perceive the integration of technology in the classroom, the corresponding objective might be to investigate specific experiences and attitudes of teachers toward technological tools. By aligning objectives with research questions, researchers can design a coherent and focused study that systematically addresses the key issues.

Designing the Research Framework

Choosing an Appropriate Qualitative Research Approach

Choosing an appropriate qualitative research approach is crucial for designing a robust research framework. The choice of approach depends on the nature of the research questions, the context of the study, and the type of data needed. Common qualitative research approaches include case study, ethnography, and grounded theory.

Case Study

A case study approach involves an in-depth examination of a single case or a small number of cases within a real-world context. This approach is particularly useful for exploring complex phenomena and gaining a deep understanding of specific instances (Yin, 2016). For example, a case study might investigate how a particular school implements inclusive education practices.

Ethnography

Ethnography focuses on studying cultures and communities by immersing the researcher in the daily lives of the participants. This approach aims to provide a detailed and nuanced understanding of social practices, behaviors, and interactions within a specific cultural context (Hammersley & Atkinson, 2019). An ethnographic study might explore the cultural practices of a remote village or the social dynamics within a corporate environment.

Grounded Theory

Grounded theory involves generating theory from data collected during the research process. This approach is iterative, with data collection and analysis occurring simultaneously to develop theoretical insights (Charmaz, 2014). Grounded theory is particularly useful for studies aiming to build new theories or concepts based on empirical evidence.

Outlining the Research Process and Stages

Outlining the research process and stages involves mapping out the key steps and phases of the study. This outline provides a structured plan for conducting the research, ensuring that all necessary activities are accounted for and sequenced appropriately (Creswell & Poth, 2018). Key stages typically include:

1. **Literature Review:** Conducting a thorough review of existing research to contextualize the study and identify gaps or areas for further investigation.
2. **Research Design:** Developing the research framework, including the selection of the qualitative approach and the formulation of research questions and objectives.
3. **Data Collection:** Implementing data collection methods such as interviews, focus groups, or observations.
4. **Data Analysis:** Analyzing the collected data using appropriate qualitative techniques to identify patterns, themes, and insights.
5. **Interpretation and Reporting:** Interpreting the findings in relation to the research questions and objectives and presenting the results in a clear and coherent manner.

Developing Data Collection Methods

Selecting Suitable Data Collection Techniques

Selecting suitable data collection techniques is essential for gathering rich and relevant data. Common qualitative data collection methods include interviews, focus groups, and observations.

Interviews

Interviews involve direct, one-on-one interactions between the researcher and the participant. They are useful for exploring participants' perspectives, experiences, and insights in depth. Interviews can be structured, semi-structured, or unstructured, depending on the research objectives. Structured interviews follow a predefined set of questions, while semi-structured interviews allow for more flexibility and probing. Unstructured interviews are more open-ended, enabling participants to guide the conversation.

Focus Groups

Focus groups involve group discussions with multiple participants, facilitated by the researcher. This method is effective for exploring collective views, generating interactive discussions, and identifying common themes. Focus groups can reveal diverse perspectives and the dynamics of group interactions, providing a richer understanding of the research problem.

Observations

Observations involve systematically watching and recording behaviors and interactions in natural settings. This method provides direct insights into how people behave and interact in specific contexts, offering valuable contextual and behavioral data. Observations can be participant or non-participant, depending on whether the researcher is actively involved in the setting or remains a passive observer.

Designing Data Collection Instruments

Designing data collection instruments involves creating tools that facilitate the systematic and consistent gathering of data. Common instruments include interview guides and observation protocols.

Interview Guides

An interview guide is a structured outline of topics and questions to be covered during an interview. It ensures that all relevant issues are addressed while allowing for flexibility and probing based on participants' responses. A well-designed interview guide includes open-ended questions that encourage detailed responses, prompts for probing deeper into specific areas, and a logical flow that facilitates a natural conversation.

Observation Protocols

An observation protocol is a structured format for recording observations in a systematic manner. It typically includes sections for noting the time, setting, participants, activities, and any relevant contextual information. A detailed observation protocol helps to ensure consistency and accuracy in recording observations, making it easier to analyze the data later.

Establishing a Timeline and Budget

Creating a Realistic Timeline for Each Stage of the Research

Establishing a realistic timeline is essential for ensuring that the research project stays on track and is completed within the desired timeframe. A timeline outlines the key milestones and deadlines for each stage of the research process, providing a structured plan for managing time effectively (Creswell & Poth, 2018). Key considerations in creating a timeline include:

1. **Literature Review:** Allowing sufficient time for a comprehensive review of existing research.
2. **Research Design:** Planning time for developing the research framework and obtaining any necessary approvals or permissions.
3. **Data Collection:** Scheduling time for recruiting participants, conducting interviews or observations, and transcribing data.
4. **Data Analysis:** Allocating time for coding and analyzing the data, identifying themes, and interpreting findings.
5. **Interpretation and Reporting:** Planning time for writing and revising the research report, including time for feedback and revisions.

Estimating and Managing the Budget for the Research Project

Estimating and managing the budget is crucial for ensuring that the research project is financially feasible and sustainable. A detailed budget outlines all anticipated costs and allocates resources appropriately, helping to avoid overspending and financial shortfalls (Mertens, 2014). Key budget considerations include:

1. **Personnel Costs:** Salaries or stipends for research assistants, transcribers, and any other personnel involved in the project.
2. **Travel and Fieldwork Expenses:** Costs for travel, accommodation, and subsistence for fieldwork or participant recruitment.
3. **Materials and Supplies:** Costs for any materials or supplies needed for data collection, such as recording equipment, transcription software, or printed materials.
4. **Data Management:** Costs for data storage, software licenses, and any other expenses related to managing and analyzing data.
5. **Dissemination and Reporting:** Costs for printing, publishing, or presenting the research findings, including conference fees or journal submission fees.

Effective budget management involves regularly monitoring expenses, making adjustments as needed, and ensuring that all costs are accounted for. By creating a detailed and realistic budget, researchers can ensure that their project is financially viable and that they can achieve their research objectives within the available resources.

Planning and structuring a qualitative research project involves setting clear and achievable research objectives, designing an appropriate research framework, developing robust data collection methods, and establishing a realistic timeline and budget. By carefully considering each of these elements, researchers can ensure that their study is well-organized, methodologically sound, and capable of producing valuable and credible findings.

DEVELOPMENT OF A QUALITATIVE RESEARCH PLAN

Developing a qualitative research plan is a meticulous process that requires careful consideration of various components to ensure a coherent and comprehensive approach to investigating the research problem. This section outlines the critical components of a research plan, the structure and content of a qualitative research proposal, and strategies for reviewing and refining the research plan.

Components of a Research Plan

Title and Introduction

The title of a research plan should succinctly reflect the main focus of the study. It needs to be clear, specific, and informative, providing potential readers with a good understanding of the research topic. The introduction sets the stage for the research, offering a brief overview of the problem under investigation, the significance of the study, and the research objectives. It should engage the reader, highlight the importance of the research, and provide context by situating the study within the broader field.

Literature Review

The literature review is a critical component of the research plan that synthesizes existing research related to the topic. It provides a theoretical framework, identifies gaps in the current knowledge, and justifies the need for the proposed study. A well-conducted literature review demonstrates the researcher's familiarity with the field and situates the new research within the existing body of knowledge. The review should be comprehensive yet focused, critically evaluating and summarizing relevant studies.

Research Questions and Hypotheses

Clearly articulated research questions and hypotheses guide the direction of the study. In qualitative research, research questions are often open-ended, aiming to explore processes, experiences, and meanings rather than testing specific hypotheses. These questions should be specific, researchable, and aligned with the study's objectives. Hypotheses, while more common in quantitative research, can also be included in qualitative studies to propose potential outcomes based on the literature review and theoretical framework.

Methodology

The methodology section outlines the research design and approach. It details the qualitative methods that will be employed, such as case studies, ethnography, or grounded theory. This section should include a description of the research setting, participant selection criteria, sampling strategies, and data collection methods (e.g., interviews, focus groups, observations). A well-defined methodology ensures the study's rigor and allows for replication or further exploration by other researchers.

Data Analysis Plan

The data analysis plan describes how the collected data will be processed and interpreted. It should outline the techniques and tools that will be used for analyzing qualitative data, such as coding, thematic analysis, or narrative analysis. The plan should also address how the data will be managed, including transcription, storage, and ensuring the confidentiality of participants. This section demonstrates the researcher's preparedness to systematically analyze and make sense of the data.

Ethical Considerations

Ethical considerations are paramount in qualitative research. This section should address issues such as informed consent, confidentiality, and the protection of participants from harm. Researchers must explain how they will obtain consent, ensure anonymity, and handle sensitive information. Ethical considerations also include the researcher's reflexivity, acknowledging their potential biases and the impact of their positionality on the research process.

Expected Outcomes and Contributions

The research plan should conclude with a discussion of the expected outcomes and contributions of the study. This section outlines the anticipated findings and their potential implications for theory, practice, and future research. It should highlight the study's significance and how it will contribute to the existing body of knowledge. Articulating the expected contributions helps to underscore the importance and relevance of the research.

Writing the Research Proposal

Structure and Content of a Qualitative Research Proposal

A well-structured research proposal is essential for clearly communicating the research plan to stakeholders such as funding bodies, ethics committees, and academic supervisors. The structure typically includes the following sections:

1. **Title Page:** Includes the title of the research, the researcher's name, and institutional affiliation.
2. **Abstract:** A brief summary of the research proposal, including the research problem, objectives, methodology, and expected outcomes.
3. **Introduction:** Provides background information, the significance of the study, and the research objectives.
4. **Literature Review:** Synthesizes relevant literature and identifies gaps that the study aims to address.
5. **Research Questions/Hypotheses:** Clearly states the research questions and, if applicable, the hypotheses.
6. **Methodology:** Details the research design, participant selection, data collection methods, and data analysis plan.
7. **Ethical Considerations:** Discusses how ethical issues will be addressed.
8. **Expected Outcomes and Contributions:** Outlines the anticipated findings and their implications.
9. **Timeline:** Provides a schedule for the research activities.
10. **Budget (if applicable):** Estimates the financial resources required for the study.

Tips for Writing a Compelling and Coherent Proposal

Writing a compelling and coherent research proposal involves clarity, coherence, and persuasiveness. Here are some tips:

- **Be Clear and Concise:** Use clear and concise language to articulate the research plan. Avoid jargon and complex sentences.
- **Logical Flow:** Ensure that the proposal has a logical flow, with each section building on the previous one. Transitions between sections should be smooth.
- **Evidence-Based:** Support your arguments with evidence from the literature. Demonstrating a thorough understanding of the field adds credibility to your proposal.
- **Engage the Reader:** Write in an engaging manner that captures the reader's interest. Highlight the significance and potential impact of your research.

- **Revise and Edit:** Carefully revise and edit your proposal to eliminate errors and improve clarity. Consider seeking feedback from peers or mentors.

Reviewing and Refining the Research Plan

Seeking Feedback from Peers and Advisors

Seeking feedback from peers and advisors is an essential step in refining a research plan. Feedback provides valuable insights and perspectives that the researcher may have overlooked. It can help identify weaknesses, suggest improvements, and validate the research approach. Engaging with a community of scholars through discussions, presentations, and peer reviews enhances the quality and rigor of the research plan.

Revising the Plan Based on Feedback and Practical Considerations

Revising the research plan based on feedback involves critically evaluating the suggestions and making necessary adjustments. This process ensures that the research plan is robust and feasible. Practical considerations, such as resource availability, timeline constraints, and ethical issues, should also be factored into the revisions. Flexibility and adaptability are crucial in refining the research plan to address any unforeseen challenges or changes in circumstances.

Developing a qualitative research plan is a comprehensive process that involves careful planning, structured writing, and iterative refinement. The components of a research plan, including the title and introduction, literature review, research questions, methodology, data analysis plan, ethical considerations, and expected outcomes, provide a solid foundation for conducting rigorous qualitative research. Writing a compelling research proposal and seeking feedback from peers and advisors further enhance the quality and coherence of the research plan. By following these steps, researchers can design and execute meaningful qualitative studies that contribute valuable insights to their fields.

ETHICAL ASPECTS OF QUALITATIVE RESEARCH IN BUSINESS

Ethical considerations in qualitative research are pivotal in ensuring the integrity, credibility, and social responsibility of the research process. This is particularly important in business research contexts, where the findings can have significant implications for various stakeholders, including companies, employees, customers, and the broader community. This section outlines the importance of ethics in qualitative research, focusing on key areas such as informed consent, confidentiality and anonymity, addressing ethical dilemmas, and compliance with institutional and legal requirements.

Importance of Ethics in Qualitative Research

Ethical Principles in Qualitative Research

Qualitative research often involves close interactions between researchers and participants, making ethical principles critical to the research process. Core ethical principles include respect for persons, beneficence, and justice (Beauchamp & Childress, 2013). Respect for persons entails recognizing the autonomy of participants and obtaining their informed consent. Beneficence involves maximizing benefits and minimizing harm to participants, while justice requires equitable treatment of all participants and ensuring that the burdens and benefits of research are distributed fairly.

Relevance of Ethics in Business Research Contexts

In business research, ethical considerations are particularly relevant due to the potential impact of findings on organizational practices and policies. Ethical research practices help build trust with participants, enhance the credibility of the research, and ensure that the results are respected and utilized appropriately by stakeholders. For instance, ethical breaches in research can damage a company's reputation, lead to legal repercussions, and negatively affect employees and customers (Bell & Bryman, 2007). Therefore, maintaining high ethical standards is crucial in business research to uphold the integrity and social responsibility of the research process.

Informed Consent

Designing Informed Consent Processes

Informed consent is a cornerstone of ethical research, ensuring that participants are fully aware of the nature, purpose, and potential risks of the study before agreeing to participate. The process of obtaining informed consent should be thorough and transparent, providing participants with all necessary information in a clear and comprehensible manner. Researchers should design consent forms that outline the study's objectives, procedures, potential risks, benefits, and the participants' rights, including the right to withdraw at any time without penalty.

Ensuring Participants' Understanding and Voluntary Participation

Ensuring that participants genuinely understand the information provided and that their participation is voluntary is crucial. Researchers should use plain language and avoid technical jargon when explaining the study. They should also provide opportunities for participants to ask questions and seek clarification. To confirm understanding, researchers can ask participants to summarize the information in their own words. Voluntary participation means that participants should not feel coerced or unduly influenced to take part in the study.

Confidentiality and Anonymity

Safeguarding Participants' Identities and Data

Protecting the confidentiality and anonymity of participants is essential in qualitative research. This involves safeguarding any personal information collected during the study and ensuring that participants cannot be identified in the published results. Researchers should implement robust data protection measures, such as secure data storage and restricted access to sensitive information. Anonymizing data, by removing or disguising identifying details, further protects participants' identities.

Strategies for Maintaining Confidentiality

Several strategies can help maintain confidentiality in qualitative research. Using pseudonyms or codes for participants during data collection and analysis prevents the association of data with specific individuals. Researchers should also be cautious when sharing data with colleagues or presenting findings, ensuring that identifying information is not inadvertently disclosed. Additionally, researchers must comply with data protection regulations, such as the General Data Protection Regulation (GDPR) in the European Union, which mandates strict protocols for handling personal data (European Commission, 2018).

Addressing Ethical Dilemmas

Common Ethical Challenges in Qualitative Research

Qualitative researchers frequently encounter ethical dilemmas, such as balancing the need for in-depth data with respect for participants' privacy, or managing conflicts of interest. For instance, a researcher might discover information that suggests unethical behavior within a company, posing a dilemma between maintaining confidentiality and reporting the issue. These challenges require careful consideration and a nuanced approach to ethical decision-making.

Approaches to Resolving Ethical Dilemmas

To address ethical dilemmas, researchers should rely on ethical guidelines and frameworks provided by professional associations and institutional review boards (IRBs). Engaging in reflexive practice, where researchers continually reflect on their values, biases, and the impact of their research, can help navigate complex ethical situations. Seeking advice from peers, mentors, and ethics committees can also provide valuable perspectives and guidance. Ultimately, researchers should aim to act in a manner that upholds the ethical principles of respect, beneficence, and justice.

Compliance with Institutional and Legal Requirements

Adhering to Institutional Review Board (IRB) Guidelines

Compliance with IRB guidelines is a fundamental aspect of conducting ethical research. IRBs review research proposals to ensure that they meet ethical standards and protect participants' rights and welfare. Researchers must submit detailed protocols outlining their research methods, informed consent procedures, and data protection measures. Adhering to IRB guidelines not only ensures ethical compliance but also enhances the credibility and legitimacy of the research (Emanuel et al., 2000).

Understanding and Complying with Relevant Legal Regulations

Researchers must also be aware of and comply with legal regulations governing research, such as data protection laws and ethical standards specific to their field. In business research, this includes adhering to corporate governance standards and industry-specific ethical codes. Understanding these regulations is crucial for avoiding legal repercussions and ensuring that the research is conducted responsibly and ethically. Researchers should stay informed about any changes in relevant laws and continuously update their practices to remain compliant.

Ethical considerations in qualitative research are paramount, especially in business contexts where the implications of research findings can significantly impact various stakeholders. Key ethical principles, such as respect for persons, beneficence, and justice, guide researchers in designing and conducting studies that uphold the rights and dignity of participants. Ensuring informed consent, safeguarding confidentiality and anonymity, addressing ethical dilemmas, and complying with institutional and legal requirements are critical components of ethical research practice. By adhering to these principles and practices, researchers can conduct robust and socially responsible qualitative research that contributes valuable insights to the field of business.

CONCLUSION

The development of a well-designed qualitative research plan is essential for conducting robust, credible, and ethically sound research in the field of business. This chapter has explored several critical aspects that contribute to the formulation and execution of such a plan.

Recap of Key Points Discussed in the Chapter

The chapter began by emphasizing the importance of ethics in qualitative research, especially within business contexts where the implications of research findings can significantly impact stakeholders. Ethical principles such as respect for persons, beneficence, and justice were highlighted as fundamental guidelines for ensuring the integrity and social responsibility of the research process.

Informed consent was discussed as a cornerstone of ethical research, requiring thorough and transparent processes to ensure participants are fully aware of the study's nature, purpose, and potential risks before agreeing to participate. The necessity of ensuring participants' understanding and voluntary participation was underscored.

Confidentiality and anonymity were identified as crucial for protecting participants' identities and data. Strategies such as using pseudonyms, secure data storage, and compliance with data protection regulations were recommended to safeguard confidentiality.

The chapter addressed common ethical challenges in qualitative research, recognizing that researchers often face dilemmas that require careful ethical decision-making. Approaches such as relying on ethical guidelines, engaging in reflexive practice, and seeking advice from peers and mentors were suggested for resolving these dilemmas.

Finally, compliance with institutional and legal requirements was discussed as an integral aspect of conducting ethical research. Adhering to Institutional Review Board (IRB) guidelines and understanding relevant legal regulations were highlighted as necessary steps for ensuring ethical compliance and enhancing the credibility of the research.

Emphasis on the Importance of a Well-Designed Research Plan

A well-designed research plan serves as the foundation for successful qualitative research. It ensures that the research process is systematic, coherent, and ethically sound. A comprehensive research plan includes clearly articulated research questions, a detailed methodology, robust data analysis strategies, and thorough ethical considerations.

By meticulously planning each aspect of the research, researchers can anticipate potential challenges, address ethical issues proactively, and ensure that their findings are credible and valuable. A well-designed research plan also facilitates effective communication of the research objectives, processes, and outcomes to stakeholders, enhancing the transparency and legitimacy of the study.

Encouragement for Researchers to Be Diligent and Ethical in Their Research Design Process

In conclusion, it is imperative for researchers to be diligent and ethical in their research design process. The integrity of qualitative research hinges on the researcher's commitment to ethical principles and meticulous planning. Researchers should continually reflect on their values, biases, and the potential impact of their research, striving to conduct studies that respect participants' rights, maximize benefits, and contribute meaningfully to the field.

Ethical diligence in research design not only protects participants but also enhances the credibility and societal relevance of the research. By adhering to high ethical standards and rigorously planning their studies, researchers can ensure that their work stands up to scrutiny and makes a positive impact on business practices and policies.

In summary, the development of a qualitative research plan is a complex but essential task that requires careful attention to ethical principles, methodological rigor, and practical considerations. Researchers are encouraged to approach this process with diligence, integrity, and a commitment to ethical research practices, thereby contributing valuable insights and advancing knowledge in the field of business.

REFERENCES

Beauchamp, T. L., & Childress, J. F. (2013). *Principles of biomedical ethics* (7th ed.). Oxford University Press.

Bell, E., & Bryman, A. (2007). The ethics of management research: An exploratory content analysis. *British Journal of Management, 18*(1), 63–77. DOI: 10.1111/j.1467-8551.2006.00487.x

Charmaz, K. (2014). *Constructing grounded theory* (2nd ed.). SAGE Publications.

Creswell, J. W., & Poth, C. N. (2018). *Qualitative inquiry and research design: Choosing among five approaches* (4th ed.). SAGE Publications.

Emanuel, E. J., Wendler, D., & Grady, C. (2000). What makes clinical research ethical? *Journal of the American Medical Association, 283*(20), 2701–2711. DOI: 10.1001/jama.283.20.2701 PMID: 10819955

European Commission. (2018). *General Data Protection Regulation (GDPR)*. Retrieved from https://ec.europa.eu/info/law/law-topic/data-protection_en

Hammersley, M., & Atkinson, P. (2019). *Ethnography: Principles in practice* (4th ed.). Routledge. DOI: 10.4324/9781315146027

Lincoln, Y. S., & Guba, E. G. (1985). *Naturalistic inquiry*. SAGE Publications. DOI: 10.1016/0147-1767(85)90062-8

Mason, M. (2010). Sample size and saturation in PhD studies using qualitative interviews. *Forum Qualitative Sozialforschung/Forum: Qualitative. Social Research, 11*(3).

Maxwell, J. A. (2013). *Qualitative research design: An interactive approach* (3rd ed.). SAGE Publications.

Merriam, S. B., & Tisdell, E. J. (2016). *Qualitative research: A guide to design and implementation* (4th ed.). Jossey-Bass.

Mertens, D. M. (2014). *Research and evaluation in education and psychology: Integrating diversity with quantitative, qualitative, and mixed methods* (4th ed.). SAGE Publications.

Patton, M. Q. (2015). *Qualitative research & evaluation methods: Integrating theory and practice* (4th ed.). SAGE Publications.

Ravitch, S. M., & Riggan, M. (2017). *Reason & rigor: How conceptual frameworks guide research* (2nd ed.). SAGE Publications.

Yin, R. K. (2016). *Case study research and applications: Design and methods* (6th ed.). SAGE Publications.

Chapter 3
Selection of Participants in Qualitative Research

Débora Araújo Severiano
Universidade Presbiteriana Mackenzie, Brazil

Ana Lucia Fontes de Souza Vasconcelos
Universidade Federal de Pernambuco, Brazil

Octavio Ribeiro Mendonça Neto
https://orcid.org/0000-0002-6123-6733
Universidade Presbiteriana Mackenzie, Brazil

ABSTRACT

This chapter, according to the authors, aims to provide a comprehensive and practical overview of participant selection in qualitative research. It explores the importance of careful participant selection to ensure relevant and representative data, highlighting purposive and snowball sampling strategies. It discusses the ethical challenges of research, including informed consent and confidentiality, and emphasizes the need for flexibility as new data and contexts emerge. Additionally, it addresses the use of digital technologies and social media for participant recruitment, promoting the inclusion of diverse and underrepresented voices to enrich the understanding of phenomena. The chapter also suggests that novice and experienced researchers consider small or pilot contexts to explore different sampling methods, preparing to adjust their approaches as new insights and interdisciplinary collaborations emerge.

DOI: 10.4018/979-8-3373-4432-4.ch003

1. SELECTION OF PARTICIPANTS IN QUALITATIVE RESEARCH

1.1 Objective

The objective of this chapter is to provide a comprehensive and practical overview of the selection of participants in qualitative research. In this sense, the theoretical and methodological approaches involved are discussed, presenting practical examples and guidelines for researchers in different fields of study.

1.2 Introduction

Qualitative research is a methodological approach used in the social sciences and other disciplines that seek to understand the complexity of human behaviors, beliefs, and social interactions within their natural environments and contexts. Unlike quantitative research, which focuses on the quantification of data and statistical techniques for analysis, qualitative research emphasizes the depth and richness of the collected data, allowing for an in-depth exploration of phenomena from the perspectives and experiences of the participants. This approach enables a detailed understanding of the subjective and contextual aspects that influence the reality being studied. (Creswell, 2007; Denzin & Lincoln, 2011)

One of the most critical aspects that requires special attention in qualitative research is the selection of participants. Careful selection of participants ensures that the most relevant and diverse voices are considered, providing a comprehensive and authentic view of the phenomenon under investigation (Marshall & Rossman, 2014). Selecting the appropriate participants enriches the quality of the collected data and strengthens the validity and credibility of the results. A judicious selection ensures that the study achieves its objectives by capturing a diverse range of perspectives that reflect the complexity of the phenomenon being studied. (Patton, 2014).

1.3 Selection criteria

The clear definition of selection criteria is an important step in conducting qualitative research. These criteria determine who will be included or excluded from the study, directly influencing the quality and relevance of the collected data (Miles et al., 1994).

Inclusion criteria are the characteristics that participants must have to be considered eligible to participate in the study. These criteria ensure that the selected individuals possess the necessary experiences or attributes to provide valid and relevant data. On the other hand, exclusion criteria are the characteristics or conditions that disqualify potential participants. These criteria help avoid the collection of data that may distort the results or that would not significantly contribute to the objectives of the research (Morse & Richards, 2002; Creswell, 2007).

Table 1 summarizes some practical examples in different contexts.

Table 1. Examples of inclusion and exclusion criteria.

Study	Example of Inclusion Criteria	Example of Exclusion Criteria
Mental Health	Individuals diagnosed with clinical depression in the past six months.	Individuals currently in acute mental health crisis or those who do not consent to recorded interviews.
Educational Study on Teaching Methods	Teachers with at least five years of experience in elementary education.	Educators who have recently changed subjects or schools and may lack consistent context.
Sociological Research on Immigration	Immigrants who arrived in the new country in the past five years and are aged between 18 and 45.	Immigrants in legal irregularity who prefer not to participate due to safety concerns.
Market Study on Technology Consumption Habits	Smartphone users aged 20 to 35 who regularly make online purchases.	Users who have not accessed the internet in the last month.

Source: Prepared by the author.

These examples highlight how selection criteria can be adapted to meet the specific needs of a study, ensuring the collection of relevant and representative data. Furthermore, they illustrate the importance of flexibly adjusting the criteria as new information and contexts emerge during the course of the research.

In qualitative research, the flexibility and adaptation of selection criteria throughout the investigation process help capture the complexity and dynamics of the phenomena being studied. This continuous adjustment of criteria allows the study to remain relevant and comprehensive while responding to new discoveries and emerging directions. (Strauss & Corbin, 1998; Charmaz, 2006).

Flexibility is important for the following aspects:

- **Responses to Initial Discoveries:** At the beginning of a qualitative research study, researchers may encounter unexpected information or trends that require a reconsideration of the inclusion and exclusion criteria. Being flexible allows for the incorporation of valuable new insights that can enhance the richness of the data and the understanding of the phenomenon being studied.
- **Adaptation to Context and Variabilities:** Social phenomena and human interactions are complex and dynamic. Adapting the selection criteria may be necessary to accommodate these variabilities and ensure that the research continues to reflect the lived reality of the participants.

1.4 Sampling in Qualitative Research

Sampling in qualitative research differs significantly from quantitative research in that it focuses on obtaining in-depth, detailed insights into a phenomenon rather than generalizing findings to a larger population (Creswell, 2007).

The choice of the sample is therefore more subjective and driven by the purpose of the study and the nature of the data that the researcher intends to collect. Qualitative sampling strategies include purposive sampling, theoretical sampling, and snowball sampling, each with its own particularities and applications (Patton, 2014).

a) Purposive sampling

Purposive sampling is a central strategy in qualitative research, where participants are selected based on specific characteristics that best enrich the study. The researcher looks for individuals who have experiences, perspectives, or knowledge that are highly relevant to the research question. (Patton, 2014).

This approach ensures that the data collected is rich and relevant, allowing for a deeper understanding of the phenomenon under study. For example, in a study on the experiences of parents of autistic children, the researcher would deliberately choose parents who live with this reality, rather than randomly selecting (Palinkas et al., 2015).

b) Theoretical sampling

Primarily used in research developed under the Grounded Theory perspective, theoretical sampling involves the continuous selection of participants based on emerging categories during data analysis. The process is iterative, allowing the researcher to modify the selection of participants as new theories or patterns emerge. This can lead to a study where the initial sample is small but expands as necessary to saturate theoretical categories. For example, if certain unexpected teaching strategies emerge during initial research on pedagogical practices, the study may expand to include more educators who use these practices. (Strauss & Corbin, 1998; Charmaz, 2006).

c) Snowball sampling

This technique is predominant when researching hard-to-reach populations or when initial contacts with the target group are limited. In snowball sampling, existing participants recruit future subjects from among their acquaintances. This approach is particularly useful in studies involving marginalized communities or contexts where individuals may be invisible or hesitant to identify themselves. For example, in a study on undocumented immigrants, initial participants, once trust is established, may refer other members of their communities who have relevant experiences to share (Goodman, 1961).

1.5 Comparison between probability and non-probability sampling

The fundamental distinction between qualitative and quantitative sampling lies in their probabilistic and non-probabilistic approaches, respectively. In probabilistic sampling, common in quantitative studies, each member of a defined population has a known and non-zero chance of being selected. This allows for the generalization of results to the entire population. In contrast, non-probabilistic sampling, typical of qualitative research, does not apply this idea of probability, as the focus is on the depth and richness of the collected data rather than its statistical representativeness. (Creswell, 2007).

The choice of sampling methodology reflects the fundamentally different objectives of qualitative research, which aims to explore and understand complex phenomena rather than predict or measure specific variables within a larger population (Patton, 2014).

The value of non-probability sampling lies in its ability to reveal nuances and details that may be lost in purely quantitative approaches. While quantitative research may focus on generalizations and patterns, the qualitative approach through non-probability sampling can capture subtle details and perspectives that are essential for a deep understanding of the phenomenon being studied (Flick, 2022).

1.6 Sample Size

In qualitative research, determining the sample size is about finding a balance between having enough participants to ensure that the phenomenon is well understood and avoiding the collection of redundant data (Guest et al., 2006).

Unlike quantitative research, where the sample size is often calculated according to rigorous statistical principles, qualitative research is guided by principles of sufficiency and data saturation (Creswell, 2007).

The ideal sample size in qualitative research is not established by mathematical formulas, but by careful judgments based on the study's objectives, data collection methods, and the nature of the phenomenon being investigated. The sample should be sufficiently comprehensive to capture the variability of experiences and perspectives while also managing practical resources such as time and access to participants (Mason, 2010).

The researcher should evaluate the scope of the study and the specific objectives, also considering the depth and richness of the necessary data. Data saturation is a key concept in determining the sample size for qualitative studies. Saturation is reached when the collection of new data no longer yields significant new insights or enhances understanding of the phenomenon under study. During this stage, patterns and categories begin to repeat, indicating that the researcher has gathered enough data on the topic (Creswell, 2007; Patton, 2014).

Throughout the qualitative research process, the researcher must be prepared to adjust the sample size as necessary. Initially, a preliminary estimate of the sample size may be made, but it is common for this estimate to evolve as data collection and analysis progress (Charmaz, 2006; Creswell, 2007; Patton, 2014).

Flexibility is crucial, as new themes emerge, and the researcher may need to include more participants to explore these areas or cease data collection if saturation is reached sooner than expected. This adaptability helps ensure that the study remains focused and relevant, aligning the sampling with the dynamic objectives of the investigative process (Patton, 2014; Patricia et al., 2015).

1.7 Diversity and Representation

The diversity and representativeness of participants are fundamental aspects of qualitative research, as they ensure that multiple perspectives on the phenomenon being studied are included. The heterogeneity of the sample enriches the analysis and lends legitimacy to the results by capturing the inherent complexity of social, cultural, and behavioral issues (Miles et al., 1994).

The diversity of participants in a qualitative study allows for the reflection of a variety of experiences and backgrounds associated with the phenomenon under investigation. Including participants from different backgrounds, genders, ages, ethnicities, and socioeconomic contexts enables the researcher to examine how these factors influence perceptions and behaviors. Diversity fosters a more robust and inclusive analysis, resulting in insights that are relevant and applicable across various contexts, while also avoiding oversimplifications or biased conclusions (Denzin & Lincoln, 2011).

To ensure representativeness in qualitative research, researchers can adopt various strategies. First, clear inclusion criteria that align with the study's objectives must be defined to ensure the incorporation of different relevant subgroups (Miles et al., 1994; Creswell, 2007; Denzin & Lincoln, 2011).

Targeted sampling techniques, such as purposive sampling and snowball sampling, are useful for identifying and recruiting participants who bring diverse perspectives. Additionally, involving intermediaries or community leaders can facilitate access to underrepresented populations. It is also vital for the researcher to maintain an open and flexible attitude, willing to adjust methods to include emerging and challenging voices. While diversity enhances the depth of research, it also involves significant challenges and ethical considerations. One of the main challenges is reaching and recruiting participants from communities or groups that are difficult to access, which may require additional time and resources (Creswell, 2007; Denzin & Lincoln, 2011).

When selecting participants, it is essential to respect ethical principles such as informed consent, confidentiality, and respect for the rights of the involved subjects. Researchers should be sensitive to power dynamics that may affect an individual's willingness or ability to participate. Furthermore, it is important to ensure that representation does not marginalize or stigmatize minority or underrepresented groups, approaching the study in a way that respects their dignity and contributes to beneficial outcomes (Patton, 2014).

1.8 Recruitment of Participants

Recruiting participants is a fundamental step in conducting qualitative research, as it determines the quality and relevance of the collected data (Heckathorn, 1997). Successful recruitment helps obtain a diverse and representative sample and facilitates an effective and ethical data collection process.

There are several effective approaches to recruiting participants in qualitative research. One common method is direct recruitment, which involves approaching potential participants through personal invitations, phone calls, or emails. Additionally, placing advertisements in relevant locations, such as on university bulletin boards, health clinics, or workplaces, can attract interested individuals (Heckathorn, 1997; Creswell, 2007). Partnerships with organizations or community groups are also valuable strategies, allowing these entities to introduce the study and recommend eligible participants within their networks (Creswell, 2007; Denzin & Lincoln, 2011).

Social media has emerged as a powerful tool for recruiting participants, enabling access to a vast and diverse audience quickly and cost-effectively. Specific online groups, hashtags, and paid advertisements can target specific interest audiences (Patton, 2014).

Partnerships with institutions such as schools, non-governmental organizations (NGOs), and community centers are also effective, as these institutions often have access to specific target populations and can facilitate reliable introductions and recruitment. Respecting and understanding the dynamics and norms of communities is essential when attempting to recruit participants from well-defined or culturally distinct groups. Participant recruitment should always be conducted with attention to ethical issues. Ensuring that the process is transparent and respectful is fundamental (Denzin & Giardina, 2016).

Potential participants should be fully informed about the nature, objectives, and possible implications of the study before agreeing to participate, always promoting informed consent. It is essential to ensure the confidentiality and anonymity of participants, safeguarding their identities and information (Denzin & Lincoln, 2011; Patton, 2014).

In practice, researchers should also consider the logistical and resource impact during recruitment, ensuring accessibility and convenience for participants, which includes considering schedules, locations, and formats that respect their needs and limitations (Creswell, 2007; Patton, 2014).

1.9 Ethical Considerations

Ethical considerations are indispensable in any research, especially in qualitative studies, where direct involvement with participants can raise complex issues related to privacy, consent, and data use. Protecting the rights and well-being of participants is a priority and must be carefully managed at all stages of the research process (Orb et al., 2001).

Informed consent is a central ethical principle that ensures participants fully understand the nature, purpose, and procedures of the study before agreeing to participate. The process involves clearly presenting information about what participation will entail, including any potential risks or discomforts, the participants' rights to withdraw at any time without penalty, and the methods of communicating the research results (Orb et al., 2001; Denzin & Lincoln, 2011; Patton, 2014).

Researchers must provide this information in an understandable manner and ensure that participants have the opportunity to ask questions and receive satisfactory answers. Documenting consent, either in writing or through recordings, is common practice and confirms that consent was obtained in an ethical and transparent manner. Protecting the confidentiality and privacy of participants is another essential factor in building trust and ensuring the ethical use of shared information. Researchers should commit to not disclosing identities or personal information that could identify participants unless explicitly consented to. Strategically, this may include the use of pseudonyms, data coding, or anonymous collection (Orb et al., 2001; Creswell, 2007; Patton, 2014).

During the dissemination of results, precautions must be taken to ensure that the narrative or presented data do not allow for the indirect identification of participants. Protecting privacy is a continuous commitment that persists beyond the data collection phase, encompassing the storage and disposal of data (Orb et al., 2001;Creswell, 2007; Patton, 2014).

Managing sensitive data ethically requires special care due to the delicate nature of the information that may be collected in qualitative research. Some areas of study, such as mental health, abuse, or discrimination, often involve data that, if neglected, can cause harm to participants. Researchers should establish clear guidelines for how this data will be handled, securely stored, and who will have access to it (Orb et al., 2001; Denzin & Lincoln, 2011; Sanjari et al., 2014).

Minimizing risk is essential, and the implementation of robust security measures, such as data encryption and restricted access controls, is fundamental. Additionally, developing a clear plan for the disposal of data after the study is completed—which should be communicated to participants—is a recommended practice for the ethical management of sensitive data (Orb et al., 2001; Flick, 2022).

CONCLUSION

For novice researchers, it is recommended that they start by gaining a solid understanding of the ethical and methodological foundations in participant selection, ensuring that the principles of informed consent and confidentiality are prioritized. Exploring different sampling methods in small practical or pilot contexts can provide valuable learning about their applicability and challenges (Orb et al., 2001).

For experienced researchers, delving into innovative and technological approaches to recruitment can expand the possibilities for engagement and access to new study groups. Both novices and experienced researchers should be open to continuous learning and the adaptation of methods as new insights emerge during the research. Additionally, promoting interdisciplinary collaborations and maintaining ongoing dialogue with the academic community about emerging practices can enrich their methodological and ethical approaches in the long term (Creswell, 2007; Patton, 2014).

In a constantly evolving methodological landscape, future trends in participant selection for qualitative research should incorporate technological advancements and new inclusion strategies (Denzin & Giardina, 2016).

With the help of digital tools, the use of social media platforms for recruitment and engagement with virtual communities is expanding, providing access to more diverse and geographically dispersed groups. Furthermore, the growing emphasis on interdisciplinary research may require a more integrated approach in sample selection that captures multiple aspects of human experiences (Patton, 2014).

Participatory and collaborative methods are also emerging, where participants play an active role not only as data sources but also as co-creators of knowledge, redefining traditional recruitment of subjects. Another innovative aspect is the increasingly conscious inclusion of marginalized and underrepresented voices to produce more complete and empathetic narratives (Charmaz, 2006).

REFERENCES

Charmaz, K. (2006). Constructing grounded theory: A practical guide through qualitative analysis. *Sage (Atlanta, Ga.).*

Creswell, J. W. (2007). Qualitative inquiry and research design : Choosing among five traditions. *Sage (Atlanta, Ga.).*

Denzin, N. K., & Giardina, M. D. (2016). *Qualitative inquiry through a critical lens*. Routledge. DOI: 10.4324/9781315545943

Denzin, N. K., & Lincoln, Y. S. (2011). *The Sage handbook of qualitative research.* Sage.

Flick, U. (2022). *An introduction to qualitative research.*

Goodman, L. A. (1961). Snowball sampling. The Annals of Mathematical Statistics. *Annals of Mathematical Statistics, 32*(1), 148–170. DOI: 10.1214/aoms/1177705148

Guest, G., Bunce, A., & Johnson, L. (2006). How Many Interviews Are Enough?: An Experiment with Data Saturation and Variability. *Field Methods, 18*(1), 59–82. DOI: 10.1177/1525822X05279903

Heckathorn, D. D. (1997). Respondent-driven sampling-A new approach to the study of hidden populations. *Social Problems, 44*(2), 174–199. DOI: 10.2307/3096941

Mason, M. (2010). *Sample Size and Saturation in PhD Studies Using Qualitative Interviews*. https://www.qualitative-research.net/

Miles, M. B., Huberman, A. M., & Saldaña, J. (1994). *Qualitative Data Analysis A Methods Sourcebook Edition.*

Morse, J. M., & Richards, Lyn. (2002). Readme first for a user's guide to qualitative methods. *Sage (Atlanta, Ga.).*

Orb, A., Eisenhauer, L., & Wynaden, D. (2001). Ethics in qualitative research. *Journal of Nursing Scholarship, 33*(1), 93–96. DOI: 10.1111/j.1547-5069.2001.00093.x PMID: 11253591

Palinkas, L. A., Horwitz, S. M., Green, C. A., Wisdon, J. P., Duan, N., & Hoagwood, K. (2015). Purposeful sampling for qualitative data collection and analysis in mixed method implementation research. *Administration and Policy in Mental Health, 42*(5), 533–544. DOI: 10.1007/s10488-013-0528-y PMID: 24193818

Patricia, I., Ph, D., & Ness, L. R. (2015). *Are We There Yet? Data Saturation in Qualitative Research.* Walden Faculty and Staff Publications. https://scholarworks.waldenu.edu/facpubs/455

Patton, M. Q. (2014). *Qualitative research & evaluation methods: Integrating theory and practice.* Sage publications.

Sanjari, M., Bahramnezhad, F., Khoshnava Fomani, F., Sho-Ghi, M., & Cheraghi, M. A. (2014). Ethical challenges of researchers in qualitative studies: the necessity to develop a specific guideline. In *J Med Ethics Hist Med* (Vol. 7).

Strauss, A., & Corbin, J. (1998). *Basics of qualitative research techniques.*

Chapter 4
The Role of Technology in Modern Qualitative Research:
Tools, Software, and Emerging Trends

Mustafa Kayyali
https://orcid.org/0000-0003-3300-262X
Maaref University of Applied Sciences, Syria

ABSTRACT

The integration of technology into qualitative research has transformed traditional methods, bringing researchers fresh tools and approaches for data gathering, analysis, and presentation. This chapter addresses the crucial role of technology in modern qualitative research, with an emphasis on tools, software, and rising trends that are altering the discipline. It begins by studying the growth of qualitative research methods and the progressive usage of technology. The presentation then covers major technology breakthroughs, including transcribing software, qualitative data analysis tools, and digital collaboration platforms. Emerging technologies such as artificial intelligence (AI), virtual reality (VR), and cloud-based research environments are explored for their potential to advance qualitative inquiry. Alongside these benefits, ethical questions and challenges linked with technology use in qualitative research are also addressed.

DOI: 10.4018/979-8-3373-4432-4.ch004

INTRODUCTION

Qualitative research has long been acknowledged for its potential to deliver rich, in-depth insights into complicated human behaviors, experiences, and social phenomena (Lim, 2024). Unlike quantitative research, which often focuses on numerical data and statistical analysis, qualitative research is concentrated on investigating the meanings, motivations, and settings underlying individual or group activities. Through methods such as interviews, focus groups, participant observation, and content analysis, qualitative researchers have historically tried to capture the complexity and nuances of social life (Mohajan, 2018). In recent years, however, the landscape of qualitative research has been significantly revolutionized by the arrival of technology. The introduction of digital tools and software has transformed how researchers collect, analyze, and interpret qualitative data. What was formerly a time-consuming and frequently labor-intensive process is now more streamlined, efficient, and, in many respects, more smart. From advanced transcribing software to powerful qualitative data analysis (QDA) tools, technology has introduced new possibilities for qualitative researchers to deepen their investigations and enhance their findings (Olapane, 2021). The impact of technology on qualitative research extends beyond the sheer convenience of data storage and processing. It has transformed the entire research process, from data gathering to delivering outcomes. Digital tools have made it possible for researchers to conduct studies remotely, reach a bigger and more diverse range of participants, and access data in ways that were previously imagined. The integration of technologies such as artificial intelligence (AI), machine learning, and natural language processing into qualitative research is pushing the boundaries of what can be achieved, enabling researchers to analyze vast amounts of textual or visual data with unprecedented speed and accuracy (Girju, 2023). While technology offers numerous advantages, it also introduces new obstacles and ethical considerations. The reliance on digital tools brings up difficulties surrounding data security, privacy, and the potential for bias in algorithm-driven analysis. Moreover, as new technologies continue to emerge, researchers must remain knowledgeable about how to properly integrate these tools into their methodological frameworks without compromising the quality of their research.

This chapter intends to analyze the changing role of technology in qualitative research, with a special focus on tools, software, and emerging trends that are shaping the field. We will investigate how technology is being employed in many stages of qualitative research, from data collection and transcription to analysis and reporting. By studying these technological innovations, we hope to provide a complete knowledge of how they are revolutionizing qualitative research, particularly within the framework of business studies. The chapter is structured as follows. First, we will analyze the historical evolution of qualitative research methodologies, tracking the move from conventional procedures to the adoption of technology in research practice. Next, we will discuss major technology breakthroughs in qualitative research, focusing on software tools that assist with data collecting, processing, and visualization. We will also analyze the rising trends in technology-enhanced qualitative research, including the use of AI, virtual reality, and cloud-based platforms. Ethical considerations will be addressed in a dedicated section, highlighting the need to sustain research integrity in the digital age. To anchor our discussion in real-world applications, we will present numerous case studies from the business sector, highlighting how technology is being leveraged to address complicated research concerns. Finally, we will end by commenting on the future of technology in qualitative research, describing the difficulties and opportunities that lie ahead.

THE EVOLUTION OF QUALITATIVE RESEARCH METHODS

Qualitative research has a long and rich history, evolving over the decades from a rather obscure and non-mainstream approach to a powerful and fundamental method in social sciences and business research (Tshabangu & Madondo, 2020). Initially, qualitative methods were considered subjective and occasionally untrustworthy, with a large dependence on the researcher's interpretation and a more intuitive approach to data processing. However, over the years, qualitative research has undergone a transition, spurred by breakthroughs in theory, technique, and technology. Early qualitative research methods, such as ethnography and case studies, evolved as tools for researchers to investigate social phenomena in-depth, providing a deeper knowledge of human behavior and society systems. These methods were anchored in the premise that to properly grasp a phenomenon, one had to immerse oneself within it—observing, experiencing, and interpreting it in context. In the 19th and early 20th centuries, individuals like Max Weber and Emile Durkheim shaped the foundations of qualitative research by proposing methodologies to examine human experience through precise, subjective data (Yılmaz, 2024). They helped build a paradigm for understanding complex social systems, cultures, and organizations.

With this early framework set, qualitative research methodologies began to take shape more distinctly in the 20th century. One of the most prominent early movements in qualitative research was symbolic interactionism, pioneered by scientists like George Herbert Mead and Herbert Blumer (Quist-Adade, 2019). They stressed the role of symbols and language in creating human connection, and their work gave rise to methodologies like participant observation and in-depth interviews, which are still key instruments in qualitative research today. These methods were developed to acquire data that captured the complexity of human experience, extending beyond simplistic quantitative metrics to gain deeper insights into social reality. As qualitative research gained importance, so too did the approaches used to collect and analyze data. One of the most notable advances was the advent of grounded theory in the 1960s (McCall & Edwards, 2021). Grounded theory, established by Barney Glaser and Anselm Strauss, revolutionized qualitative research by offering a systematic approach to data collecting and theory construction. It introduced the idea of building theory directly from the data rather than applying preexisting theoretical frameworks, enabling researchers to identify patterns and linkages spontaneously. This technique signaled a change towards a more structured and rigorous approach to qualitative research, offering researchers tools for evaluating and interpreting huge amounts of unstructured qualitative data.

Another major development during this period was the establishment of phenomenology as a distinct qualitative technique. Inspired by the work of Edmund Husserl and later refined by researchers such as Maurice Merleau-Ponty and Alfred Schutz, phenomenology focuses on comprehending the lived experiences of individuals from their viewpoints (Deep, 2020). By using in-depth interviews and careful observation, phenomenologists aimed to reveal the essence of human experience and provide vivid explanations of how people make sense of their worlds. This method brought a dimension of depth to qualitative research, emphasizing the subjective character of human experience and the necessity of understanding individuals' perspectives and interpretations. Alongside phenomenology, story inquiry evolved as another significant qualitative method (Nigar, 2020). Narrative inquiry stresses the storytelling aspect of human experience, noting that people typically make meaning of their lives through tales. By studying these narratives, researchers gain insight into how individuals construct meaning and identity, which is especially important in fields like business, where understanding personal and organizational stories can provide rich insights into decision-making, leadership, and organizational culture. Narrative inquiry has become a useful tool for understanding the intricacies of human behavior and has been employed extensively in numerous areas.

While these qualitative methodologies were reshaping the landscape of research, the 1980s and 1990s saw a huge shift in how qualitative research was regarded and used (Morse, 2020). With the rise of critical theory and postmodern ideas, qualitative research began to take on a more reflexive and critical posture. Scholars like Michel Foucault and Pierre Bourdieu questioned established power structures and narratives, asking academics to investigate how information is formed and who holds the power in producing that knowledge. This theoretical change gave rise to new approaches such as discourse analysis, which focuses on the ways language is used to generate social realities and power dynamics. The increased understanding of the subjectivity of both the researcher and the researched prompted qualitative researchers to look more critically at the role of the researcher's perspective in the research process (Dodgson, 2019). This led to the creation of more transparent and self-reflexive techniques, which emphasize the necessity for researchers to assess their own biases, preconceptions, and positionality in connection to the research. These advancements served to establish qualitative research as a more rigorous and thoughtful discipline, one that appreciates subjectivity and context while preserving a commitment to systematic inquiry.

With the turn of the 21st century, technological improvements began to impact qualitative research in important ways (Kraus et al, 2019). The digital age brought about new means for data collecting and analysis, such as online surveys, social media analysis, and the usage of digital audio and video capture. The internet opens up new possibilities for reaching participants in rural or hard-to-reach regions, enabling researchers to access more diverse and worldwide populations. This enhanced accessibility and diversity in research subjects brought fresh insights into social behavior and organizational dynamics, notably in the business domain. The digital transformation of qualitative research also extended to data analysis. Traditional qualitative data analysis approaches, such as thematic analysis and content analysis, were updated to accommodate larger and more complex datasets (Vaismoradi & Snelgrove, 2019). Software technologies like NVivo, Atlas.ti, and MAXQDA allowed academics to easily code and analyze huge amounts of qualitative data (de Wildt, 2020). These technologies featured advanced capabilities like as text mining, sentiment analysis, and automatic coding, allowing for more efficient processing and deeper analysis of data. The creation of these software tools substantially boosted the rigor and efficiency of qualitative research, enabling researchers to examine data more methodically while yet keeping the richness of qualitative inquiry. As qualitative research has evolved, it has increasingly adopted mixed-methods approaches (Dawadi et al, 2021). These approaches incorporate both qualitative and quantitative methods to create a more holistic view of research topics. Mixed-methods research enables the depth and complexity of qualitative data to be complemented by the objectivity and generalizability of quantitative data. This combination has shown to be especially beneficial in business research, where both qualitative insights into consumer behavior and quantitative market data are vital for comprehending the complexity of consumer decision-making and organizational strategy.

Today, qualitative research methodologies continue to evolve as they respond to the ever-changing demands of study disciplines, including business, health, and education (Gupta et al, 2021). The increased availability of big data, for example, gives new opportunities for qualitative researchers to study large-scale qualitative datasets, such as open-ended survey replies, social media interactions, and customer feedback. As data science tools such as machine learning and natural language processing are merged into qualitative research, new pathways for analysis and interpretation are opening up. Researchers may now employ algorithms to find patterns in enormous amounts of unstructured data, yielding insights that were previously unreachable. The growth of qualitative research methodologies has been distinguished by a steady adaptation to changing sociological, theoretical, and technical circumstances. From its early days as a primarily descriptive and subjective approach, qualitative research has matured into a sophisticated and rigorous methodology that combines a wide range of methods and tools. Today, qualitative research continues to expand, embracing digital tools and mixed methodologies approaches, while being founded on the values of deep understanding, context, and human experience. As the discipline advances forward, the integration of classic qualitative methodologies with developing technologies will surely affect the future of research, enabling new approaches to investigate and understand the intricacies of human behavior, organizational dynamics, and social phenomena.

KEY TECHNOLOGIES SHAPING MODERN QUALITATIVE RESEARCH

Over the past few decades, the landscape of qualitative research has been changed by significant improvements in technology (Thunberg & Arnell, 2022). From the methods used to collect data to how researchers evaluate and interpret that data, technology has provided new possibilities, making the research process more efficient, sophisticated, and inclusive. These technology advances have had a major impact on how qualitative research is conducted across numerous fields, including business, social sciences, education, and healthcare. As the instruments for qualitative research continue to advance, they are not only boosting the depth of analysis but also broadening the scope and accessibility of research. One of the most significant technological breakthroughs impacting qualitative research is the development of digital instruments for data collecting (Tracy, 2024). Traditionally, qualitative data collecting methods like interviews, focus groups, and ethnographic observations relied significantly on manual processes—researchers would take notes, record audio, and occasionally transcribe lengthy hours of interviews by hand. While these procedures have been effective for millennia, they can be time-consuming and prone to human error. The emergence of digital recording technologies, such as high-quality audio recorders and digital cameras, has eased data collection, allowing researchers to focus more on the interaction with participants rather than the mechanics of gathering information.

With the rise of digital platforms, the scope of qualitative data collecting has also grown beyond traditional face-to-face contact (Gill & Baillie, 2018). Online surveys, video conferencing, and social media platforms have become useful sources of qualitative data. For example, researchers can now conduct interviews or focus groups utilizing platforms like Zoom, Skype, or Google Meet, erasing geographical boundaries that traditionally limited participation (Howlett, 2022). Online discussion forums, social media posts, and even online reviews give a variety of qualitative data that might offer insight into consumer behavior, public opinion, or societal trends. This transition to online platforms has not only enhanced the accessibility of research for participants, but has also made it easier for researchers to contact various populations from different regions of the world. This democratization of research is particularly beneficial in a worldwide world where knowing a wide range of perspectives is essential.

Beyond digital data gathering, the introduction of qualitative data analysis software has greatly impacted the way qualitative data is processed (Mezmir, 2020). Software such as NVivo, ATLAS.ti, MAXQDA, and Dedoose give strong capabilities for organizing, coding, and analyzing enormous volumes of textual, audio, and visual data (Olson, 2021). Before these tools, qualitative researchers were limited by the manual processes of coding data by hand and organizing it in physical files, which might become daunting, especially for huge datasets. Now, these software applications allow for the systematic management of data, enabling researchers to go through vast volumes of interview transcripts, focus group discussions, or social media posts with ease. The usefulness of these technologies has grown dramatically in recent years. NVivo, for example, now offers a range of qualitative and mixed-methods analysis techniques, such as content analysis, theme analysis, and sentiment analysis (Elliott, 2022). It also interfaces with other tools, such as survey software, and has the flexibility to manage both qualitative and quantitative data in the same project. The use of such technologies helps academics to detect patterns in data, extract relevant themes, and develop insights that are often missed with human analysis. More importantly, these technologies enable transparency and reproducibility in research, ensuring that the processes done in the data analysis process are apparent and can be examined or repeated by others.

Improvements in machine learning and artificial intelligence (AI) have opened up new opportunities for qualitative data analysis (Longo, 2019). Natural language processing (NLP), a subset of AI, is particularly crucial for processing and analyzing huge textual datasets. With NLP tools, researchers may analyze complicated language patterns in a far more effective way than ever before. For instance, sentiment analysis techniques can help determine the emotional tone behind words used in interviews, social media posts, or consumer reviews. This is particularly important in business research, as knowing client sentiment is key to influencing marketing tactics, upgrading products, or enhancing customer service. Machine learning algorithms can also be used to categorize and tag qualitative data more rapidly and accurately (Ebrahim et al, 2020). While traditional coding approaches require researchers to manually assign categories or topics to distinct areas of data, machine learning algorithms can learn from the data itself and uncover patterns that human researchers would overlook. This allows for a more extensive examination and can reveal insights into enormous datasets that would otherwise be difficult to manage. For example, in qualitative business research, AI-powered technologies can assist researchers in recognizing emerging trends from customer feedback, social media conversations, or product evaluations, delivering real-time insights into market dynamics.

Another area where technology is making an influence is in the field of transcribing. Transcription software has gone a long way in recent years, with tools like Otter.ai and Trint providing automatic transcription services that can translate spoken words into text with high levels of accuracy (Gaber et al, 2020). These technologies use voice recognition technology to transcribe interviews, focus groups, and other oral data, saving researchers a significant amount of time and effort. Although transcribing software is not flawless and still requires human monitoring to assure correctness, it represents a huge leap forward in expediting the qualitative data processing process. With these technologies, researchers may focus on evaluating the content of the interviews rather than spending hours physically transcribing.

In addition to transcription tools, data visualization technologies are increasingly playing a vital part in qualitative research (Ntsobi et al, 2024). Visualization tools help researchers to depict complex data in a more digestible and interpretable format, making it easier to communicate findings to stakeholders, clients, or the broader research community. For instance, qualitative data analysis tools like NVivo may generate visual representations of data, such as word clouds, graphs, and network maps, that assist researchers uncover patterns and connections between themes. These visual tools make the data more accessible and intelligible, allowing researchers to explain their findings more engagingly and successfully. The integration of digital technologies with qualitative research has also altered how researchers retain and disseminate their findings. Cloud computing has transformed the way qualitative data is stored and accessible. Cloud-based storage systems like Dropbox, Google Drive, and OneDrive allow researchers to store their data securely and access it from anywhere, as long as they have an internet connection (Ramya & Rajeswari, 2023). This flexibility not only makes cooperation easier—especially for academics working in teams across different locations—but it also allows for safer storage of sensitive data, as cloud services often incorporate robust security features such as encryption and password protection.

Cloud-based technologies enable real-time collaboration on qualitative research initiatives (Cresswell et al, 2022). Using platforms like Google Docs or Microsoft Teams, researchers can work together on analyzing data, coding transcripts, or creating reports, regardless of their physical location (Choukaier, 2024). This capacity to work smoothly in real-time is especially vital for multi-disciplinary and international research teams, as team members often bring diverse knowledge to the table. The continued development of immersive technologies, such as virtual reality

(VR) and augmented reality (AR), holds promise for future qualitative research applications (Baikulova & Suderevskaia, 2019). VR and AR offer new ways to collect data through simulation and real-time interaction, generating more dynamic environments for users. For example, VR can imitate real-world scenarios, allowing researchers to watch how individuals respond in controlled yet genuine contexts. This can be particularly beneficial in disciplines like business, psychology, and healthcare, where researchers need to study how individuals make decisions or interact with their surroundings in a more realistic context. As the discipline of qualitative research continues to expand, it's apparent that technology will play an increasingly essential role. Researchers currently have access to a large array of digital tools and platforms that make it easier to acquire, analyze, and interpret qualitative data (Lobe et al, 2020). These tools not only enhance the efficiency and rigor of qualitative research but also bring up new options for examining complicated human behavior, social phenomena, and organizational dynamics. From digital transcription tools to AI-powered analysis, technology is transforming qualitative research, making it more accessible, efficient, and insightful than ever before. As these technologies continue to improve, the future of qualitative research seems tremendously exciting, with new methodologies and tools continuously emerging to assist researchers in confronting the challenges of an increasingly digitized and linked world.

SOFTWARE SOLUTIONS FOR DATA COLLECTION AND ANALYSIS

The evolution of technology has completely altered how data is collected and evaluated in qualitative research (Hesse et al, 2019). Software tools have developed as crucial companions for researchers, helping them to rapidly gather, analyze, and understand huge amounts of qualitative data (Tracy, 2024). Whether researchers are conducting in-depth interviews, focus groups, ethnographic studies, or monitoring social media trends, software solutions provide both structure and flexibility, allowing for deeper insights and more robust conclusions. As technology progresses, so too does the capability of these tools, which are increasingly smart and diverse, capable of handling the complexities and nuances inherent in qualitative data. One of the key benefits of employing software for data collecting and analysis is the efficiency it gives (Wang et al, 2018). Traditional techniques of data collecting, such as manual note-taking during interviews or focus groups, can be time-consuming and prone to inaccuracy. Transcribing interviews by hand, classifying vast volumes of textual material, and systematically organizing conclusions all involve significant effort. Software solutions ease these pressures by automating many of the repetitive activities connected with data gathering and analysis, freeing researchers to focus on interpretation and theory-building (Übellacker, 2024). These tools not only save time but also reduce the possibility of human error, which is especially significant when dealing with enormous datasets. When it comes to data gathering, the digital era has introduced a wide array of instruments that make it easier to record and preserve data. Many software solutions are built to work fluidly across many platforms, enabling researchers to collect data from different sources, whether it's through online surveys, digital interviews, or social media platforms. One of the most popular software options for qualitative data collection is Qualtrics (Weber, 2021). This robust platform is frequently used for developing and distributing surveys, collecting replies, and organizing data. Qualtrics helps researchers to adapt surveys to match the specific demands of their research and acquire data from varied demographics quickly and effectively. The platform also contains powerful analytics capabilities that assist researchers in interpreting the responses, making it an all-in-one solution for qualitative data collecting and analysis.

Another tool that has made considerable progress in qualitative data collection is Dedoose (Sampson & Wong, 2023). This program is particularly valuable for researchers undertaking mixed-methods studies, as it allows them to examine both qualitative and quantitative data within the same platform. Dedoose is noted for its intuitive interface, which makes it easy to manage massive datasets and undertake extensive analysis. It is used to code and categorize text, audio, and video files, and it gives visual analytics tools, like as bar graphs and pie charts, to assist researchers uncover patterns and trends in their data. What sets Dedoose distinct is its focus on cooperation, allowing numerous academics to work together on the same subject, regardless of location (Lieber et al, 2021). This is particularly valuable for research teams dispersed across different areas or countries, as it promotes real-time collaboration and assures that all team members can access and contribute to the data analysis process. For researchers conducting interviews or focus groups, platforms like Otter.ai have changed the process of transcription. Otter.ai employs advanced speech recognition technology to transcribe spoken words into text rapidly and correctly (Seyedi et al, 2023). While transcription software has been around for some time, Otter.ai delivers a high level of precision and ease of use, making it a go-to choice for qualitative researchers. The software is capable of transcribing audio from numerous sources, such as Zoom meetings, recorded interviews, and voice notes, and it can even distinguish different speakers within the recording. This feature is vital for researchers who need to safeguard the integrity of their data and ensure accurate representation of all voices involved in a discourse. Once data is collected, the challenge switches to evaluating it. Qualitative data analysis software has become indispensable for academics who need to code enormous amounts of text, detect themes, and make sense of complicated, unstructured data. NVivo is one of the most extensively used software tools for qualitative data analysis, and it is a favorite among academics in a range of disciplines. NVivo is a package of tools meant to aid researchers with activities such as coding, categorizing, and visualizing qualitative data (Mortelmans, 2019). It supports a wide range of file kinds, including text, music, video, and even social network data.

NVivo's flexibility to handle many forms of data makes it a perfect solution for researchers who need to analyze interviews, focus group discussions, and other sorts of unstructured data simultaneously (Ireland, 2020). The software allows researchers to readily find recurring patterns and themes, classify data into categories, and even show the relationships between different codes and concepts. NVivo includes extensive tools for constructing visual representations of data. For example, researchers can generate word clouds, maps, and charts that highlight the frequency of certain terms or themes across their dataset. These visual tools are invaluable for recognizing trends, making data more digestible, and presenting findings more compellingly. Furthermore, NVivo's capacity to allow cooperation makes it a good alternative for research teams working on huge projects. Multiple team members can work on the same project concurrently, making real-time adjustments and sharing their thoughts without having to send files back and forth. This collaboration aspect is particularly useful for qualitative research initiatives that involve varied viewpoints and require input from multiple stakeholders.

Another prominent qualitative data analysis tool is ATLAS.ti (Kalpokaite & Radivojevic, 2020). This software is notable for its versatility, allowing researchers to evaluate a range of data formats, including text, audio, and video. ATLAS.ti is often used by academics in the social sciences, education, and business, and it includes a wide array of tools for coding, categorizing, and visualizing data. Like NVivo, ATLAS.ti enables team-based collaboration, ensuring that multiple researchers can contribute to the analysis process in real-time (Niedbalski & Ślęzak, 2023). One of the main aspects of ATLAS.ti is its powerful coding capabilities. Researchers can design complicated code systems to categorize their data, and the program enables hierarchical coding, meaning that codes can be divided into categories, subcategories, and sub-subcategories (Rädiker & Kuckartz, 2020). This structure is particularly beneficial when working with enormous datasets that demand a high degree of organization. For academics interested in a cheaper choice, MAXQDA is another qualitative data analysis program that includes powerful functionality for data management and analysis. MAXQDA supports a wide range of data kinds, including text, audio, video, and graphics, and it is noted for its user-friendly interface (Liu & Eagan, 2021). The software allows researchers to code and categorize their data, make memos to capture discoveries during the study process and generate visualizations to uncover trends and linkages. MAXQDA also has a major focus on mixed-methods research, enabling researchers to blend qualitative and quantitative data in a single analysis (Marjaei et al, 2019). This makes it a popular choice for researchers conducting studies that require both forms of data to provide a more comprehensive picture of the research issue.

The introduction of cloud-based technologies has also had a huge impact on qualitative research. Cloud-based solutions like Google Drive and Dropbox allow researchers to save, share, and collaborate on their data remotely (Andriani et al, 2019). These platforms make it easier for research teams to work together, especially when team members are in various places. Cloud-based solutions also provide a level of security that is often stronger than traditional file storage systems, making it easier to preserve sensitive research data. As technology continues to progress, software solutions for qualitative research are getting more integrated and more capable. Researchers now have access to tools that not only automate many aspects of data collecting and processing but also reveal crucial insights that were earlier difficult to uncover (Stieglitz et al, 2018). Whether it's through the use of AI-powered transcription software, advanced coding tools, or cloud-based collaboration platforms, technology is playing an increasingly essential role in determining the future of qualitative research. These tools have altered the research process, making it faster, more efficient, and more accessible to scholars throughout the world. As these technologies continue to improve, the potential for qualitative research to deliver even deeper and more extensive insights into human behavior, social trends, and organizational dynamics will only expand, creating new options for researchers in every field.

EMERGING TRENDS IN TECHNOLOGY-ENHANCED QUALITATIVE RESEARCH

In recent years, qualitative research has been undergoing a dramatic transition, driven mostly by the rapid expansion of technology. What was formerly a field that depended mostly on traditional procedures like face-to-face interviews, focus groups, and manual coding of data, has now been transformed by a range of cutting-edge technologies. These tools and platforms have not only made the research process more efficient but have also opened up new opportunities for analyzing complex data, boosting cooperation, and gaining deeper insights into human behavior, social dynamics, and organizational processes. As qualitative researchers continue to investigate and embrace new technological solutions, it's apparent that technology is transforming the future of the discipline in profound ways. One of the most prominent developing developments in technology-enhanced qualitative research is the use of artificial intelligence (AI) and machine learning (Novawan et al, 2024). These technologies have begun to play a key role in data collecting, analysis, and interpretation. AI-powered technologies are capable of automating many of the labor-intensive operations that researchers typically had to conduct manually, such as transcribing audio or video data, coding text, and recognizing patterns within vast datasets (Zhang, 2023). What's particularly fascinating about AI is its capacity to analyze qualitative data at scale, which was historically a huge hurdle for researchers. For example, AI can swiftly evaluate thousands of interview transcripts or focus group conversations, discovering common themes or keywords, and even revealing insights that researchers may not have initially considered. This form of automation not only saves time but also lowers the potential for human bias in the analysis process.

Natural Language Processing (NLP), a subset of AI, is one of the primary technologies being utilized to augment qualitative research (Guetterman et al, 2018). NLP allows computers to perceive, interpret, and generate human language, which is invaluable for evaluating textual data. Researchers can utilize NLP technologies to perform sentiment analysis, discover significant phrases, and extract valuable insights from enormous amounts of text (Cheligeer et al, 2023). For instance, researchers conducting studies on social media platforms can use NLP to evaluate huge amounts of user-generated content, such as posts, comments, and tweets, to uncover trends, attitudes, and patterns in real-time. This has opened up new options for qualitative research, particularly in subjects like marketing, sociology, and political science, where understanding public opinion and consumer behavior is vital.

Another trend that is gaining momentum is the use of virtual and augmented reality (VR and AR) technologies in qualitative research (Baikulova & Suderevskaia, 2019). VR and AR have typically been associated with gaming or entertainment, but they are increasingly being employed in research settings to create immersive environments for data collecting. For instance, VR can be used to recreate real-world events in which participants can interact with a virtual environment, providing researchers with a more accurate understanding of how people react in various contexts. This can be particularly valuable for research involving human behavior, such as those linked to decision-making, education, or healthcare. VR can also be used to conduct virtual focus groups or interviews, allowing researchers to connect with participants more dynamically and interactively, even when they are located in different areas of the world (Wu et al, 2022). AR has the potential to enhance the way qualitative data is collected and processed. AR overlays digital information onto the actual environment, allowing people to engage with both physical and digital objects concurrently. In a research context, AR can be used to develop interactive surveys, where participants are asked to reply to questions or perform tasks while engaging with digital objects inside their physical environment. This technique not only makes the data collection process more interesting for participants but also offers new options for collecting data in complicated situations, such as metropolitan spaces, museums, or healthcare settings.

Social media platforms and online communities are also playing an increasingly essential role in qualitative research (Kozinets, 2019). In the past, qualitative researchers generally collected data from in-person interviews or focus groups, but today, much of the interaction and participation happens online. Researchers are now turning to platforms like Twitter, Facebook, Instagram, and Reddit to collect qualitative data, using social media material to evaluate public mood, trends, and behaviors. Social media analytics tools are rapidly being used to watch and analyze online interactions, offering real-time data on everything from consumer preferences to political beliefs. One of the primary advantages of using social media as a data source is the capacity to acquire vast volumes of data from varied demographics. It also allows researchers the chance to investigate the dynamics of online communities, such as how groups form around certain hobbies, ideas, or causes. The shift toward internet platforms for conducting interviews and focus groups is another key trend in qualitative research. While face-to-face contact is vital for certain forms of qualitative research, many researchers are now employing video conferencing systems like Zoom, Skype, or Microsoft Teams to conduct interviews remotely (Archibald et al, 2019). This has been particularly crucial during the COVID-19 pandemic when in-person research became increasingly problematic. Remote interviews have the advantage of being more flexible and convenient for both researchers and participants (Mirick & Wladkowski, 2019). They also provide for greater geographical reach, enabling researchers to access people from other regions or even countries. In addition to this, online platforms offer unique capabilities, such as automatic recording, transcription services, and digital whiteboards, which enhance the data collection process.

The integration of cloud-based tools in qualitative research has further broadened the possibilities for cooperation and data exchange. With cloud systems like Google Drive, Dropbox, and Microsoft OneDrive, researchers can save, organize, and exchange data in real-time, regardless of where they are situated (Ramya & Rajeswari, 2023). This has made collaboration on major research projects extremely easy, particularly for teams located across multiple geographic areas. It also guarantees that data is securely saved and immediately available to all team members, eliminating the danger of losing crucial research materials. Cloud platforms have also made it easier for researchers to handle massive datasets, as they can access and analyze their data from numerous devices, whether they are working in the field, at home, or in the office (Langmead & Nellore, 2018). The increased emphasis on data visualization is another trend that is affecting the way qualitative researchers report and interpret their findings. Data visualization technologies allow researchers to portray qualitative data in visual formats, such as graphs, charts, and maps, making complex material easier to grasp and analyze (Grant, 2018). These methods are particularly beneficial when dealing with enormous datasets, as they allow researchers to uncover patterns and links more quickly. In qualitative research, visual data can also assist researchers explain their findings to stakeholders, making it simpler to express crucial insights in a way that is both interesting and accessible. Platforms like Tableau and Power BI allow researchers to construct interactive dashboards that provide a more dynamic perspective of the data, enabling users to explore different elements of the research and interact with the data in real-time (Martucci, 2024).

Lastly, the rise of big data is having a tremendous impact on qualitative research (Gupta & Rani, 2019). While qualitative research has historically concentrated on smaller, more focused datasets, the rise of big data has offered new opportunities for academics to study huge amounts of qualitative information. Using the use of new data analytics tools, researchers can now examine massive volumes of text, audio, and video data, discovering insights that would have been difficult, if not impossible, to identify using previous approaches. Big data tools are also making it easier to incorporate qualitative and quantitative data, enabling researchers to undertake more comprehensive studies that provide a fuller, more nuanced understanding of their research issues. The growing trends in technology-enhanced qualitative research are transforming the way researchers approach data collecting, analysis, and interpretation (Nicolaou, 2021). AI, machine learning, virtual and augmented reality, social media analytics, remote interviewing tools, cloud-based platforms, data visualization, and big data are just some of the technical breakthroughs that are transforming the field. As technology continues to improve, the future of qualitative research will be distinguished by even greater prospects for efficiency, cooperation, and insight generation. These developments are not only boosting the quality of qualitative research but are also making it more accessible, enabling researchers to engage with broader and more diverse populations, collect richer data, and find deeper insights into human behavior and social dynamics.

ETHICAL CONSIDERATIONS IN TECHNOLOGY USE FOR QUALITATIVE RESEARCH

As the incorporation of technology into qualitative research continues to develop, researchers are provided with unique opportunities to gather, analyze, and interpret data in once inconceivable ways (Bryda & Costa, 2023). These technological breakthroughs, from artificial intelligence to virtual reality, offer new pathways for acquiring insights into human behavior, culture, and social dynamics. However, with these opportunities come substantial ethical challenges that must be addressed to respect the rights, dignity, and privacy of participants. Ethical considerations in the use of technology in qualitative research are crucial to ensuring that technology is utilized ethically, transparently, and in a way that aligns with established research ethics norms. One of the major ethical difficulties when employing technology in qualitative research is providing informed permission (Wangmo et al, 2019). In classic qualitative research methods, obtaining informed permission is a basic process, where researchers describe the goal of the study, the type of data being collected, and how that data will be utilized. However, when technology is included, particularly when dealing with remote data-collecting methods such as video conferencing, online surveys, or social media platforms, the dynamics of informed consent can become more complex. Participants may not fully comprehend how their data will be handled, especially if it is obtained using AI-driven techniques or big data analytics.

Participants in virtual focus groups or online interviews may not be aware that their facial expressions or tones of voice are being evaluated by AI algorithms to assess sentiment or emotional responses. Similarly, when using social media data, the distinctions between public and private information can blur, and it becomes more difficult for participants to manage how their data is utilized. Researchers must be proactive in ensuring that participants are completely informed of how their information will be used and that they express explicit consent for both the collection and analysis of their data. This is particularly crucial when using modern technologies such as AI, because the extent and depth of analysis may not be immediately obvious to participants. The issue of privacy is another key ethical dilemma when employing technology in qualitative research. In an increasingly digitized environment, personal data is often acquired without individuals' express understanding or agreement. Whether through social media platforms, mobile apps, or online surveys, massive volumes of personal information are collected and kept. Researchers must be cautious in securing this data, ensuring that it is anonymized and securely maintained and that participants' privacy is not compromised at any point in the research process.

For instance, while collecting qualitative data using social media, researchers must be careful of the fact that many users may not understand the outcomes of revealing personal information on public platforms (Taylor & Pagliari, 2018). Even though the data may be publically available, the ethical obligation still falls with the researcher to guarantee that the data is utilized properly and that participants' identities and personal details are not disclosed. Additionally, when adopting technologies like VR or AR, researchers must evaluate the possibility of invasion of privacy, especially when participants are participating in immersive environments (O'Hagan et al, 2023). Researchers must guarantee that these tools do not expose sensitive or personal information without specific agreement. Another ethical issue connected to the use of technology in qualitative research is the potential for data tampering and misunderstanding (Mills, 2018). Technologies like AI and machine learning algorithms can process massive amounts of data, but they are only as good as the data fed into them (Tyagi & Chahal, 2020). Researchers must be vigilant about how they design their studies and interpret the data to prevent unwittingly introducing bias or distorting outcomes. For instance, if AI technologies are used to code qualitative data, there is a risk that the algorithms may misinterpret the meaning of specific words or phrases, leading to erroneous findings. This is particularly significant when evaluating sensitive themes, such as mental health, discrimination, or trauma, where the intricacies of language are critical to understanding the context.

The employment of AI in qualitative research raises difficulties concerning responsibility (Cheng et al, 2021). If an AI tool misinterprets or erroneously categorizes data, who is liable for the error?

While AI can deliver valuable insights, it is ultimately up to the researcher to guarantee that the findings are accurate, dependable, and ethically sound (Lim, 2024). This obligation goes not only to the design and implementation of the research but also to the interpretation of the data. Researchers must be clear about the tools they are employing and the limits of those instruments. Ethical research mandates that academics critically analyze the outputs of AI and other technologies and guarantee that their conclusions are based on the lived experiences and voices of the participants. In addition to these considerations, the use of technology in qualitative research brings ethical challenges relating to accessibility and inclusivity. While technology can democratize research by making it more accessible to various populations, it can also create hurdles for particular groups of participants (Seskir et al, 2023). For example, some persons may not have access to essential equipment, such as smartphones or dependable internet connections, to participate in online surveys or virtual interviews. Others may lack the technical abilities required to browse digital platforms or interact with sophisticated technology like VR or AR (Ullah et al, 2018). Researchers must be conscious of these hurdles and take steps to ensure that their studies are inclusive and that all potential participants have an equal opportunity to participate. This can involve providing alternative means to participate in the research or offering technical support to people who may struggle with the technology.

Equally crucial is the ethical issue of the digital imprint left by participants when engaged in technology-driven qualitative research (Mogaji et al, 2024). With the development of online research methods, such as email interviews, social media data collecting, and virtual focus groups, there is an inherent risk of leaving traces of personal information across the internet. These digital footprints can be traced, evaluated, and potentially exploited by third parties, making it necessary for researchers to ensure that participants' digital privacy is secured. This can be especially hard when data is obtained from public platforms or digital environments where the limits between private and public domains are often blurry. Researchers must take steps to anonymize data and eliminate identifiable details wherever possible.

The growing expansion of big data and the usage of cloud computing in qualitative research create further ethical challenges. Big data can afford researchers access to an unparalleled variety of information, but it also raises issues about permission and the possibility of over-surveillance. When huge datasets are collected, particularly from online platforms, it becomes increasingly difficult to track individual permission, especially when data is pooled from several sources. Researchers must take care to guarantee that participants' privacy is protected and that their data is not being used for reasons beyond the scope of the original consent (Xu et al, 2020).

Another worry is the influence that the employment of advanced technologies might have on the participants themselves (Brougham & Haar, 2018). While technologies such as VR and AR can create immersive experiences for participants, they also have the potential to be overwhelming or unsettling. For example, VR simulations intended to examine emotional or behavioral responses could provoke uncomfortable memories or sensations in certain participants, especially when studying sensitive themes such as trauma or addiction. Researchers must be attentive in monitoring the well-being of participants and provide appropriate support if any emotional or psychological disturbance develops. Ethical behavior mandates that researchers prioritize the safety and well-being of participants at all times (Olson, 2023). As technology continues to advance, the ethical landscape in qualitative research must also adapt. Researchers must be educated about the newest breakthroughs in technology and regularly analyze the ethical implications of new tools and procedures. This needs a continual commitment to ethical behavior and a desire to engage in open discourse about the difficulties and obligations that come with employing technology in research. By developing a culture of ethical awareness and accountability, researchers may ensure that technology enhances the quality of qualitative research while respecting the rights and dignity of participants.

The use of technology in qualitative research gives many benefits, but it also presents a range of ethical concerns. From obtaining informed permission and protecting privacy to resolving issues of data manipulation, accessibility, and participant well-being, researchers must be careful in their ethical considerations. By addressing technology with a dedication to transparency, responsibility, and respect for participants, researchers can pass these hurdles and use technology to enhance the depth, accuracy, and impact of qualitative research. As technology continues to grow, it will be vital for researchers to stay agile, ethical, and proactive in addressing the many ethical difficulties that arise in this constantly changing context.

CASE STUDIES: APPLICATIONS OF TECHNOLOGY IN BUSINESS QUALITATIVE RESEARCH

In recent years, organizations throughout the world have increasingly turned to technology to better their research methodologies and obtain deeper insights into customer behavior, market trends, and organizational dynamics. Technology, when integrated into business qualitative research, can revolutionize how data is acquired, analyzed, and understood. By delivering more efficient, cost-effective, and thorough research tools, technology can unleash new perspectives and empower businesses to make better-informed decisions. To comprehend the full potential of technology in business qualitative research, let's look at numerous case studies that highlight its real-world applications.

One noteworthy example is the use of sentiment analysis in consumer research (Hartmann et al, 2023). Sentiment analysis, powered by artificial intelligence (AI) and natural language processing (NLP), is increasingly being utilized by enterprises to assess how customers feel about their products or services. In a case study by a worldwide retail corporation, sentiment analysis was applied to examine hundreds of social media postings, customer reviews, and forum discussions. By processing this unstructured data, the organization was able to obtain insights into consumer emotions, views, and attitudes toward their brand. This system allowed the organization to obtain real-time input on new introductions, identify growing patterns, and spot possible consumer issues before they escalated. The qualitative data obtained from these sources was crucial in developing the company's marketing tactics, product development, and customer service initiatives.

Another case study involves the use of virtual reality (VR) for examining consumer behavior in a simulated retail setting (Meißner et al, 2020). A multinational electronics manufacturer launched a VR-based study to assess how customers engage with their store layout and product displays. In this immersive environment, people could stroll through a virtual store and interact with objects as if they were physically present. Researchers could then observe their behavior, such as which things drew their interest, how long they spent on various displays, and how they made purchasing decisions. The qualitative data obtained through this technology provided extensive insights into consumer preferences, buying behaviors, and decision-making processes (Akter et al, 2019). This research finally helped the company restructure its shop layouts and optimize product placement to enhance sales and customer happiness. In a more organizational setting, corporations have also employed AI and machine learning algorithms to analyze interview transcripts and focus group conversations. In one case study undertaken by a renowned management consultancy, AI techniques were applied to assess qualitative data from employee interviews addressing organizational culture. The firm was working with a large client that was experiencing a significant transformation, and knowing the company's culture was important to establishing a successful change management approach. By utilizing machine learning techniques to code and analyze the transcripts, the researchers were able to discover major themes and patterns that arose across the interviews. These themes helped the consultancy comprehend employees' opinions of leadership, communication, and collaboration within the firm. This technology not only sped up the study process but also found insights that may have been ignored in traditional, manual analysis.

A third case study brings us to the domain of market research, where technology is increasingly being employed to assist in-depth customer insights (Kumar et al, 2021). A famous market research business joined with a mobile app developer to collect data on consumer behavior in real-time. The software was designed to track users' shopping patterns, including the products they viewed, how long they spent browsing, and the variables that influenced their purchasing decisions. This data was then evaluated to find patterns, preferences, and pain areas in the customer journey. By merging this real-time qualitative data with other sources of information, such as customer surveys and internet reviews, the firm was able to give the client a comprehensive knowledge of their target market. The app also allowed for continuous data collecting, meaning the research was continuing, generating significant insights that could be utilized to change marketing strategies, improve product offers, and adapt user experiences. In the field of human resources and organizational development, qualitative research methodologies are also being utilized to strengthen leadership development programs. A major telecoms business deployed an AI-driven platform to analyze employee feedback from leadership assessments, performance reviews, and 360-degree feedback surveys (Chakraborty et al, 2024). The platform gave extensive data on common patterns that were forming across multiple departments and teams, delivering a sophisticated view of the leadership attributes that employees valued the most. This data was then utilized to inform the company's leadership development programs, ensuring that they were aligned with the expectations and needs of their staff. The use of technology makes it easier to discover trends and patterns in employee feedback, leading to more targeted and successful leadership development programs.

Another significant application of technology in business qualitative research is through internet ethnography, particularly in comprehending global customer behavior (Moisander et al, 2020). A renowned luxury fashion brand conducted online ethnography research to discover how its target demographic was interacting with the brand across multiple geographies. The research team employed complex data mining tools to evaluate internet content, including blogs, social media posts, and video reviews, to capture consumers' true perspectives about the brand. The technology allowed the researchers to examine enormous amounts of qualitative data from multiple cultural contexts, providing a more holistic view of how consumers in different markets regarded the brand. This insight was crucial in helping the business alter its marketing strategy to better resonate with varied consumer segments, boosting both engagement and loyalty. As organizations grow increasingly worldwide, another application of technology in qualitative research is in the study of cross-cultural consumer behavior. One famous case study featured a multinational food company using video interviews to get insights from clients in different countries. Using a cloud-based video conferencing tool, researchers conducted virtual focus groups with participants from several countries to discuss their food preferences, cultural influences, and shopping habits. The technology allowed the researchers to perform in-depth qualitative studies without the requirement for overseas travel, which would have been costly and time-consuming. The video data acquired was evaluated to reveal cross-cultural variances in attitudes toward food, which ultimately guided the company's product development and marketing tactics for different countries.

Technology has been a game-changer in the field of qualitative research by enabling real-time data analysis and cooperation (Green Jr, 2022). One multinational management consulting firm employed cloud-based qualitative research software to simplify the analysis of interviews and focus group data. Researchers from diverse locations were able to work in real-time, coding and analyzing the data concurrently. This program allowed the team to swiftly discover developing themes and patterns, which could then be shared with clients for prompt action. The efficiency of this technology not only lowered the time needed for data analysis but also increased the accuracy of the conclusions by allowing multiple academics to cooperate on the same dataset. This study demonstrates how cloud-based tools are revolutionizing qualitative research by making it more collaborative, transparent, and efficient.

These case studies highlight the numerous and revolutionary ways that technology is being utilized in qualitative research in business. From sentiment analysis and virtual reality to AI-driven interview coding and mobile data gathering, technology is transforming how businesses gather and interpret qualitative data. By giving new opportunities for in-depth analysis, real-time insights, and cross-cultural understanding, technology is boosting the quality of qualitative research and helping businesses to make more informed, strategic decisions. As technology continues to grow, even more inventive applications will probably arise, further altering the landscape of business qualitative research and offering new opportunities for firms to communicate with their customers, employees, and stakeholders.

CHALLENGES AND FUTURE DIRECTIONS

As technology continues to transform the landscape of business qualitative research, it brings with it both substantial potential and noticeable obstacles (Volberda et al, 2021). While improvements in technology have made qualitative research more efficient, accessible, and far-reaching, the intricacies and limitations of these tools and approaches must be carefully negotiated. Moving forward, the role of technology in qualitative research will expand, providing new difficulties that will require researchers, organizations, and stakeholders to adapt and innovate. One of the major issues faced by businesses when adding technology into qualitative research is the issue of data protection and security (Liu et al, 2020). As digital platforms and technologies become increasingly crucial to the collecting and analysis of qualitative data, the risks connected with data breaches and unauthorized access expand. This is particularly critical when dealing with sensitive data such as consumer feedback, staff interviews, or focus group discussions when participants expect their comments to stay confidential. The incorporation of AI, machine learning, and cloud-based platforms into qualitative research methods raises the potential for data to be accessed, altered, or exploited. As businesses collect more data online, they must also invest in robust security measures, implement strict data protection protocols, and ensure that their research methodologies comply with data privacy regulations, such as the General Data Protection Regulation (GDPR) in the European Union or the California Consumer Privacy Act (CCPA) in the United States (Park, 2019). Failing to address these concerns can result in serious consequences, including reputational harm, legal penalties, and loss of consumer trust.

Alongside the privacy and security concerns, another challenge resides in the complexity and technical skill required to apply new technological tools for qualitative research (Carter et al, 2021). While techniques like sentiment analysis, machine learning, and natural language processing (NLP) can dramatically improve the speed and efficiency of data gathering and analysis, they also demand a comprehensive understanding of both the technology itself and the study objectives. Many firms may lack the specific knowledge needed to effectively harness new technologies, which can lead to misinterpretations or erroneous outcomes. For instance, machine learning algorithms, although strong, are only as successful as the data they are trained on. If the data is biased, incomplete, or badly curated, the insights generated may not be reliable. Additionally, organizations may struggle to find experienced people who can not only work with these new tools but also apply them to the special needs of qualitative research. This lack of experience can restrict the full potential of these technologies and limit their applicability in business research. Another important concern is the issue of data overload (Bawden & Robinson, 2020). With the rise of digital platforms and online data sources, organizations now have access to huge amounts of unstructured data from social media, consumer reviews, online forums, and other digital channels. While this data gives an unprecedented potential to understand customer opinion and behavior, it also comes with the complexity of storing and interpreting these vast databases. The sheer volume of qualitative data can overwhelm traditional research methodologies, making it difficult for organizations to extract significant insights. Even with advanced technology such as AI and machine learning, the process of cleaning, organizing, and understanding such massive quantities of data may be time-consuming and resource-intensive. Researchers must establish ways to filter out unnecessary data, prioritize essential findings, and guarantee that their analyses remain focused on the research objectives rather than becoming lost in the noise of an ever-expanding data landscape.

The problem of assuring accuracy and validity in qualitative research becomes much more evident when dealing with digital instruments (Lim, 2024). Unlike traditional qualitative research methods, such as face-to-face interviews or focus groups, technology-driven research approaches often rely on data that may not necessarily reflect the whole context in which participants contribute their feedback (Zou et al, 2023). For example, sentiment analysis algorithms can assess text-based data, but they generally struggle to effectively understand variations in tone, irony, or sarcasm. Similarly, automatic transcribing technologies, while incredibly efficient, may overlook essential contextual details or misinterpret participants' words. As a result, businesses must be cautious in confirming the authenticity of the data they collect and the insights they generate. This issue becomes more crucial when the data are utilized to inform vital company decisions or impact strategic orientations. While technology might make qualitative research more accessible, it can also expand the divide between organizations with the resources to invest in cutting-edge techniques and those without. Small and medium-sized firms (SMEs) sometimes struggle to stay up with the latest technologies due to budgetary restrictions, insufficient technical competence, and a lack of infrastructure (Wellem, 2022). This creates an unequal playing field where larger organizations, with higher finances and access to specialized people, can dominate the research space, while smaller businesses may be left behind. As technology becomes more vital to the research process, policymakers, researchers, and business leaders need to develop methods to democratize access to these tools and ensure that smaller players can also profit from the advancements in technology.

As we look toward the future of technology-enhanced qualitative research, many major approaches are developing that may assist overcome some of the issues highlighted above. First, we should anticipate seeing further breakthroughs in the integration of artificial intelligence and machine learning into qualitative research approaches. These technologies can dramatically increase the efficiency and accuracy of data analysis by automating laborious operations, discovering patterns that may not be immediately visible, and delivering deeper insights into complex datasets. The future of AI in qualitative research is anticipated to involve technologies that can better grasp human emotions, context, and nuance, leading to more accurate sentiment analysis and increasing the overall quality of the research process. However, the development of these technologies must go hand-in-hand with ethical norms and best practices to guarantee that they are used ethically and transparently.

Another major development is the increased emphasis on data visualization tools that assist researchers and business leaders make sense of complex qualitative data (O'Halloran et al, 2018). As businesses receive more and more data from digital sources, the ability to present this data in a clear and visually engaging manner will become increasingly critical. Advanced data visualization tools may translate raw data into intuitive dashboards, charts, and infographics that make it easier for stakeholders to understand the insights gained from qualitative research. By making qualitative data more available and actionable, these tools have the potential to enable better decision-making and foster collaboration between academics, executives, and other stakeholders inside the business. As remote work and digital communication grow more popular, we are expected to see an increased usage of virtual platforms and tools for doing qualitative research. Virtual focus groups, online interviews, and even immersive technology like virtual reality (VR) may become increasingly widespread as firms continue to adapt to a changing global landscape (Mathysen & Glorieux, 2021). These tools not only make it easier to reach a broad and geographically distributed sample but also create opportunities for businesses to engage with participants in novel and relevant ways. Virtual platforms can also make research more inclusive by eliminating barriers to participation for persons who may have previously been excluded from traditional face-to-face studies owing to geographical, economic, or physical constraints.

Future innovations in qualitative research will likely entail a larger concentration on mixed-methods techniques (Hendren et al, 2023). By combining both qualitative and quantitative methodologies, organizations can acquire a more comprehensive knowledge of the issues they are examining. Technology can play a crucial role in easing the integration of different methodologies, allowing for more seamless data collection and analysis (Ajegbile et al, 2024). For example, organizations might mix AI-driven sentiment analysis with survey data to create a more comprehensive view of consumer sentiments or habits. This integration of qualitative and quantitative data will help firms generate more accurate models and predictions, making it feasible to create more successful plans and solutions. The hurdles of incorporating technology into qualitative research are important but not insurmountable (Smith et al, 2024). As technology continues to grow, organizations and researchers must remain adaptable, continually addressing concerns related to data privacy, security, and accuracy, while also adopting new tools and methodologies that promise to improve the research process. The future of business qualitative research will be defined by ongoing breakthroughs in artificial intelligence, machine learning, data visualization, and virtual platforms, among other technologies. As these tools continue to grow, they will offer new chances to solve existing difficulties and improve the quality, efficiency, and impact of qualitative business research. Ultimately, the success of technology-enhanced qualitative research will depend on its appropriate and inventive usage, ensuring that it serves as a valuable tool for businesses to understand their customers, employees, and markets.

CONCLUSION

As we complete this investigation of the role of technology in modern qualitative research, it is evident that we are at the vanguard of a fundamental shift in research methodology. Technology has fundamentally altered the landscape of qualitative research, providing unparalleled options for improved data gathering, processing, and interpretation. The rapid evolution of tools and software has led to more efficient and effective qualitative research processes, while also allowing researchers new approaches to handle challenging research topics. From strong data collection platforms and transcription software to sophisticated qualitative data analysis tools, technology is transforming the way we conduct research and interpret social phenomena. The integration of artificial intelligence (AI), machine learning, natural language processing, and other digital advancements has already had a major impact on qualitative research. These techniques have enabled researchers to evaluate huge quantities of textual and visual data with better accuracy, speed, and depth. For example, AI-powered software may now spot patterns in qualitative data that may have otherwise gone unnoticed, generating fresh insights and opening doors to more sophisticated analysis. This is particularly significant in business research, as qualitative data often incorporates interviews, consumer feedback, market reports, and other forms of unstructured data. By integrating AI and machine learning, researchers may infer meaning from these complex data sources in ways that were traditionally time-prohibitive or physically impossible. The use of digital tools also extends to the fundamental process of data gathering, providing for increased flexibility, accessibility, and variety in study participants. Digital platforms enable researchers to conduct studies remotely, breaking down geographical barriers and reaching people who may have previously been excluded. Furthermore, digital technologies allow researchers to acquire data in a more broad range of formats, such as audio, video, or social media posts, thereby presenting a larger and more diversified data set for analysis. In corporate contexts, this enhanced reach is significant since it allows researchers to obtain viewpoints from worldwide or otherwise hard-to-reach populations.

Despite these advantages, it is vital to realize that the usage of technology in qualitative research also presents new obstacles. One of the biggest issues is the question of ethics, particularly addressing data privacy, informed permission, and the potential for prejudice. As more qualitative data is obtained through digital means, preserving participants' personal information and ensuring the integrity of the data become crucial. Additionally, the increased dependence on automated tools for data analysis raises worries about the possibility of algorithmic bias, where the systems themselves may accidentally provide slanted conclusions based on the data they process. To mitigate these dangers, researchers must be diligent and proactive in addressing ethical issues throughout the research process, from data collection to analysis and reporting. Another problem arises in the constant requirement for researchers to stay aware of technological changes. The speed of technical progress is tremendous, and new tools and procedures are always being produced. Researchers must therefore maintain a commitment to continual learning, ensuring that they are conversant with the latest software and tools and understand how to integrate these technologies into their research designs without sacrificing the rigor and reproducibility of their investigations. Additionally, it is crucial for researchers to critically analyze the technical tools they utilize to ensure that these tools correspond with their research questions, objectives, and theoretical frameworks. Technology should not be considered as a panacea for research challenges but rather as a facilitator that can enhance, supplement, and refine established research procedures.

Looking to the future, the use of technology in qualitative research is expected to become even more crucial. As new technologies such as virtual reality (VR), augmented reality (AR), and blockchain continue to emerge, they will certainly open up even more options for doing innovative and disruptive qualitative research. VR, for example, has the opportunity for immersive ethnographic study, where researchers can explore and engage with virtual worlds that match real-world contexts. This could lead to totally new forms of qualitative investigation, particularly in sectors such as business and consumer behavior. Similarly, AR might allow researchers to overlay digital data onto real-world situations, offering rich, context-specific insights into how people engage with products, venues, or brands. In business research, these developing technologies could alter the way we study consumer experiences, organizational behavior, and market trends. For instance, VR-based focus groups could enable firms to study consumer reactions in simulated surroundings, delivering deeper insights into decision-making processes. Likewise, AR might be utilized in customer experience studies, where researchers track how customers interact with products or services more interactively and engagingly. The application of blockchain technology in qualitative research is another intriguing prospect. Blockchain could enhance the transparency and traceability of data, ensuring that the research process is auditable and safe. This is particularly relevant when working with sensitive or private data in corporate situations when concerns of data ownership, confidentiality, and intellectual property are critical. By employing blockchain, researchers can construct a secure, immutable record of data gathering and analysis, providing more trust in the integrity of the research findings.

Despite these encouraging improvements, the broad implementation of these technologies in qualitative research will require continual discussion, collaboration, and critical evaluation. Researchers, institutions, and policymakers must engage in debates about the best practices for integrating modern technologies into qualitative research while ensuring that ethical standards are maintained. Research training programs will need to embrace computer literacy, educating the next generation of qualitative researchers with the ability to navigate this increasingly digital terrain. The function of technology in qualitative research is no longer a peripheral consideration but a key component of modern research practice. The tools and software available today have altered how academics acquire, analyze, and interpret data, opening up new possibilities for inquiry and discovery. However, as with any big shift in study methods, some obstacles must be addressed. Ethical difficulties, the potential for prejudice, and the requirement for continual technological literacy are significant difficulties that researchers must navigate. Nevertheless, the ongoing expansion of technology presents intriguing prospects for qualitative research, particularly within the corporate setting. By adopting these innovations, researchers can not only boost the quality of their work but also contribute to a deeper and more complex knowledge of human behavior, organizational dynamics, and market pressures. As we look to the future, the integration of technology and qualitative research holds the potential to transform our view of the world and offer fresh insights into the complex interplay of social, cultural, and economic phenomena.

REFERENCES

Ajegbile, M. D., Olaboye, J. A., Maha, C. C., & Tamunobarafiri, G. (2024). Integrating business analytics in healthcare: Enhancing patient outcomes through data-driven decision making. *World J Biol Pharm Health Sci, 19*(1), 243–250. DOI: 10.30574/wjbphs.2024.19.1.0436

Akter, S., Bandara, R., Hani, U., Wamba, S. F., Foropon, C., & Papadopoulos, T. (2019). Analytics-based decision-making for service systems: A qualitative study and agenda for future research. *International Journal of Information Management, 48*, 85–95. DOI: 10.1016/j.ijinfomgt.2019.01.020

Andriani, G., Godoy, E., Koslovski, G., Obelheiro, R., & Pillon, M. (2019). An architecture for synchronising cloud file storage and organisation repositories. *International Journal of Parallel, Emergent and Distributed Systems : IJPEDS, 34*(5), 538–555. DOI: 10.1080/17445760.2017.1422500

Archibald, M. M., Ambagtsheer, R. C., Casey, M. G., & Lawless, M. (2019). Using zoom videoconferencing for qualitative data collection: Perceptions and experiences of researchers and participants. *International Journal of Qualitative Methods, 18*, 1609406919874596. DOI: 10.1177/1609406919874596

. Baikulova, M., & Suderevskaia, E. (2019). Futures Research: The Application of VR and AR Emerging Technologies in New Media: Future of Immersive Visuals Online.

. Bawden, D., & Robinson, L. (2020). Information overload: An overview.

Brougham, D., & Haar, J. (2018). Smart technology, artificial intelligence, robotics, and algorithms (STARA): Employees' perceptions of our future workplace. *Journal of Management & Organization, 24*(2), 239–257. DOI: 10.1017/jmo.2016.55

Bryda, G., & Costa, A. P. (2023). Qualitative research in digital era: Innovations, methodologies and collaborations. *Social Sciences (Basel, Switzerland), 12*(10), 570. DOI: 10.3390/socsci12100570

Carter, S. M., Shih, P., Williams, J., Degeling, C., & Mooney-Somers, J. (2021). Conducting qualitative research online: Challenges and solutions. *Patient, 14*(6), 711–718. DOI: 10.1007/s40271-021-00528-w PMID: 34114170

Chakraborty, T., Sharada, V. S., & Sharma, J. (2024). Data-Driven Decisions for Employee Motivation: Strategies for Implementing Technology in Performance Reviews in a Tech Landscape. In Employee Performance Management for Improved Workplace Motivation (pp. 311-338). IGI Global.

Cheligeer, C., Yang, L., Nandi, T., Doktorchik, C., Quan, H., Zeng, Y., & Singh, S. (2023). Natural language processing (NLP) aided qualitative method in health research. *Journal of Integrated Design & Process Science*, *27*(1), 41–58. DOI: 10.3233/JID-220013

Cheng, L., Varshney, K. R., & Liu, H. (2021). Socially responsible ai algorithms: Issues, purposes, and challenges. *Journal of Artificial Intelligence Research*, *71*, 1137–1181. DOI: 10.1613/jair.1.12814

Choukaier, D. (2024). Enhancing English As A Foreign Language (EFL) Instruction Through Digital Teaching Platforms: Analyzing The Impact Of Microsoft Teams, Zoom, And Google Meet On Communication And Participation. *Educational Administration: Theory and Practice*, *30*(6), 2404–2418.

Cresswell, K., Domínguez Hernández, A., Williams, R., & Sheikh, A. (2022). Key challenges and opportunities for cloud technology in health care: Semistructured interview study. *JMIR Human Factors*, *9*(1), e31246. DOI: 10.2196/31246 PMID: 34989688

Dawadi, S., Shrestha, S., & Giri, R. A. (2021). Mixed-methods research: A discussion on its types, challenges, and criticisms. *Journal of Practical Studies in Education*, *2*(2), 25–36. DOI: 10.46809/jpse.v2i2.20

de Wildt, L. (2020). Qualitative data analysis software. In *The SAGE Encyclopedia of the Sociology of Religion* (pp. 643–644). SAGE Publications, Inc.

Deep, B. (2020). Lived experience and the idea of the social in Alfred Schutz: A phenomenological study of contemporary relevance. *Journal of Indian Council of Philosophical Research*, *37*(3), 361–381. DOI: 10.1007/s40961-020-00211-9

Dodgson, J. E. (2019). Reflexivity in qualitative research. *Journal of Human Lactation*, *35*(2), 220–222. DOI: 10.1177/0890334419830990 PMID: 30849272

Ebrahim, S. A., Poshtan, J., Jamali, S. M., & Ebrahim, N. A. (2020). Quantitative and qualitative analysis of time-series classification using deep learning. *IEEE Access : Practical Innovations, Open Solutions*, *8*, 90202–90215. DOI: 10.1109/ACCESS.2020.2993538

Elliott, J. (2022). The craft of using NVivo12 to analyze open-ended questions: An approach to mixed methods analysis. *The Qualitative Report*. Advance online publication. DOI: 10.46743/2160-3715/2022.5460

Gaber, M., Pastor, G. C., & Omer, A. (2020). Speech-to-Text technology as a documentation tool for interpreters: A new approach to compiling an ad hoc corpus and extracting terminology from video-recorded speeches. TRANS: revista de traductología, (24), 263-281.

Gill, P., & Baillie, J. (2018). Interviews and focus groups in qualitative research: An update for the digital age. *British Dental Journal, 225*(7), 668–672. DOI: 10.1038/sj.bdj.2018.815 PMID: 30287965

Girju, R. (2023, October). Understanding Lived Experience: Bridging Artificial Intelligence and Natural Language Processing with Humanities and Social Sciences. []). IOP Publishing.]. *IOP Conference Series. Materials Science and Engineering, 1292*(1), 012020. DOI: 10.1088/1757-899X/1292/1/012020

Grant, R. (2018). *Data visualization: Charts, maps, and interactive graphics*. Chapman and Hall/CRC. DOI: 10.1201/9781315201351

Green Jr, C. (2022). Industry 4.0 Technologies and Biotechnology Biosensors a Game Changer in US Supply Chain Food Safety: An Exploratory Study. Capella University.

Guetterman, T. C., Chang, T., DeJonckheere, M., Basu, T., Scruggs, E., & Vydiswaran, V. V. (2018). Augmenting qualitative text analysis with natural language processing: Methodological study. *Journal of Medical Internet Research, 20*(6), e231. DOI: 10.2196/jmir.9702 PMID: 29959110

Gupta, D., & Rani, R. (2019). A study of big data evolution and research challenges. *Journal of Information Science, 45*(3), 322–340. DOI: 10.1177/0165551518789880

Gupta, J., Mitra, B., & Garg, K. (2021). Technological disruptions reshaping management education: A qualitative study of stakeholder perspectives. *International Journal of Management Education, 15*(2), 179–198.

Hartmann, J., Heitmann, M., Siebert, C., & Schamp, C. (2023). More than a feeling: Accuracy and application of sentiment analysis. *International Journal of Research in Marketing, 40*(1), 75–87. DOI: 10.1016/j.ijresmar.2022.05.005

Hendren, K., Newcomer, K., Pandey, S. K., Smith, M., & Sumner, N. (2023). How qualitative research methods can be leveraged to strengthen mixed methods research in public policy and public administration? *Public Administration Review, 83*(3), 468–485. DOI: 10.1111/puar.13528

Hesse, A., Glenna, L., Hinrichs, C., Chiles, R., & Sachs, C. (2019). Qualitative research ethics in the big data era. *The American Behavioral Scientist, 63*(5), 560–583. DOI: 10.1177/0002764218805806

Howlett, M. (2022). Looking at the 'field'through a Zoom lens: Methodological reflections on conducting online research during a global pandemic. *Qualitative Research, 22*(3), 387–402. DOI: 10.1177/1468794120985691 PMID: 35663097

Ireland, E. (2020). Strategies Big Data Analytics Specialists Need to Improve the Analytical Methods Used for Classifying Structured and Unstructured Data (Doctoral dissertation, Colorado Technical University).

Kalpokaite, N., & Radivojevic, I. (2020). Teaching qualitative data analysis software online: A comparison of face-to-face and e-learning ATLAS. ti courses. *International Journal of Research & Method in Education, 43*(3), 296–310. DOI: 10.1080/1743727X.2019.1687666

Kozinets, R. (2019). Netnography: The essential guide to qualitative social media research.

Kraus, S., Palmer, C., Kailer, N., Kallinger, F. L., & Spitzer, J. (2019). Digital entrepreneurship: A research agenda on new business models for the twenty-first century. *International Journal of Entrepreneurial Behaviour & Research, 25*(2), 353–375.

Kumar, V., Ramachandran, D., & Kumar, B. (2021). Influence of new-age technologies on marketing: A research agenda. *Journal of Business Research, 125*, 864–877. DOI: 10.1016/j.jbusres.2020.01.007

Langmead, B., & Nellore, A. (2018). Cloud computing for genomic data analysis and collaboration. *Nature Reviews. Genetics, 19*(4), 208–219. DOI: 10.1038/nrg.2017.113 PMID: 29379135

Lieber, E., Salmona, M., & Kaczynski, D. (2021). Introduction to Dedoose for mixed analysis. In *The Routledge Reviewer's Guide to Mixed Methods Analysis* (pp. 319–330). Routledge. DOI: 10.4324/9780203729434-27

Lim, W. M. (2024). What is qualitative research? An overview and guidelines. *Australasian Marketing Journal*, •••, 14413582241264619.

Liu, J., & Eagan, J. (2021). ADQDA: a cross-device affinity diagramming tool for fluid and holistic qualitative data analysis. Proceedings of the ACM on Human-Computer Interaction, 5(ISS), 1-19. DOI: 10.1145/3488534

Liu, N., Nikitas, A., & Parkinson, S. (2020). Exploring expert perceptions about the cyber security and privacy of Connected and Autonomous Vehicles: A thematic analysis approach. *Transportation Research Part F: Traffic Psychology and Behaviour, 75*, 66–86. DOI: 10.1016/j.trf.2020.09.019

Lobe, B., Morgan, D., & Hoffman, K. A. (2020). Qualitative data collection in an era of social distancing. *International Journal of Qualitative Methods, 19*, 1609406920937875. DOI: 10.1177/1609406920937875

Longo, L. (2019, September). Empowering qualitative research methods in education with artificial intelligence. In *World conference on qualitative research* (pp. 1-21). Cham: Springer International Publishing.

. Marjaei, S., Yazdi, F. A., & Chandrashekara, M. (2019). MAXQDA and its Application to LIS Research. Library Philosophy and Practice, 1-9.

Martucci, C. (2024). Qualitative evaluation of Power BI and its application in an Aerospace Industry cluster (Doctoral dissertation, Politecnico di Torino).

Mathysen, D., & Glorieux, I. (2021). Integrating virtual reality in qualitative research methods: Making a case for the VR-assisted interview. *Methodological Innovations, 14*(2), 20597991211030778. DOI: 10.1177/20597991211030778

McCall, C., & Edwards, C. (2021). New perspectives for implementing grounded theory. *Studies in Engineering Education, 1*(2), 93–107. DOI: 10.21061/see.49

Meißner, M., Pfeiffer, J., Peukert, C., Dietrich, H., & Pfeiffer, T. (2020). How virtual reality affects consumer choice. *Journal of Business Research, 117*, 219–231. DOI: 10.1016/j.jbusres.2020.06.004

. Mezmir, E. A. (2020). Qualitative data analysis: An overview of data reduction, data display, and interpretation. Research on humanities and social sciences, 10(21), 15-27.

Mills, K. A. (2018). What are the threats and potentials of big data for qualitative research? *Qualitative Research, 18*(6), 591–603. DOI: 10.1177/1468794117743465

Mirick, R. G., & Wladkowski, S. P. (2019). Skype in qualitative interviews: Participant and researcher perspectives. *The Qualitative Report, 24*(12), 3061–3072. DOI: 10.46743/2160-3715/2019.3632

Mogaji, E., Jain, V., & Sharma, H. (2024). *Qualitative Research for a Digital World: A Practical Guide*. SAGE Publications Limited.

. Mohajan, H. K. (2018). Qualitative research methodology in social sciences and related subjects. Journal of economic development, environment and people, 7(1), 23-48.

Moisander, J., Närvänen, E., & Valtonen, A. (2020). Interpretive marketing research: Using ethnography in strategic market development. In *Marketing management* (pp. 237–253). Routledge. DOI: 10.4324/9780203710807-19

Morse, J. (2020). The changing face of qualitative inquiry. *International Journal of Qualitative Methods, 19*, 1609406920909938. DOI: 10.1177/1609406920909938

Mortelmans, D. (2019). Analyzing qualitative data using NVivo. The Palgrave handbook of methods for media policy research, 435-450. DOI: 10.1007/978-3-030-16065-4_25

Nicolaou, C. (2021). Qualitative methods research through the Internet Applications and Services: The contribution of audiovisual media technology as technology-enhanced research. *International Research in Higher Education, 6*(1), 1–14. DOI: 10.5430/irhe.v6n1p1

Niedbalski, J., & Ślęzak, I. (2023, January). NVivo as a tool for supporting teamwork in the context of qualitative research conducted remotely-opportunities, limitations, and practical tips. In *World conference on qualitative research* (pp. 38-59). Cham: Springer International Publishing. DOI: 10.1007/978-3-031-31346-2_3

Nigar, N. (2020). Hermeneutic phenomenological narrative enquiry: A qualitative study design. *Theory and Practice in Language Studies, 10*(1), 10–18. DOI: 10.17507/tpls.1001.02

Novawan, A., Walker, S. A., & Ikeda, O. (2024). The new face of technology-enhanced language learning (TELL) with artificial intelligence (AI): Teacher perspectives, practices, and challenges. *Journal of English in Academic and Professional Communication, 10*(1), 1–18. DOI: 10.25047/jeapco.v9i1.3754

Ntsobi, M. P., Costa, A. P., Kasperiuniene, J., Brandão, C., & Ribeiro, J. (2024, January). Digital Tools and Techniques in Qualitative Research: Digital Skills and Research Optimisation. In World Conference on Qualitative Research (pp. 1-25). Cham: Springer Nature Switzerland.

O'Hagan, J., Saeghe, P., Gugenheimer, J., Medeiros, D., Marky, K., Khamis, M., & McGill, M. (2023). Privacy-enhancing technology and everyday augmented reality: Understanding bystanders' varying needs for awareness and consent. *Proceedings of the ACM on Interactive, Mobile, Wearable and Ubiquitous Technologies, 6*(4), 1–35. DOI: 10.1145/3569501

O'Halloran, K. L., Tan, S., Pham, D. S., Bateman, J., & Vande Moere, A. (2018). A digital mixed methods research design: Integrating multimodal analysis with data mining and information visualization for big data analytics. *Journal of Mixed Methods Research, 12*(1), 11–30. DOI: 10.1177/1558689816651015

Olapane, E. C. (2021). An in-depth exploration on the praxis of computer-assisted qualitative data analysis software (CAQDAS). *Journal of Humanities and Social Sciences Studies*, *3*(11), 57–78. DOI: 10.32996/jhsss.2021.3.11.5

. Olson, K. (2021). Assessing the use of Qualitative Data Analysis Software (QDAS) by Art Historians and Archaeologists.

Olson, R. E. (2023). Emotions in human research ethics guidelines: Beyond risk, harm and pathology. *Qualitative Research*, *23*(3), 526–544. DOI: 10.1177/14687941211039965

Park, G. (2019). The changing wind of data privacy law: A comparative study of the European Union's General Data Protection Regulation and the 2018 California Consumer Privacy Act. *UC Irvine L. Rev.*, *10*, 1455.

Quist-Adade, C. (2019). *Symbolic interactionism: The basics*. Vernon Press.

Rädiker, S., & Kuckartz, U. (2020). *Focused analysis of qualitative interviews with MAXQDA*. MAXQDA Press.

. Ramya, R., & Rajeswari, G. (2023). CLOUD TECHNOLOGIES IN EDUCATIONAL RESEARCH. EDUCATION 5.0: REVOLUTIONIZING LEARNING FOR THE FUTURE, 90.

Sampson, C., & Wong, L. S. (2023). Using Qualitative Data Analysis Software to Help Explore Critical Research Questions: A Tool, Not a Replacement. In Handbook of Critical Education Research (pp. 637-654). Routledge.

Seskir, Z. C., Umbrello, S., Coenen, C., & Vermaas, P. E. (2023). Democratization of quantum technologies. *Quantum Science and Technology*, *8*(2), 024005. DOI: 10.1088/2058-9565/acb6ae

Seyedi, S., Griner, E., Corbin, L., Jiang, Z., Roberts, K., Iacobelli, L., Milloy, A., Boazak, M., Bahrami Rad, A., Abbasi, A., Cotes, R. O., & Clifford, G. D. (2023). Using HIPAA (health insurance portability and accountability act)–compliant transcription services for virtual psychiatric interviews: Pilot comparison study. *JMIR Mental Health*, *10*, e48517. DOI: 10.2196/48517 PMID: 37906217

Smith, A. H., Touchett, H., Chen, P., Fletcher, T., Arney, J., Hogan, J., Wassef, M., Cloitre, M., & Lindsay, J. A. (2024). Patient satisfaction with a coach-guided, technology-based mental health treatment: Qualitative interview study and theme analysis. *JMIR Mental Health*, *11*(1), e50977. DOI: 10.2196/50977 PMID: 38306167

Stieglitz, S., Mirbabaie, M., Ross, B., & Neuberger, C. (2018). Social media analytics–Challenges in topic discovery, data collection, and data preparation. *International Journal of Information Management, 39*, 156–168. DOI: 10.1016/j.ijinfomgt.2017.12.002

Taylor, J., & Pagliari, C. (2018). Mining social media data: How are research sponsors and researchers addressing the ethical challenges? *Research Ethics, 14*(2), 1–39. DOI: 10.1177/1747016117738559

Thunberg, S., & Arnell, L. (2022). Pioneering the use of technologies in qualitative research–A research review of the use of digital interviews. *International Journal of Social Research Methodology, 25*(6), 757–768. DOI: 10.1080/13645579.2021.1935565

Tracy, S. J. (2024). *Qualitative research methods: Collecting evidence, crafting analysis, communicating impact*. John Wiley & Sons.

Tshabangu, I., Ba, S., & Madondo, S. M. (Eds.). (2020). *Approaches and processes of social science research*. IGI Global.

Tyagi, A. K., & Chahal, P. (2020). Artificial intelligence and machine learning algorithms. In *Challenges and applications for implementing machine learning in computer vision* (pp. 188–219). IGI Global. DOI: 10.4018/978-1-7998-0182-5.ch008

Übellacker, T. (2024). AcademiaOS: Automating Grounded Theory Development in Qualitative Research with Large Language Models. arXiv preprint arXiv:2403.08844.

Ullah, F., Sepasgozar, S. M., & Wang, C. (2018). A systematic review of smart real estate technology: Drivers of, and barriers to, the use of digital disruptive technologies and online platforms. *Sustainability (Basel), 10*(9), 3142. DOI: 10.3390/su10093142

. Vaismoradi, M., & Snelgrove, S. (2019). Theme in qualitative content analysis and thematic analysis.

Volberda, H. W., Khanagha, S., Baden-Fuller, C., Mihalache, O. R., & Birkinshaw, J. (2021). Strategizing in a digital world: Overcoming cognitive barriers, reconfiguring routines and introducing new organizational forms. *Long Range Planning, 54*(5), 102110. DOI: 10.1016/j.lrp.2021.102110

Wang, Y., Kung, L., & Byrd, T. A. (2018). Big data analytics: Understanding its capabilities and potential benefits for healthcare organizations. *Technological Forecasting and Social Change, 126*, 3–13. DOI: 10.1016/j.techfore.2015.12.019

Wangmo, T., Lipps, M., Kressig, R. W., & Ienca, M. (2019). Ethical concerns with the use of intelligent assistive technology: Findings from a qualitative study with professional stakeholders. *BMC Medical Ethics*, *20*(1), 1–11. DOI: 10.1186/s12910-019-0437-z PMID: 31856798

Weber, S. (2021). A step-by-step procedure to implement discrete choice experiments in Qualtrics. *Social Science Computer Review*, *39*(5), 903–921. DOI: 10.1177/0894439319885317

Wellem, A. (2022). Budgeting and cost control challenges in small and medium enterprises construction projects in Cape Town (Doctoral dissertation, Cape Peninsula University of Technology).

Wu, Y. F., & Kim, E. Y. (2022). Users' perceptions of technological features in augmented reality (AR) and virtual reality (VR) in fashion retailing: A qualitative content analysis. *Mobile Information Systems*, *2022*(1), 3080280. DOI: 10.1155/2022/3080280

Xu, A., Baysari, M. T., Stocker, S. L., Leow, L. J., Day, R. O., & Carland, J. E. (2020). Researchers' views on, and experiences with, the requirement to obtain informed consent in research involving human participants: A qualitative study. *BMC Medical Ethics*, *21*(1), 1–11. DOI: 10.1186/s12910-020-00538-7 PMID: 33008387

. Yılmaz, A. O. (2024). Political Ontological Analysis of Classical French Sociological Thought: Auguste Comte and Émile Durkheim.

Zhang, Y. (2023). *Use of artificial intelligence (AI) in historical records transcription: Opportunities, challenges, and future directions*. McGill University.

Zou, P. X., & Xu, X. (2023). *Research methodology and strategy: theory and practice*. John Wiley & Sons. DOI: 10.1002/9781394190256

Chapter 5
Evaluating Qualitative Data Analysis Tools: A Comparative Study of NVivo and ATLAS.ti

Jeganathan Gomathi Sankar
https://orcid.org/0000-0002-5077-5109
BSSS Institute of Advanced Studies, India

Dina Darwish
https://orcid.org/0000-0002-6708-649X
Ahram Canadian University, Egypt

P. Valan
https://orcid.org/0009-0007-0638-1325
Vellore Institute of Technology, India

ABSTRACT

This chapter provides a comparative analysis of NVivo and ATLAS.ti, two prominent software tools for qualitative data analysis. It evaluates the core functionalities, user interfaces, and analytical capabilities of both platforms, focusing on their strengths and limitations. The chapter details how each tool supports data management, coding, and analysis, and considers their impact on research efficiency and effectiveness. Key factors such as usability, cost, integration with other research tools, and user support are discussed to assist researchers in selecting the most suitable software for their qualitative research needs. The analysis aims to offer practical insights and guidance for optimizing the qualitative research process

DOI: 10.4018/979-8-3373-4432-4.ch005

INTRODUCTION

The use of qualitative data analysis software (QDAS) packages has been heralded as a hopeful and troubling development for the qualitative research community from its inception 25 years ago. In the 1980s, when the first generation of QDAS systems was developed, researchers hypothesized that these tools may advance qualitative research in many important ways. The original strategy entailed enabling more complex, adaptable, exhaustive, and thorough coding methodologies, therefore extending paper-based methods for categorizing, identifying, and assessing data "beyond the feasible manual limits" (Richards & Richards, 1994, p. 29), a more direct, reliable, and effective method of obtaining coded data for analysis (Richards & Richards, 1994; Richards, 2002) and improved evaluation of the qualitative data supporting new theoretical claims and assumptions (Hesse-Biber et al., 1991; Hesse-Biber, 2022). The second option was that by enabling studies that "would be impossible to carry out manually," QDAS may enhance qualitative analysis and interpretation (Richards & Richards, 1991, p. 319), As an example, it can facilitate the "manually impossible "linking of text, analysis, and non-text resources (graphics, audio, and video) in a single analytical environment outside the mind's eye." (Dohan & Sanchez-Jankowski, 1998, p. 484). The third possibility was that QDAS could facilitate more transparent analytical processes, which would make it easier to demonstrate validity, rigor, and trustworthiness (Morison & Moir, 1998) and improve knowledge of its applicability, limitations, and practical application (Blismas & Dainty, 2003). By doing so, others may be able to avoid frequent blunders (Sin, 2007), and the qualitative research community may be able to better understand how to advise new QDAS users on how to steer clear of typical difficulties and issues in qualitative research (Bassett, 2004).

Concerns about the possible impact of QDAS on qualitative research have been raised since its introduction. One of the most enduring concerns is that these instruments may impose a limiting framework on research practices since they are designed for particular research approaches, such grounded theory (Holbrook & Butcher, 1996, p. 60). Also, researchers may find it more difficult to modify their studies to satisfy the particular requirements of their projects due to the technological limitations imposed by QDAS (Woods et al., 2016). This limitation can take many different forms. For example, researchers may prioritize program requirements (Gilbert, 2002), adjust study designs to match the software's capabilities (Gilbert et al., 2014), or use analysis techniques only because the software makes them easier (Woods et al., 2016). Also, the dependence on software might obscure the understanding of researchers' methodological procedures (Bryman & Burgess, 1994; MacMillan & Koenig, 2004). These worries emphasize the need for a critical and reflective approach to QDAS use, making sure that researchers continue to be conscious of the ways in which these instruments might influence qualitative research methods (Brown, 2002; Woods et al., 2016).

The two main methodological strands that steer social science research are qualitative and quantitative. Usually, these methods' corresponding techniques are grouped together. Numerous statistical tools are used in quantitative analysis, including well-known programs like LISREL, R, SPSS, and AMOS. On the other hand, software packages referred to as QDA (Qualitative Data Analysis), CAQDAS (Computer-Assisted Qualitative Data Analysis Software), or QDAS (Qualitative Data Analysis Software) are becoming more and more popular in qualitative research. These tools support researchers in the analysis of information gathered from sources including documents, field notes, interviews, focus groups, and open-ended survey questions. According to Woods et al. (2016), QDAS enables a range of study approaches and comprehensive analysis of both textual and visual data. Additionally, mixed-methods research, which combines quantitative and qualitative methodologies, commonly uses these tools. Before QDAS was created, content analysis in qualitative research was frequently done using programs like Excel or SPSS. The main purpose of these previous applications was to assist users in finding differences or correlations between variables and frequency (Oliveira, 2023). On the other hand, QDAS provides a number of sophisticated features that improve research capacities, such as the capacity to generate word clouds, construct models, retrieve data, and automatically code words. The ease of qualitative analysis has been greatly enhanced by this development. Even with QDAS's user-friendly interfaces, these tools' efficacy mostly depends on how well-versed the researcher is in analytic techniques. The program itself as well as the researcher's analytical skills and knowledge limit the efficiency that QDAS can deliver (Oliveira et al., 2013).

Prior to the development of QDAS, Excel or SPSS were often used by researchers for qualitative studies, especially those that included content analysis. The main purpose of these software tools was to help users find differences or correlations between variables and frequencies. On the other hand, QDAS offers a number of sophisticated capabilities that go beyond these fundamental functions and greatly improve convenience for researchers. These features include the ability to create word clouds, develop models, retrieve data, and code automatically based on phrases. Even while QDAS's interfaces are meant to be user-friendly, many of its fundamental ideas follow the same reasoning. Rather than a researcher's existing software experience, the success of these tools mostly depends on the researcher's analytical understanding (Oliveira et al., 2013). Researchers' interest in examining different software tools has increased as the adoption of software for qualitative data analysis has expanded (Woods et al., 2016). Nonetheless, there have long been continuing epistemological discussions over the application of qualitative software in social research. Some academics voiced worries about computer software's influence on qualitative research in the early 1990s, arguing that the tools may be used to manipulate the researcher as well as the study (Gilbert et al., 2014). A major concern is that researchers could modify their data to suit the features of the software rather than the requirements of their studies (Woods et al., 2016). Further, the researcher and the data may become separated by the usage of software, which might impose strict restrictions on the analytical process (Bourdon, 2002). Nevertheless, there is minimal reason for mistrust by the 2020s when using qualitative data analysis tools becomes necessary. According to Peters and Wester (2007), researchers have benefited greatly from technological improvements in the last 20 years that have made it easier for them to undertake analysis. Among the well-known tools for qualitative data analysis (QDA) are Quirkos, QDA Miner, NVivo, Dedoose, Elan, F4 Analyze, and Transana (Paulus & Lester, 2020). But preferences for these programs differ among disciplines, nations, and eras. Although there are other brands of QDAS, Hoover and Koerber (2011) highlight that NVivo, Atlas.ti, and MAXQDA are the three most popular brands. ATLAS.ti 5.0 and NVivo 2.0 are acknowledged as some of the most useful QDA tools in communication research, according to an assessment of the literature and papers on the Web of Science (WoS). Using these tools, researchers may create classifications that represent testable models of the conceptual framework underpinning the data, as well as correlate codes or labels with text, audio, picture, or video segments and look for trends in these codes. Because of their adaptability, they may be used in a wide variety of research applications (Gilbert et al., 2014; Woods et al., 2016; Oliveira, 2023).

This study aims to do a thorough comparative examination of two popular QDA tools: ATLAS.ti and NVivo. The present review aims to evaluate the practical consequences of applying various software packages in qualitative research, while also highlighting their unique strengths, limits, and features. The goals involve assessing the main characteristics of both products, including their capacities for data input and export, their coding features, their analytical tools, and their visualization possibilities. To evaluate usability, the user interface and total user experience will be analyzed in order to identify the more accessible and user-friendly tool. The study will also look into the quality and accessibility of user manuals, training materials, and technical support for ATLAS.ti and NVivo. Value for money and pricing models will be compared, with consideration given to individual and institutional licensing choices, in a cost-effectiveness study. Researchers who have used both technologies will provide comments on their user experiences, with an emphasis on overall efficacy and satisfaction. Along with identifying specific case studies that highlight the advantages and disadvantages of each QDA tool, the study will also investigate the many research applications of both tools. In light of the results, the study's final goal is to offer researchers useful advice on how to choose and apply QDA technologies that are appropriate for their particular research requirements.

OVERVIEW OF NVIVO

Known for being one of the best tools for QDA, NVivo was created especially to help researchers organize and analyze qualitative data more effectively. Because of the way it is designed, it may be used by researchers in a wide range of fields, which makes it a very useful tool for qualitative research methods (Alam, 2021). Fundamentally, NVivo provides an extensive toolkit that facilitates the research process. These tools provide a variety of features that make organizing, coding, and analyzing data easier. Text, audio, video, and picture data may all be imported by users, enabling a multifaceted approach to qualitative research. With the use of this capabilities, researchers may examine a wide range of rich data sources and get insights that they would miss using more conventional techniques. Numerous qualitative research approaches, such as content analysis, theme analysis, and grounded theory, are supported by NVivo (Dalkin et., 2021). NVivo enables researchers to apply rigorous analytical frameworks to their data and ensures that the results are legitimate and trustworthy by offering tools specifically designed for these approaches. Users may methodically identify and categorize data segments using the software's coding features, which improves the analytical process and encourages deeper interaction with the content. Through its visualization capabilities, which let researchers construct graphical representations of their data and conclusions, NVivo also improves the study experience. These visuals, which include word clouds, charts, and models, help convey complicated data to a variety of audiences while also assisting in the identification of patterns and trends (Dhakal, 2022; Niedbalski & Ślęzak, 2023). NVivo is an essential tool for qualitative research as it gives researchers the tools they need to efficiently handle and evaluate complicated data. Researchers may concentrate on gaining significant insights that can advance their fields of study and guide practice thanks to its intuitive interface and robust analytical tools (Allsop et al., 2022).

Key Features and Functionalities

NVivo has sophisticated coding features that enable users to methodically classify and identify data segments. This capability, which enables researchers to divide massive amounts of data into digestible chunks, is essential for qualitative research. In order to support an organized analysis process, users can generate codes that stand in for themes, concepts, or categories produced from the data. Because coding is flexible, it may be arranged hierarchically such that more specialized sub-codes can be included into larger themes. This creates a clear foundation for comprehending complicated data sets. The program is adaptable to varied research procedures since it supports a range of analytical approaches (Niedbalski & Ślęzak,

2023). For example, while content analysis concentrates on measuring certain features of the data, such the frequency of themes or concepts, thematic analysis allows researchers to find and understand patterns across qualitative data. Additionally supported is the grounded theory technique, which enables researchers to create theories based on the data itself, strengthening the breadth and precision of their studies (Limna, 2023). NVivo's capacity to adjust to various methods guarantees that it can fulfill the requirements of a wide range of research initiatives. NVivo has strong visualization features in addition to coding that are intended to improve the analysis process. These resources, which include word clouds, charts, and models, let researchers see intricate connections and patterns in the data (Niedbalski & Ślęzak, 2023). For instance, word clouds may provide instant insight into important themes by highlighting the most frequently used phrases, and charts can show how codes are distributed throughout various data sources. These illustrations support the analytical process and function as powerful communication tools for presenting results to diverse audiences. Efficient data querying is another feature of NVivo that lets researchers do intricate searches across large data sets. Users may quickly and correctly get specified information by refining their searches using numerous parameters (Kraiwanit et al., 2023). Large qualitative data sets, where manually sorting through the data might be time-consuming, benefit greatly from this feature. NVivo makes it easier for researchers to get specific data, which improves their capacity to interact with the data in-depth and extract insightful information.

User Interface and Usability

NVivo's user interface is well-thought-out to guarantee simplicity of use, with a logical structure that suits both inexperienced and seasoned users. Because of its simple layout, users may become used to the program fast, which lowers the learning curve that is sometimes associated with sophisticated data analysis tools (Lovelace et al., 2020). The dashboard acts as a central hub, giving a summary of ongoing initiatives and facilitating quick access to different data sets and analytical resources. By using a more efficient approach to workflow, researchers are free to concentrate on their investigation instead of wasting time looking for functions or data. Intuitive drag-and-drop features that streamline coding and data management further increase usability. Users need little effort to apply codes to certain text segments or transfer data files into the proper categories (Soehardi et al., 2021). By enabling researchers to interact more freely with their data and fostering a more dynamic analytic experience, this user-friendly tool increases productivity. For customers looking to get the most out of the program, NVivo also offers a wealth of tutorials and support materials. These materials address a broad spectrum of subjects, from sophisticated analytical methods to fundamental features (Elliott-Mainwaring, 2021).

Step-by-step instructions, video tutorials, and community forums are available to users, creating a helpful learning and troubleshooting environment. This dedication to user assistance promotes continued interaction with the program, which in turn improves the caliber of qualitative research outputs, while also enabling researchers to make the most of NVivo's features (Elliott-Mainwaring, 2021; Almusallam, 2021).

Types of Data Supported

NVivo's adaptability for qualitative research is further enhanced by its ability to handle a wide range of data formats. For researchers who often gather a variety of data to fully comprehend their study issues, this skill is crucial. Text data is supported by NVivo, including open-ended survey questions and transcripts from focus groups and interviews (Lovelace et al., 2020; Niedbalski & Ślęzak, 2023). These text-based resources are essential to qualitative research because they provide a thorough examination of the viewpoints, experiences, and feelings of participants. NVivo makes it easier for researchers to analyze such rich textual material, allowing them to detect themes, patterns, and insights that are essential for understanding qualitative findings (Kraiwanit et al., 2023). NVivo facilitates the integration of multimedia data, such as audio files, videos, and pictures, in addition to text data. This feature allows interaction with data in many formats, extending the reach of qualitative analysis beyond regular text. For example, it is possible to transcribe and analyze interview audio recordings in conjunction with the original recordings, allowing for the capture of subtleties in tone and intonation that would not be apparent in written text. In a similar vein, video data may offer insightful background, including nonverbal signs and interactions, which are critical for comprehending intricate social dynamics (Almusallam, 2021; Limna, 2023). NVivo facilitates a more thorough analysis that captures the complexity of qualitative research by supporting these different kinds of data. The software's capacity to manage multimedia data makes creative analytical methods easier to implement (Soehardi et al., 2021). Researchers can create a rich, integrated dataset by connecting audio or visual data to certain text portions. Researchers can gain better insights by cross-referencing several data kinds, which improves their overall comprehension of the study issue (Dalkin et., 2021; Allsop et al., 2022).

When doing mixed-methods research, where qualitative and quantitative data are combined to create a more comprehensive picture of the study issue, NVivo's support for a variety of data sources is very helpful. NVivo allows researchers to examine both qualitative and quantitative survey data, allowing for a more thorough interpretation of the results (Alam, 2021). The process of triangulation—where several data sources join together to offer confirming evidence—is made possible by this integration, which increases the validity of the study findings. Because of its adaptability, NVivo may be used in a variety of research environments, including social, business, and academic ones (Soehardi et al., 2021). Researchers in a variety of disciplines, including sociology, education, health, and market research, may take use of NVivo's features to carry out in-depth qualitative analyses that are in line with their particular study goals. NVivo promotes a comprehensive approach to qualitative data analysis by supporting a broad range of data kinds and formats, allowing for the investigation of complicated problems from several perspectives (Soehardi et al., 2021; Kraiwanit et al., 2023). Because NVivo can handle a wide range of data sources, including video and text, it is an invaluable tool for qualitative researchers. Because of its capabilities, in-depth and nuanced analyses are made possible, leading to a greater comprehension of the study setting and participant experiences. NVivo improves the qualitative research process by combining various data formats, allowing researchers to get insightful knowledge that guides their conclusions and advances their fields of study (Lovelace et al., 2020; Kraiwanit et al., 2023; Niedbalski & Ślęzak, 2023).

OVERVIEW OF ATLAS.TI

A sophisticated QDA tool called ATLAS.ti was created to help researchers manage and analyze complicated qualitative data. ATLAS.ti is a popular tool in many fields, such as market analysis, health research, and social sciences, because of its strong features and intuitive design. The program is appropriate for extensive qualitative research since it allows users to input a variety of data kinds, such as text, audio, video, and photos. Because of its adaptability, researchers may work with a wide range of data types, which improves their capacity to examine complex and varied data (Smit & Scherman, 2021). By supporting a hierarchical data structure, ATLAS.ti enables users to develop projects that illustrate the connections between various data segments and codes. The ability to code is one of ATLAS.ti's main advantages. Users can systematically organize and categorize data by assigning codes to particular data segments (Kalpokaite & Radivojevic, 2020; Gupta, 2023). With the software's support for many coding techniques, such as open, axial, and selective coding, researchers may modify their strategy according to the analytical framework they are using. This methodical coding procedure adds to the richness of qualitative analysis by assisting in the discovery of themes and patterns. Strong visualization capabilities are another feature that ATLAS.ti provides to improve data understanding (Soratto et al., 2020). To provide a better understanding of the intricate linkages included in the data, researchers might utilize networks and diagrams to show the links between codes and concepts. These graphic aids are useful for analysis as well as for effectively communicating results to academic or stakeholder audiences. Because of the software's sophisticated querying features, researchers may do in-depth searches and extract certain data from their data sets (Smit & Scherman, 2021). Large projects, where manually sorting through data might be time-consuming, benefit greatly from this feature. ATLAS.ti facilitates effective data retrieval, which improves research efficiency and encourages more in-depth interaction with the data. ATLAS.ti offers a wealth of online resources, documentation, and tutorials to help users navigate its functionalities (Gupta, 2023). In order to ensure that users can fully utilize the software's capabilities and apply it successfully in their analyses, these resources are beneficial for both inexperienced and seasoned researchers. With a wide range of capabilities that make managing, categorizing, and analyzing various data kinds easier, ATLAS.ti is an effective tool for qualitative researchers. Its adaptability, together with its sophisticated visualization and querying features, enables researchers to carry out in-depth and insightful studies, which eventually improves their comprehension of intricate qualitative phenomena (Smit & Scherman, 2021; Gupta, 2023).

Key Features and Functionalities

Several features and functions that improve ATLAS.ti's usefulness as a tool for QDA are included. The foremost advantage of ATLAS.ti is its capacity to import and handle a wide range of data formats, such as text documents, audio files, photos, videos, and webpages (Friese, 2019; Soratto et al., 2020). This adaptability makes it possible for researchers to work with several formats on a same project, which makes it easier to analyze qualitative data more thoroughly. Particularly impressive are ATLAS.ti's coding and classification abilities. The program allows users to methodically assign codes to certain data segments through the use of a hierarchical coding framework that supports a variety of coding strategies, including axial, selective, and open coding (Ronzani et al., 2020). Enhancing the analytical process, this structured method helps uncover themes, patterns, and concepts. ATLAS.ti increases the quality of qualitative research by offering an organized framework for coding. Further assisting with data interpretation are the visualization tools included in ATLAS.ti. According to Rambaree (2013) researchers can depict linkages between codes, concepts, and data segments through the creation of networks and diagrams. These graphic depictions help to communicate study findings more effectively and provide a greater grasp of intricate connections. Visualizing data relationships facilitates analysis and improves the way results are presented to a range of audiences (Soratto et al., 2020). Another important aspect of ATLAS.ti is its advanced querying and retrieval capabilities. Researchers may do comprehensive searches throughout their data sets with the program, which makes it possible to get certain information based on a variety of criteria (Kuckartz, 2019). This feature is especially helpful for large projects, where it might be difficult to navigate through vast amounts of data. The effectiveness of the querying process facilitates data analysis and improves accessibility, encouraging a deeper interaction with the content. Reflective analysis is further supported by ATLAS.ti's memoing and annotating tools. Memos may be created by researchers and connected to certain codes or portions of data, which enables them to record ideas, conclusions, and interpretations as they go (Paulus, 2017). This feature is useful for subsequent reflection and analysis in addition to helping to record the researcher's developing understanding. The program facilitates cooperation and enhances collective analytical efforts by allowing numerous users to collaborate on the same topic using collaborative capabilities (Kuckartz & Rädiker, 2022). Researchers can collaborate and gain common understanding by exchanging codes, remarks, and data segments. As well, users may build thorough documentation of their findings with ATLAS.ti's tools for exporting data and creating reports (Friese, 2019). Since it enables researchers to combine code lists and visualizations into coherent reports, this capability is crucial for communicating findings to academic or stakeholder audiences. Both inexperienced and seasoned researchers will find ATLAS.ti's user-friendly interface beneficial. Data administration and coding procedures are made easier by the layout's drag-and-drop functionality and simple organization (Smit, 2021). By putting a strong emphasis on usability, researchers may interact with their data more successfully and enjoy a better overall study experience.

User Interface and Usability

ATLAS.ti's user interface is intended to make navigating around easier and improve the user experience in general. Its well-thought-out design accommodates both inexperienced and seasoned researchers. According to Friese (2019), the dashboard offers a summary of active projects and facilitates rapid access to various data sets, scripts, and analytic tools. By using this simplified method, users may spend more time concentrating on their study because it takes less effort to find certain traits. The user-friendly design of ATLAS.ti is one of its most noteworthy features. Drag-and-drop features in the UI make it easy for users to effectively assign codes and arrange data (Smit, 2021). This feature lowers the barrier to entry and promotes efficient engagement with the data, making it especially helpful for researchers who might not be familiar with QDA software. Moreover, ATLAS.ti offers thorough tutorials and support materials to assist users in realizing the full potential of the program. According to Kucharz (2019), these tools provide researchers with confidence as they negotiate the intricacies of qualitative data analysis by providing assistance on a range of features. Such resources are crucial for creating a fruitful study atmosphere, especially for those who might be unfamiliar with qualitative approaches. The configurable interface of ATLAS.ti is another noteworthy feature. According to Ronzani et al. (2020), users have the ability to customize the layout to suit their tastes, resulting in a more individualized experience that complements specific processes. By allowing researchers to organize tools and resources in a way that best meets their analytical needs, this flexibility improves usability. The usefulness of the program is additionally enhanced by its responsive design. According to Friese (2019), ATLAS.ti is compatible with several platforms, such as tablets and smartphones, making it easier to access projects while on the road. This flexibility ensures that the analysis process stays fluid and is especially useful for researchers who may need to study data in various situations or environments.

Types of Data Supported

ATLAS.ti is an extremely flexible program for QDA that can handle a wide range of data formats. This makes it a significant resource for academics working in many fields. This skill makes it possible to conduct thorough studies that fully capture the richness and complexity of qualitative data, which is crucial for comprehending social interactions, cultural phenomena, and human behavior (Gulsia & Yadav, 2023). Textual data is one of the main data kinds that ATLAS.ti supports. Numerous formats, such as transcripts from focus groups, interviews, and open-ended survey questions, may be imported and analyzed by researchers. This capability makes it possible to analyze language, themes, and narratives that come up in qualitative research in great depth. Researchers might gain deeper insights into participants' experiences and opinions by classifying and categorizing these writings to reveal patterns and trends (Friese, 2019; Kalpokaite & Radivojevic, 2020). The program's ability to handle vast amounts of text documents makes it even more useful for doing long qualitative research. In addition to textual data, researchers may analyze audio recordings with ATLAS.ti, which allows them to examine conversations, interviews, and other spoken information. By linking individual audio segments to pertinent codes, users of the program are able to record not just spoken material but also subtleties of tone, pitch, and emotional expression. Understanding the nuances of communication and the environment in which interactions take place requires a multidimensional approach to audio data (Gulsia & Yadav, 2023). Additionally, while they do analysis, researchers may record their comments and interpretations in notes that are connected to audio samples. Videos, graphics, and other visual data types are also supported by ATLAS.ti. Photographs and illustrations pertinent to the study can be imported by researchers, who can then associate these visual resources with certain codes and topics (Kuckartz, 2019). This feature is especially helpful in disciplines like sociology, anthropology, and cultural studies, where the use of visual objects to convey context and meaning is crucial. Through the integration of visual and auditory data analysis, scholars may get a comprehensive comprehension of their study topics. Web-based data integration is supported by ATLAS.ti due to the growing significance of digital settings. Scholars have the ability to import data from blogs, websites, online forums, and social media platforms, which makes it easier to analyze current social concerns in a digital environment (Ronzani et al., 2020). This feature is especially helpful for research on digital narratives, online communities, and how technology affects relationships and communication. The capacity to examine web-based data allows academics to investigate new patterns and phenomena that are influencing contemporary society. The efficiency with which ATLAS.ti can process massive amounts of data is another important feature. The organizing functions of the program facilitate the methodical classification and administration of various forms of data, which may be advantageous for researchers engaged in long qualitative studies. This is necessary to keep complicated studies coherent and clear, allowing researchers to browse massive data sets without becoming overwhelmed (Friese, 2019). Large-scale qualitative data management and analysis made possible by the program encourage in-depth research. ATLAS.ti's ability to integrate many data types makes it possible to approach qualitative analysis in a complete manner. Through simultaneous interaction with textual, aural, visual, and web-based data, the program facilitates a comprehensive comprehension of the phenomenon being studied. This methodology is consistent with the principles of qualitative research, which highlight the richness and complexity of the human experience and allow researchers to examine relationships between data types.

Table 1. Comparison of ATLAS.ti and NVivo

Aspect	ATLAS.ti	NVivo
User Interface	Well-organized layout; easy navigation for all users	Ribbon-style toolbar; customizable workspace
Usability	Intuitive drag-and-drop functionalities; extensive tutorials	User-friendly; strong support resources available
Types of Data Supported	Text (transcripts, documents), audio, video, images, web-based data	Text, audio, video, images, social media content
Coding Capabilities	Advanced coding features; supports thematic analysis and grounded theory	Robust coding features; flexible coding and analysis
Visualization Tools	Word clouds, charts, and models for pattern identification	Models, maps, and charts for data representation
Data Querying	Complex data querying capabilities for targeted searches	Powerful querying tools for in-depth data analysis
Analytical Approaches	Supports thematic analysis and content analysis	Supports various qualitative research methodologies
Holistic Approach	Integrates multimedia data for comprehensive analysis	Incorporates multimedia data for multi-dimensional insights

continued on following page

Table 1. Continued

Aspect	ATLAS.ti	NVivo
Ease of Use	Intuitive interface with drag-and-drop functionality; user-friendly for both novice and experienced researchers	User-friendly interface with a ribbon-style toolbar; customization options enhance usability
Functionality	Advanced coding, visualization tools, and complex querying capabilities; supports a wide range of data types	Robust coding features, powerful visualization tools, and comprehensive data querying options
Cost	Pricing may vary based on licensing options (e.g., student discounts, institutional licenses); generally considered competitive	Pricing structures vary; offers student and educational discounts; often perceived as more expensive
Support and Resources	Extensive tutorials, user forums, and support documentation available; responsive customer service	Comprehensive support resources including tutorials, user guides, and community forums; customer support rated highly
Integration and Compatibility	Compatible with various data formats and tools; seamless integration with Microsoft Office	Integrates well with multiple data sources and software; strong compatibility with Office products and web data
Performance	Efficient handling of large data sets; quick response times during analysis	Known for robust performance; can manage extensive qualitative data efficiently
Flexibility	Supports a variety of analytical approaches; adaptable to different research methodologies	Highly flexible in terms of coding and analysis; suitable for diverse qualitative research contexts
Community and User Base	Active user community that shares best practices; well-regarded in academic circles	Strong user community; widely used in academia and industry, with extensive resources shared online

Source: Primary data (Author created)

The comparative evaluation of NVivo and ATLAS.ti, as summarized in Table 1, showcases distinct strengths and functionalities tailored to various research needs.

STRENGTHS AND WEAKNESSES OF NVIVO

Benefits for Researchers

Because of its various benefits, NVivo is a helpful tool for academics evaluating qualitative data. Its broad coding abilities are one of its most impressive features. The tool makes it easier for researchers to efficiently classify and analyze qualitative data, opening the door to a variety of analytical methods including theme analysis and grounded theory. Because of its coding flexibility, which allows users to extract insightful findings from massive, complex data sets, qualitative research may go deeper (Dhakal, 2022). NVivo also provides powerful visualization tools to aid in the discovery of patterns and relationships within data. Word clouds, charts, and models are a few examples of elements that assist researchers visually represent their findings and simplify difficult information. The ability to visualize data enhances comprehension and may boost the overall impact of research findings (Phillips & Lu, 2018). One of NVivo's main benefits is its querying capabilities. Researchers may do complex data searches and get specific information precisely by using the program. This skill enables researchers to more thoroughly analyze the nuanced aspects of their data, hence deepening the investigation (Limna, 2023). Additionally, NVivo supports a wide range of data types, including text, audio, video, photographs, and social media information, which makes it feasible to conduct a holistic approach to qualitative research. Because it can accommodate a range of qualitative data sources, this flexibility improves the analysis. NVivo's intuitive user interface is a further advantage. The format is easy to navigate and provides enhanced use for both novice and expert researchers (Elliott-Mainwaring, 2021). Workspace options that are adjustable allow users to further tailor the application to their own needs. When paired with an abundance of helpful resources—like tutorials, discussion boards, and user manuals—NVivo empowers researchers to get the most out of the software (Soehardi et al., 2021; Almusallam, 2021; Limna, 2023). Nevertheless, NVivo has a number of shortcomings. Cost is one of the key difficulties, as it may be fairly pricey in compared to other tools for evaluating qualitative data. Smaller colleges or independent researchers with limited resources would find this pricing structure unaffordable. Moreover, even with NVivo's user-friendly interface, some users may find that mastering all of its features requires a significant time investment. Due to the complexity of several elements, there could be a steep learning curve for beginners (Allsop et al., 2022). Resource intensity is also another drawback. A large amount of memory and processing power may be needed for NVivo when dealing with large data sets. Its need may restrict its effectiveness on older or less capable machines, hindering future study. Some customers have also reported intermittent technical issues, such bugs or malfunctions, which might annoy users and impede the analysis process (Dalkin et., 2021). NVivo's collaborative features may not be as robust as those of certain other platforms, despite the fact that it supports solo usage effectively. For study teams that rely on efficient communication and information sharing, this restriction might provide challenges (Kraiwanit et al., 2023).

Limitations and Challenges

Despite its robust capabilities and widespread use in qualitative research, NVivo has flaws and challenges. One of the biggest challenges is the steep learning curve that must be overcome in order to master the software. NVivo's advanced features may be frightening to novice users due to their complexity, despite the program's user-friendly interface. Researchers sometimes need to invest a large amount of time and effort in training in order to fully utilize the capabilities of the program. This might provide a particular challenge for those who are either unfamiliar with digital technologies or are transitioning from manual, traditional methods of qualitative analysis (Allsop et al., 2022). One other significant limitation is the cost of the program. NVivo is one of the priciest qualitative data analysis tools available, therefore independent researchers, students, and small enterprises on a limited budget can find it too expensive. The cost structure, which also includes quarterly updates and maintenance costs, may deter potential customers from making the first purchase. This may make it harder for users to utilize the software, which would limit the opportunity for more in-depth qualitative analysis—especially in environments with few resources. Technical limitations are another issue that NVivo users must deal with (Lovelace et al., 2020). The program consumes a lot of RAM and processing power, particularly when working with large or multi-media datasets. When using older or less powerful computers, researchers may experience data loss, crashes, or slow performance, all of which can significantly impair their study. Moreover, NVivo's reliance on specific system configurations may result in incompatibilities, particularly for users of other operating systems like Mac or Linux. Despite being available for both Mac and Windows, there's a possibility that some features won't function flawlessly on all platforms, which might lead to unequal use. NVivo's limitations also extend to teamwork (Soehardi et al., 2021; Almusallam, 2021). Although the application offers certain collaborative work features, such as project sharing and team coding, these are not as seamless or well-integrated as those available in other collaborative research platforms. Managing collaborative projects may be challenging, and rigorous version control is required to prevent corruption or data loss. This might be a significant drawback for research teams that rely on smooth, real-time collaboration and data integration. Automatic analytic capabilities in NVivo, like as auto-coding, are useful but can also oversimplify complex qualitative data. It is possible for researchers to rely too much on these automated processes, which might result in the loss of nuanced findings that require a more interpretive and human approach (Elliott-Mainwaring, 2021; Kraiwanit et al., 2023). Because of this reliance on automation, the emphasis on careful analysis may inadvertently shift to more mechanical data processing, which may not be consistent with the epistemological foundations of qualitative research. NVivo also brings up privacy and data security concerns, particularly when handling sensitive or private data. Data loss or breaches are always possible, even with a program that offers safe data management and encryption. Because researchers must use best practices for data protection with attentiveness, the research process may become more challenging. While NVivo has powerful analytical tools for qualitative data, there are a number of issues with the software (Dhakal, 2022; Niedbalski & Ślęzak, 2023). Researchers may run into a variety of challenges, such as a steep learning curve, high expenses, technological requirements, and limited collaborative features. Data security concerns and the potential for an over-reliance on automated features also require careful consideration. Researchers must weigh these restrictions against the program's benefits to determine whether it is suitable for their specific research goals (Alam, 2021; Soehardi et al., 2021).

STRENGTHS AND WEAKNESSES OF ATLAS.ti

Benefits for Researchers

The robust tool suite provided by ATLAS.ti is a well-known improvement to the qualitative research process as a whole. One of its key benefits is the ability to manage and analyze complex data from a range of sources, including text, audio, video, and images. Researchers may integrate and assess different kinds of qualitative data within a single project thanks to this versatility, which also makes it feasible to triangulate data from several sources and provide a holistic view of the study material. This flexibility is also beneficial to multidisciplinary research that requires the synthesis of several data sources for in-depth analysis (Friese, 2019). The strong coding capabilities of ATLAS.ti are yet another significant advantage. The application may be used to develop sophisticated coding systems that researchers can use to systematically organize and categorize data. This feature is crucial for research with large datasets since it maintains a systematic and uniform coding procedure. Furthermore, because ATLAS.ti supports both deductive and inductive coding approaches, researchers may work with pre-existing codes or develop new ones as they go through the analytic process. This adaptability is essential for studies that require a dynamic coding method since it permits the identification of fresh themes and patterns as the inquiry progresses (Kalpokaite & Radivojevic, 2020). For seeing connections and linkages in the data, the software's visualization tools—such as concept maps and network views—are quite useful. Researchers may gain a deeper understanding of the underlying patterns and connections in the data by using these tools to illustrate their theoretical frameworks and coding structures. This visual approach enhances data interpretation and facilitates a wider audience's comprehension of complex analytical findings (Ronzani et al., 2020). ATLAS.ti also offers powerful teamwork capabilities. When several team members must work together on the same dataset, it makes project sharing and merging easier. This is especially important for multi-researcher projects. The program's capabilities for documenting changes and combining different code efforts (Kuckartz & Rädiker, 2022). This collaboration capability is especially helpful for large-scale projects or research teams that are geographically scattered, as it speeds up productivity and reduces errors that might occur when merging projects manually (Paulus & Lester, 2020). The software's intuitive interface is beneficial to both novice and expert researchers. ATLAS.ti offers a range of online courses, resources, and customer support to help users quickly become familiar with its functionality. Because the support infrastructure makes it easier to use the program's sophisticated features and decreases the initial learning curve, it is especially helpful for people who are new to the software or are switching from other qualitative data analysis tools (Rambaree, 2013). Researchers can do complex searches and extract specific data segments based on a variety of characteristics, such as codes, memoranda, or document categories, thanks to ATLAS.ti's robust query capabilities. This ability is crucial for in-depth analysis since it enables researchers to go deeper into their data and gain important insights relevant to their study issues (Soratto et al., 2020). Because ATLAS.ti can run such large queries, it is a good choice for researchers doing in-depth and complex studies. This can significantly increase the rigor and depth of qualitative analysis (Friese, 2019). ATLAS.ti provides researchers with an extensive set of tools for coding, visualizing, collaborating, and querying qualitative data. Thanks to its extensive support system, user-friendly interface, and capacity to handle a wide range of data types, it is an essential tool for both novice and seasoned qualitative researchers. The program is widely used in academia due to its significant benefits, which include its ability to facilitate collaborative research projects and to enable thorough, rigorous analysis (Gupta, 2023).

Limitations and Challenges

While ATLAS.ti is a feature-rich tool, there are a number of issues and challenges that could limit its effectiveness and usefulness in qualitative research. One of the primary problems with the program is its steep learning curve. Even though ATLAS.ti has a plethora of resources, tutorials, and customer support, new users might find it challenging to fully comprehend and utilize its sophisticated functions, especially if they haven't used qualitative data analysis software before (Kalpokaite & Radivojevic, 2020; Gupta, 2023). Researchers who are switching from more simplistic software—like NVivo, which is occasionally perceived as being more user-friendly—might find this learning curve particularly daunting (Kuckartz & Rädiker, 2022). Another major restriction affects the program's performance and system requirements. The memory and processing requirements of ATLAS.ti, especially in its latest versions, may be too much for researcher computers that are more recent or less powerful. This issue is particularly apparent when dealing with large datasets, multimedia files, or intricate coding schemes, and it may result in lower performance or even program failures (Friese, 2019). These technology issues might disrupt workflow and lead to discontent when there are tight deadlines or time-sensitive analysis. The intricacy of ATLAS.ti may also be the reason for underusing its functionalities. Many users may not utilize the program to its full potential even though it offers a wide range of sophisticated tools for data analysis, visualization, and collaboration (Paulus, 2017; Paulus & Lester, 2020). This is because they may not be aware of or comprehend these possibilities. This issue is made worse by the fact that ATLAS.ti has fewer comprehensive training materials or courses available than more widely used applications like NVivo. As a result, researchers might decide to just use the basic elements, which would limit the scope and depth of their analyses (Soratto et al., 2020). There are also compatibility and integration problems with ATLAS.ti. The software supports a wide variety of data types, although occasionally it has problems interacting with other applications and systems. For example, importing data from external sources—like survey platforms or other qualitative tools—can occasionally be time-consuming and require further formatting or human revisions. This lack of seamless interoperability may make it challenging for researchers to integrate data from multiple sources or use a variety of software tools in their research process (Paulus & Lester, 2020). The price of ATLAS.ti is a significant barrier to its general use. The program is occasionally perceived as being expensive, particularly for independent researchers or those working in developing countries where research funding is scarce. The overall cost may still be excessive even with academic and student licenses available, especially when compared to free or affordable alternatives (Friese, 2019). Some researchers may decide to employ less sophisticated instruments due to budgetary constraints, however these instruments may not offer the same level of functionality or analytical depth (Gulsia & Yadav, 2023). In addition, not all qualitative research methodologies will be compatible with the coding and data organization techniques employed by ATLAS.ti. Some scholars have criticized the program for promoting a certain rigidity in coding procedures, which may obstruct more adaptable, interpretive data processing techniques. The standardized coding system of ATLAS.ti is useful for systematic research, but it may not be the greatest option for methods that require more flexibility and adaptation, like ethnography or story analysis (Paulus & Lester, 2020). This methodological mismatch may limit the software's applicability to the full spectrum of qualitative research approaches. In the end, even though ATLAS.ti promotes teamwork, using the tool for project management can be difficult. Coordinating multiple users, synchronizing coding techniques, and merging project files without sacrificing data integrity may prove to be challenging for larger research teams (Friese, 2019). In contrast to some of its rivals, Gulsia & Yadav (2023) point out that researchers are forced to rely on external file-sharing services due to the lack of an integrated cloud-based platform for real-time collaboration, which can hinder collaboration and lead to security and version control difficulties. ATLAS.ti has numerous shortcomings even though it's a useful tool for assessing qualitative data (Smit, 2021). Obstacles for researchers can be quite challenging and include a steep learning curve, performance issues, cost, and challenges with integration and teamwork (Ronzani et al., 2020). Prospective users need to be aware of these challenges in order to make an informed decision about whether ATLAS.ti is the ideal tool for their specific research goals and circumstances (Gulsia & Yadav, 2023).

CONCLUSION

This side-by-side analysis of ATLAS.ti and NVivo provides a comprehensive look at their benefits, drawbacks, and suitability for various kinds of qualitative research. NVivo is commended for its strong coding abilities, user-friendly interface, and extensive data visualization features. Because it supports a wide range of data types, it is a versatile choice for academics dealing with a variety of datasets, including text, audio, and visual content. NVivo's ability to carry out complex searches and efficiently obtain relevant data is particularly helpful for methodical and structured qualitative research methodologies, like content and theme analysis. On the other hand, ATLAS.ti is renowned for its robust relationship-building and network view capabilities, which are highly useful for jobs demanding intricate conceptual mapping and theory formulation. With its flexible coding options and support for several data types, the platform makes it easier to conduct a thorough and sophisticated analysis of qualitative data. While ATLAS.ti's high equipment needs can pose a challenge in certain research scenarios, its complex interface and steep learning curve could be a turnoff for inexperienced users. While both programs have unique features that can significantly enhance the qualitative research process, there are also disadvantages that researchers should consider when choosing the best software for their needs. NVivo is commendable for its user-friendliness and abundance of support resources, whereas ATLAS.ti is recommended for its analytical depth and ability to handle complex data structures.

Recommendations for Researchers Based on Specific Needs

- Researchers who are new to utilizing software for qualitative data analysis or who are searching for an intuitive interface might give NVivo some thought. Its friendly community, extensive help features, and user-friendly design make it the ideal choice for novices. Because it provides easy navigation and optimized methods, it is also suitable for researchers with limited technical experience.

- For projects requiring complex data analysis, such building complex networks or exploring deep relational data, ATLAS.ti would be a better option. Deep, multi-layered insights are made easier by its flexible coding options and strong network perspective, which are crucial for grounded theory or phenomenological research.
- NVivo offers great data visualization tools, including word clouds, charts, and models, for researchers who need to present their findings in an aesthetically pleasing manner. For individuals working on projects that require regular reporting or results dissemination to a non-academic audience, NVivo's visualisation tools may be extremely useful.
- When working in groups or on extensive projects, researchers should consider the collaborative aspects of both technologies. While operating ATLAS.ti's collaborative features without external file-sharing services may be more challenging, NVivo offers robust collaboration tools, such as cloud-based sharing and synchronization. Consequently, NVivo might be a superior option for projects requiring real-time collaboration among multiple locations.
- For investigators or groups with limited funding, the cost of the program could be a deciding issue. While NVivo and ATLAS.ti offer academic pricing, it's crucial to carefully consider the entire cost of licenses as well as any potential additional costs for upgrades or maintenance. While they might not offer the same level of support and utility, alternatives that are free or less priced could also be investigated by researchers.

FUTURE TRENDS IN QUALITATIVE DATA ANALYSIS TOOLS

Tools for analyzing qualitative data will likely change in the future due to a variety of new advances that are being driven by both the evolving needs of academics and technological advancements. Machine learning (ML) and artificial intelligence (AI) technologies are being added to software for qualitative data analysis more and more. With the help of these technologies, coding and pattern recognition duties can be automated, saving time and allowing scholars to focus more on theory development and interpretation. However, the ethical implications and any biases resulting from AI-driven evaluations must be carefully considered. Another trend is the increasing significance of mixed-methods research, which combines qualitative and quantitative data. Thus, more seamless connections between statistical software and quantitative data analysis platforms should be made possible by future QDA systems, enabling researchers to perform in-depth data analysis and interpretation. This integration will support the growing need for multidisciplinary research, which looks at complex processes from multiple perspectives. Cloud-based solutions and real-time communication tools should also become increasingly common. Software for qualitative data analysis is anticipated to keep up with the COVID-19 pandemic's increased usage of online collaboration tools and remote work. With improved cloud features, researchers can ensure data security and integrity, collaborate more effectively, and manage projects across several locations. User experience and accessibility are, at last, becoming increasingly crucial. Qualitative methodologies are being used by more researchers in various domains, which is driving up demand for software that is easy to use and requires less training. Future developments of QDA tools will most likely prioritize ease of use and simplicity above analytical power in order to cater to a larger range of users. The characteristics that NVivo and ATLAS.ti offer that are beneficial for analyzing qualitative data rely on the specific needs and constraints of the research project. The requirements of their study, their level of technical proficiency, and the resources at their disposal should all be carefully considered by researchers before choosing the finest software. Future advancements in QDA tools will further enhance the research process as qualitative research advances, increasing the tools' potency and accessibility to a wider audience.

REFERENCES

Alam, M. K. (2021). A systematic qualitative case study: Questions, data collection, NVivo analysis and saturation. *Qualitative Research in Organizations and Management*, *16*(1), 1–31. DOI: 10.1108/QROM-09-2019-1825

Allsop, D. B., Chelladurai, J. M., Kimball, E. R., Marks, L. D., & Hendricks, J. J. (2022). Qualitative methods with Nvivo software: A practical guide for analyzing qualitative data. *Psych*, *4*(2), 142–159. DOI: 10.3390/psych4020013

Almusallam, I. (2021). Using NVivo for Data Analysis in Qualitative Research. In *Paper presentation], Conference: Academic Improvement Committee at English Language & Literature Department*. King Saud University.

Bassett, R. (2004). Qualitative data analysis software: Addressing the debates. *Journal of Management Systems*, *16*(4), 33–39.

Blismas, N. G., & Dainty, A. R. (2003). Computer-aided qualitative data analysis: Panacea or paradox? *Building Research and Information*, *31*(6), 455–463. DOI: 10.1080/0961321031000108816

Bourdon, S. (2002, May). The integration of qualitative data analysis software in research strategies: Resistances and possibilities. In *Forum Qualitative Sozialforschung/Forum: Qualitative. Social Research*, *3*(2), .

Brown, D. (2002, May). Going digital and staying qualitative: Some alternative strategies for digitizing the qualitative research process. In *Forum Qualitative Sozialforschung/Forum: Qualitative. Social Research*, *3*(2), .

Bryman, A., & Burgess, R. G. (Eds.). (1994). *Analyzing qualitative data* (Vol. 11). Routledge.

Butcher, L. M., & Holbrook, A. (1996). Uses of qualitative data analysis software in educational research: The literature, the hard questions and some specific research applications. *Australian Educational Researcher*, *23*(3), 55–80. DOI: 10.1007/BF03219631

Dalkin, S., Forster, N., Hodgson, P., Lhussier, M., & Carr, S. M. (2021). Using computer assisted qualitative data analysis software (CAQDAS; NVivo) to assist in the complex process of realist theory generation, refinement and testing. *International Journal of Social Research Methodology*, *24*(1), 123–134. DOI: 10.1080/13645579.2020.1803528

David, A., Ahmed, R. R., Ganesh Kumar, C., & Sankar, J. G. (2020). Consumer purchasing process of organic food product: An empirical analysis. *Journal of Management Systems. Quality - Access to Success*, *21*(177), 128–132.

David, A., Jeganathan, G. S., Sikandar, M. D., & Azam, S. A. (2023, May). Internet Users Top Concerns Ensuring Data Privacy, Security, and Protection. In *Arokiaraj David, Jeganathan Gomathi Sankar & Md Sikandar Azam, —Internet Users Top Concerns Ensuring Data Privacy, Security, and Protection , 10th International Conference on Multidisciplinary Research and Modern Education, ISBN* (No. 978-920, pp. 5-20241).

David, A., Maheswari, M., Sankar, G., Christy, F., Karuppannan, A., & Banumathi, M. (2023). Mental illness of Pregnant Women determined by Covid-19: An Empirical Analysis of Pre-and- Post Delivery. *The Saybold Report Journal*, *18*(03), 1706–1721.

David, A., & Sankar, J. G. (2024). *Generative AI and Implications for Ethics, Security, and Data Management*. IGI Global Scientific Publishing.

David, A., & Sankar, J. G. (2024). *Generative AI for Transformational Management*. IGI Global Scientific Publishing.

David, A., Sankar, J. G., Ganeshkumar, C., & Kashif, M. (2024). An Introduction to Generative AI: Crafting Worlds From Code. In *Generative AI and Implications for Ethics, Security, and Data Management* (pp. 1-16). IGI Global.

Dhakal, K. (2022). NVivo. *Journal of the Medical Library Association: JMLA*, *110*(2), 270. DOI: 10.5195/jmla.2022.1271 PMID: 35440911

Dohan, D., & Sanchez-Jankowski, M. (1998). Using computers to analyze ethnographic field data: Theoretical and practical considerations. *Annual Review of Sociology*, *24*(1), 477–498. DOI: 10.1146/annurev.soc.24.1.477

Elliott-Mainwaring, H. (2021). Exploring using NVivo software to facilitate inductive coding for thematic narrative synthesis. *British Journal of Midwifery*, *29*(11), 628–632. DOI: 10.12968/bjom.2021.29.11.628

Friese, S. (2019). Atlas. ti 8 Mac-User Manual Updated for Program Version 8.4. *ATLAS. ti, Berlin.*

Ganeshkumar, C., David, A., & Sankar, J. G. (2024). Blockchain Technology Acceptance in Agribusiness Industry. In Idrees, S. M., & Nowostawski, M. (Eds.), *Blockchain Transformations. Signals and Communication Technology*. Springer., DOI: 10.1007/978-3-031-49593-9_15

Ganeshkumar, C., David, A., Sankar, J. G., & Saginala, M. (2023). Application of Drone Technology in Agriculture: A Predictive Forecasting of Pest and Disease Incidence. In *Applying Drone Technologies and Robotics for Agricultural Sustainability* (pp. 50–81). IGI Global.

Ganeshkumar, C., Sankar, J. G., & David, A. (2023). Adoption of Big Data Analytics: Determinants and Performances Among Food Industries. [IJBIR]. *International Journal of Business Intelligence Research, 14*(1), 1–17. DOI: 10.4018/IJBIR.317419

Ganeshkumar, C., Sankar, J. G., & David, A. (2023). Impact of Artificial Intelligence on Agriculture Value Chain Performance: Agritech Perspective. In *Blockchain, IoT, and AI Technologies for Supply Chain Management* (pp. 71-98). CRC Press.

Ganeshkumar, C., Sankar, J. G., & David, A. (2025). Impact of Artificial Intelligence on Organizational Performance of Agritech Firm. In *AI-Powered Business Intelligence for Modern Organizations* (pp. 265–282). IGI Global.

Gilbert, L. S. (2002). Going the distance:'closeness' in qualitative data analysis software. *International Journal of Social Research Methodology, 5*(3), 215–228. DOI: 10.1080/13645570210146276

Gilbert, L. S., Jackson, K., & Di Gregorio, S. (2014). Tools for analyzing qualitative data: The history and relevance of qualitative data analysis software. *Handbook of research on educational communications and technology*, 221-236.

Gulsia, O., & Yadav, S. (2023). Conceptualizing network approaches for a successful farm entrepreneurship using ATLAS. ti®. *Sustainable Technology and Entrepreneurship, 2*(1), 100026. DOI: 10.1016/j.stae.2022.100026

Gupta, A. (2023, January). A Qualitative Study of Management Practices & Influence on Middle Managers' Morale in Public Sector Banks Using ATLAS. ti. In *World Conference on Qualitative Research* (pp. 18-37). Cham: Springer International Publishing.

Hesse-Biber, S. (2022). Mixed methods research designs in qualitatively driven research. *The SAGE Handbook of Qualitative Research Design*, 618-635.

Hesse-Biber, S., Dupuis, P., & Kinder, T. S. (1991). HyperRESEARCH: A computer program for the analysis of qualitative data with an emphasis on hypothesis testing and multimedia analysis. *Qualitative Sociology, 14*(4), 289–306. DOI: 10.1007/BF00989642

Jeganathan, G. S., David, A., & Ilangovan, K. (2022). Determination of Hospitality Services Quality and Customer Satisfaction–a Holserv Approach. In *AU Virtual International Conference on" Entrepreneurship & Sustainability in Digital Era" under the theme of" Challenges of Organizational & Business Management in Dynamic Digital Dimension"* (Vol. 3, No. 1, pp. 325-334).

Jeganathan, G. S., & Srinivasulu, Y. (2014). Growth Drivers of Tourism Industry in India. *International Journal of Business and Management*, *2*(11), 100.

Jeganathan, G. S., & Srinivasulu, Y. Impact of tourism on macro–economic indicators: A Comparative study of BRICS nations.

Jeganathan, G. S., & Srinivasulu, Y. The international journal of business & management.

Kalpokaite, N., & Radivojevic, I. (2020). Teaching qualitative data analysis software online: A comparison of face-to-face and e-learning ATLAS. ti courses. *International Journal of Research & Method in Education*, *43*(3), 296–310. DOI: 10.1080/1743727X.2019.1687666

Kalpokaite, N., & Radivojevic, I. (2020). Teaching qualitative data analysis software online: A comparison of face-to-face and e-learning ATLAS. ti courses. *International Journal of Research & Method in Education*, *43*(3), 296–310. DOI: 10.1080/1743727X.2019.1687666

Kraiwanit, T., Limna, P., & Siripipatthanakul, S. (2023). NVivo for social sciences and management studies: A systematic review. *Advance Knowledge for Executives*, *2*(3), 1–11.

Kuckartz, U., & Rädiker, S. (2022). Using Maxqda for integration in mixed methods research. In *The Routledge handbook for advancing integration in mixed methods research* (pp. 540–562). Routledge. DOI: 10.4324/9780429432828-41

Kuckartz, U., Rädiker, S., Kuckartz, U., & Rädiker, S. (2019). Introduction: analyzing qualitative data with software. *Analyzing qualitative data with MAXQDA: text, Audio, and video*, 1-11.

Kumar, P., Kumar, A., & Sankar, J. G. (2020). Service quality of health tourism in India: A structural equation modelling approach. *Solid State Technology*, *63*(2), 1187–1191.

Limna, P. (2023). The impact of NVivo in qualitative research: Perspectives from graduate students. *Journal of Applied Learning and Teaching*, *6*(2).

Lovelace, S., Trudel, C., Dulude, C., & King, W. J. (2020, September). Cost vs. Benefit: what does NVivo video analysis of EMR simulations add to our understanding of user experience? In *Proceedings of the International Symposium on Human Factors and Ergonomics in Health Care* (Vol. 9, No. 1, pp. 24-32). Sage CA: Los Angeles, CA: SAGE Publications. DOI: 10.1177/2327857920091056

MacMillan, K., & Koenig, T. (2004). The wow factor: Preconceptions and expectations for data analysis software in qualitative research. *Social Science Computer Review*, *22*(2), 179–186. DOI: 10.1177/0894439303262625

Morison, M., & Moir, J. (1998). The role of computer software in the analysis of qualitative data: Efficient clerk, research assistant or Trojan horse? *Journal of Advanced Nursing*, *28*(1), 106–116. DOI: 10.1046/j.1365-2648.1998.00768.x PMID: 9687137

Niedbalski, J., & Ślęzak, I. (2023, January). NVivo as a tool for supporting teamwork in the context of qualitative research conducted remotely-opportunities, limitations, and practical tips. In *World conference on qualitative research* (pp. 38-59). Cham: Springer International Publishing. DOI: 10.1007/978-3-031-31346-2_3

Oliveira, G. (2023). Developing a codebook for qualitative data analysis: Insights from a study on learning transfer between university and the workplace. *International Journal of Research & Method in Education*, *46*(3), 300–312. DOI: 10.1080/1743727X.2022.2128745

Oliveira, M., Bitencourt, C., Teixeira, E., & Santos, A. C. (2013, July). Thematic content analysis: Is there a difference between the support provided by the MAXQDA® and NVivo® software packages. In *Proceedings of the 12th European Conference on Research Methods for Business and Management Studies* (pp. 304-314).

Paulus, T., Woods, M., Atkins, D. P., & Macklin, R. (2017). The discourse of QDAS: Reporting practices of ATLAS. ti and NVivo users with implications for best practices. *International Journal of Social Research Methodology*, *20*(1), 35–47. DOI: 10.1080/13645579.2015.1102454

Paulus, T. M., & Lester, J. N. (2020). Using software to support qualitative data analysis. In *Handbook of qualitative research in education* (pp. 420–429). Edward Elgar Publishing. DOI: 10.4337/9781788977159.00048

Peters, V., & Wester, F. (2007). How qualitative data analysis software may support the qualitative analysis process. *Quality & Quantity*, *41*(5), 635–659. DOI: 10.1007/s11135-006-9016-8

Phillips, M., & Lu, J. (2018). A quick look at NVivo. *Journal of Electronic Resources Librarianship, 30*(2), 104–106. DOI: 10.1080/1941126X.2018.1465535

Rambaree, K. (2013). Three methods of qualitative data analysis using ATLAS. ti:'A Posse Ad Esse'. In *ATLAS. ti User Conference 2013, 12-14 September 2013, Technische Universität Berlin, Germany* (Vol. 1, pp. 1-15). Universitätsverlag der TU Berlin.

Richards, L. (2002). Qualitative computing—A methods revolution? *International Journal of Social Research Methodology, 5*(3), 263–276. DOI: 10.1080/13645570210146302

Richards, T., & Richards, L. (1991). The NUDIST qualitative data analysis system. *Qualitative Sociology, 14*(4), 307–324. DOI: 10.1007/BF00989643

Richards, T. J., & Richards, L. (1994). Using computers in qualitative research. *Handbook of qualitative research, 2*(1), 445-462.

Ronzani, C. M., da Costa, P. R., da Silva, L. F., Pigola, A., & de Paiva, E. M. (2020). Qualitative methods of analysis: An example of Atlas. TI™ Software usage. *Revista Gestão & Tecnologia, 20*(4), 284–311. DOI: 10.20397/2177-6652/2020.v20i4.1994

Sankar, G., & Naveenkumar, K. (2018). *Material Handling Management in Third Party Logistics Warehouse. International Journal for Research in Engineering Application & Management*. IJREAM.

Sankar, J. G. (2020). Customer perception about innovative safety food delivery during lockdown. *Journal of Contemporary Issues in Business and Government, 26*(2).

Sankar, J. G. (2020). Innovative celebrity endorsement and consumer purchase. *Journal of Contemporary Issues in Business and Government, 26*(2).

Sankar, J. G. (2020). Innovative consumer behaviour towards sportswear. *Journal of Contemporary Issues in Business and Government, 26*(2).

Sankar, J. G. (2020). Service Factors Influence on Customer Satisfaction in Innovative Ecommerce. *SANKAR, D.(2020). Service factors influence on customer satisfaction in innovative e-commerce.Journal of Contemporary Issues in Business and Government, 26*(2), 651–656.

Sankar, J. G. (2020). Service quality of e-Commerce logistics services. *Journal of Contemporary Issues in Business and Government, 26*(2).

Sankar, J. G., & KUMAR G, H. E. M. A. N. T. H. (2020). Customer perception towards electric two-wheeler innovation. *Journal of Contemporary Issues in Business and Government, 26*(2).

Sankar, J. G. (2024). AI-Driven Marketing Success Stories: A Case Note of Industry Pioneers. In *AI-Driven Marketing Research and Data Analytics* (pp. 48-66). IGI Global. DOI: 10.4018/979-8-3693-2165-2.ch003

Sankar, J. G. (2024). Travel Redefined: Future Trends in AI-Enhanced Tourism Experiences. In *AI Innovations for Travel and Tourism* (pp. 135-162). IGI Global.

Sankar, J. G. (2025). Measuring Service Quality in Generative AI Environments: A Comprehensive GAISQUAL Framework. In *Design and Validation of Research Tools and Methodologies* (pp. 317-346). IGI Global.

Sankar, J. G., & David, A. (2023). Measuring the Service Quality of Artificial Intelligence in the Tourism and Hospitality Industry. In *Handbook of Research on Innovation, Differentiation, and New Technologies in Tourism, Hotels, and Food Service* (pp. 133-155). IGI Global. DOI: 10.4018/978-1-6684-6985-9.ch007

Sankar, J. G., & David, A. (2024). A Comprehensive Examination of Mobile Augmented Reality in Tourism (MART) Adoption: Using the UTAUT2 Framework. In *Contemporary Trends in Innovative Marketing Strategies* (pp. 241-262). IGI Global.

Sankar, J. G., & David, A. (2024). Transforming the Travel Landscape: Smart Contracts in Tourism Management. In *Decentralizing the Online Experience With Web3 Technologies* (pp. 246-266). IGI Global.

Sankar, J. G., David, A., & Ganesh Kumar, C. (2022). Adaptation of Blockchain Technology In HRM. *Available atSSRN* 4515408.

Sankar, J. G., David, A., & Ramasamy, R. (2023). Assessing Water and Sanitation Service Quality at Pilgrimage Sites After COVID-19: An Empirical Investigation. In *Prospects and Challenges of Global Pilgrimage Tourism and Hospitality* (pp. 160-179). IGI Global.

Sankar, J. G., David, A., & Shrutimeena, G. (2024). From Hesitation to Integration: Understanding Digital Healthcare Adoption. In *Multi-Sector Analysis of the Digital Healthcare Industry* (pp. 1-32). IGI Global.

Sankar, J. G., David, A., & Valan, P. (2023). Examining User Understanding and Perceptions of E-Commerce Data Privacy, Security, and Protection. In *Confronting Security and Privacy Challenges in Digital Marketing* (pp. 159–185). IGI Global. DOI: 10.4018/978-1-6684-8958-1.ch008

Sankar, J. G., & Devanand, K. (2020). Social media innovation on consumer buying behaviour. *SANKAR, JG (2020). Social media innovation on consumer buying behaviour.Journal of Contemporary Issues in Business and Government, 26*(2).

Sankar, J. G., & Ilangovan, K. (2025). The Role of Smart Technology in Enhancing Sustainable Tourism Experiences. In *Integrating Architecture and Design Into Sustainable Tourism Development* (pp. 379–408). IGI Global Scientific Publishing. DOI: 10.4018/979-8-3693-7388-0.ch014

Sankar, J. G., & Kant, S. (2025). Assessing Consumer Trust and Perception in Online Food Marketplaces. In *Innovative Trends Shaping Food Marketing and Consumption* (pp. 375–394). IGI Global Scientific Publishing. DOI: 10.4018/979-8-3693-8542-5.ch015

Sankar, J. G., & Kiruba Sagar, M, V. (2020). Examining the Risk Factor Associated with UTAUT Model. *Solid State Technology*, *63*(2), 1196–1198.

Sankar, J. G., Rajeswari, M., & Dwarak, M. V. S. (2020). To Measure the Service Quality of Film and Television Technology. *Solid State Technology*, *63*(2), 1192–1195.

Sankar, J. G., & Srinivasulu, Y. (2014). A Study of Tourism Industry Impact on Macroeconomic Indicators. *GE-International Journal of Management Research*, *2*(11), 2321–1709.

Sankar, J. G., & Srinivasulu, Y. (2014). Contemporary Sustainable Development Practices in Indian Hospitality Sector: Triple Bottom Line Approach. *SSRN*, , 2. DOI: 10.2139/ssrn.4204732

Sankar, J. G., Valan, P., & Siranjeevi, M. S. (2020). Various Models to Evaluate Quality in the Service Industry. In *Digital Transformation and Innovative Services for Business and Learning* (pp. 181–194). IGI Global.

Sasikala, J., & Jeganathan, G. S. (2017). A Study on Employee Engagement in Information Technology (IT) Industry. *SSRN*, *1*(1), 233–239. DOI: 10.2139/ssrn.3536354

Simitha, S., Mugila, P., & Sankar, J. G. (2020). A Study on Customer Relationship Management in Siemens Healthineers. *Solid State Technology*, *63*(2), 1199–1203.

Sin, C. H. (2007). Using software to open up the 'Black Box'of qualitative data analysis in evaluations: The experience of a multi-site team using NUD* IST version 6. *Evaluation*, *13*(1), 110–120. DOI: 10.1177/1356389007073684

Smit, B. (2021). Introduction to ATLAS. ti for Mixed Analysis. In *The Routledge Reviewer's Guide to Mixed Methods Analysis* (pp. 331–342). Routledge. DOI: 10.4324/9780203729434-28

Smit, B., & Scherman, V. (2021). Computer-assisted qualitative data analysis software for scoping reviews: A case of ATLAS. ti. *International Journal of Qualitative Methods*, *20*, 16094069211019140. DOI: 10.1177/16094069211019140

Soehardi, F., Putri, L. D., & Dinata, M. (2021). NVivo Software Training for young researchers. *Mattawang: Jurnal Pengabdian Masyarakat, 2*(1), 8–13. DOI: 10.35877/454RI.mattawang265

Soratto, J., Pires, D. E. P. D., & Friese, S. (2020). Thematic content analysis using ATLAS. ti software: Potentialities for researchs in health. *Revista Brasileira de Enfermagem, 73*(3), e20190250. DOI: 10.1590/0034-7167-2019-0250 PMID: 32321144

SP, M. S., & Sankar, J. G. An Overview of Measurement model in Marketing Research.

Velmurugan, T. A., & Sankar, J. G. (2017). A comparative study on motivation theory with Maslow's hierarchy theory and two factor theory in organization. *Indo-Iranian Journal of Scientific Research, 1*(1), 204–208.

Woods, M., Paulus, T., Atkins, D. P., & Macklin, R. (2016). Advancing qualitative research using qualitative data analysis software (QDAS)? Reviewing potential versus practice in published studies using ATLAS. ti and NVivo, 1994–2013. *Social Science Computer Review, 34*(5), 597–617. DOI: 10.1177/0894439315596311

Chapter 6
Discourse Analysis

Joshua Onome Imoniana
University of São Paulo, Brazil

Cristiane Onome Almeida
https://orcid.org/0009-0003-7933-4168
University of São Paulo, Brazil

ABSTRACT

This chapter aims to discuss the methodological design for the adoption of discourse analysis. The methodological approach that may decide for discourse analysis is an important part of the structure of a study and; it is hinged on various methodological decisions but most importantly five, namely: the choice of the type of research whether qualitative or quantitative; describing its philosophical underpinnings that nurture the study; the approach and strategy to be adopted say discourse analysis in conducting the investigation; means used for data collection and in the language of qualitative research data building, say content analysis, interviews, imagery reviews, social media posts analysis to mention just a few; distinguishing the form of analysis; and evaluating ethical considerations. This chapter is structured as follows: Introduction, Position of a Speech in discourse analysis, Appropriateness of usage of Discourse Analysis, Act of data analysis within the contexts of Discourse Analysis and Conclusion.

1.DISCOURSE ANALYSIS

This chapter aims to discuss the methodological design for the adoption of discourse analysis. The methodological approach that may decide for discourse analysis is an important part of the structure of a study and; it is hinged on various methodological decisions but most importantly five, namely: the choice of the type

DOI: 10.4018/979-8-3373-4432-4.ch006

of research whether qualitative or quantitative; describing its philosophical underpinnings that nurture the study; the approach and strategy to be adopted say discourse analysis in conducting the investigation; means used for data collection and in the language of qualitative research data building, say content analysis, interviews, imagery reviews, social media posts analysis to mention just a few; distinguishing the form of analysis; and evaluating ethical considerations. This chapter is structured as follows: Introduction, Position of a Speech in discourse analysis, Appropriateness of usage of Discourse Analysis, Act of data analysis within the contexts of Discourse Analysis and Conclusion.

1.1. Introduction

Discourse Analysis is a qualitative strategy used in research in the areas of social sciences and humanities to investigate how language is used in data corpora narrated in texts and research papers drawing on their respective contexts. It involves evaluating the structures and patterns of information disseminated, spoken language, writing and signs, as well as visual communications in multimedia and places. The discourse analysis intends to understand how language shapes and is shaped by social, economic and political practices, identities and power relations. Additionally, it comprehends the impact and how it is impacted upon business control environments and cultural norms.

Discourse analysis can toe an interpretative perspective or a critical perspective. On one hand when it follows an interpretative, it tends to be more narrative, descriptive and analysis of wordings. On the other hand, when it is critical, it draws on power relations and more cogent reasons of possible conflicts of interest. As observed by Almeida (2020) the adoption of discourse analysis favors the analysis of imaginaries which in turn helps in the search for legitimacy that justifies the investigation of the phenomenon. Defacto, this forehand methodological decision is very important to create a basis for the design and consequent analysis of data. This is corroborated in the views of Varpio et al (2020) in the adoption of a method.

Discourse analysis is a term first proposed by Zellig Harris (1952) to describe how elements of sentences constitute connected discourse (Huan and Guan, 2020). Discourse Analysis (DA) applies to a wide range of text analysis, including conversations, interviews, media articles, official documents, and more, providing insights into the underlying social and political implications of language use of words derived from these data corpus. Words are woven from a multitude of ideological threads and serve as the fabric of all social relations in all domains (Bakhtin, 1997, p. 41).

Thus, when referring to DA, due to its multidisciplinary nature in the branches of philosophy, linguistics, sociology and psychology, diverse knowledge surfaces one should not forget to point out different epistemological positioning, otherwise, paradigms. This is what Wood and Kroger (2000) termed as the nature and role of theory, types of data for analysis, understandings and treatments of the context and how statements are justified.

Fiorin (1990) informs that DA does not aim to seek "the" true meaning of the text, nor "the" hidden meaning and "the" new and unprecedented interpretation destined to overturn all other interpretations and all other meanings. DA seeks to explain how the text says what it says and why the text says it, to explain the system of rules that presides over the constitution of meaning, as well as the order of needs that the text responds to.

There are various approaches to discourse analysis already in use by researchers in different countries. DA of French origin works with the meaning and not with the content of the text in the search for understanding the construction of discourses. It is attached to distribution in society, how they talk about those involved, about history, about the conditions of living and production. It is, therefore, worthy to describe the French approach of DA as descriptive, analytical and reflective, with a speculative and interpretative nature.

For Mussalim (2009), DA considers the context of enunciation (historical-social context) as part of the meaning of the discourse, otherwise, the meanings are historically constructed. In fact, there is no given meaning, the only true one, but several meanings that are beyond evidence (Florêncio et al., 2009, p. 65). It is not the intention of DA to label a historically constructed meaning as right or wrong.

For Maingueneau (2015, p. 47), the specific interest of DA is to relate the structuring of texts to the social places that proposes them, as its object of analysis is the enunciation device that ties together texts, resulting from the verbal and the institutional communication. Note that the subject itself is not of interest to DA, but rather the ideological place from which the subject enunciates.

History has it that Michel Pêcheux (1938-1983) in the second half of the 20th century, upon identifying the limitations of the linguistic discipline, presented studies that proposed explaining the functioning of discourse through ideology. The French philosopher seeks to reveal the ideology that the texts are destined to hide (Maingueneau, 2015, p. 19), demanding an epistemological rupture, in which theoretical questions relating to ideology and the subject intervene (Musalim, 2009).

2.2 Position of a Speech in discourse analysis

Discourse analysis may be dissociated with the text when speech therein is analyzed. In this aspect, the subject is decentralized from the analysis, since, having been challenged by the State Ideological Apparatus(s) (SIA), it is not seen as "master of its wills", nor responsible for the speech it provides. The subject is now considered as just a social position. Discourse is a form of ideological materialization, with the subject being a deposit of ideology, without its own will, and language is a process that crosses the different fields of society. See the illustration in Figure 1.

Figure 1. Origination of discourse

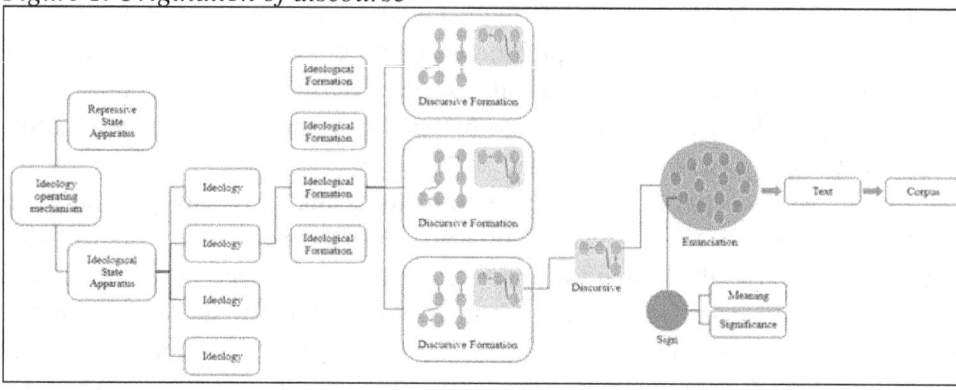

Thus, appear the heterogeneity of discourses. The notion of heterogeneity of discourse was developed throughout several works: Pêcheux and in Remontémonos de Spinoza a Foucault (1981). It is known that, beneath our words, other words resound, words of other subjects, as discourse is of the repeatable order and this repetition does not just refer to what was previously said by the subject of the speech, in the present or in the past (Indursky, 2001). See the illustration demonstrated in Figure 2. carved from the previous literatures.

Figure 2. Discourse analysis phases

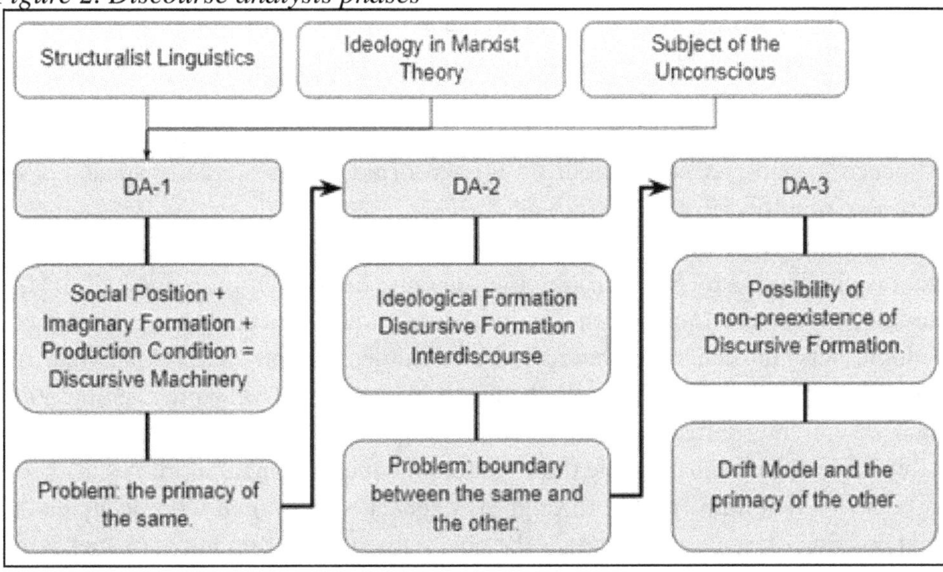

Therefore, there is no neutral or innocent speech, because, when producing it, the subject does so from a social place, from an ideological perspective and, thus, conveys values, beliefs, world views that represent the social places he occupies (Florêncio et al., 2009, p. 25-26). The DA developed by Pêcheux is characterized by revisions and changes in its basic concepts over time, being presented in three periods, as shown below in DA-1, DA-2 and DA-3. Meaning, from the perspective of DA-1, is positioned in the connection between statements and depends on the conditions of production, social position and imaginary formation. In DA-2, meaning is understood as constructed by ideological formation and resides in the space of circulation between discursive formations, that is, in interdiscourse. In turn, in DA-3, there is difficulty in establishing the meaning due to the opacity of the statements.

2.2.1 DA-1 or Automatic Discourse Analysis (ADA-1969)

The subject is not master of his will; either we have a subject who suffers the coercion of an ideological and discursive formation, or we have a subject subjected to his own unconscious nature. (Mussalim, 2009).

In this first period, the subject who believed in being the origin of what he said, the exclusive source of the meaning of his speech (ideological forgetfulness 1) and having autonomy over the form of enunciation (ideological forgetfulness 2) was treated as subject to the discursive machinery.

Discursive machine is defined by Pêcheux (1983) as conditions of stable productions. The subject speaks of a certain place (occupied position) within a social formation (Pêcheux, 1997), which is the conditions of production.

From this perspective, speeches are generated by specific discursive machines, with the subject not being the producer of meanings, nor responsible for the way the speech is said. As pointed out by Possenti (n.d.), what actually speaks is an institution, or a theory, or an ideology.

In the same vein, when using Althusser's (1980) concept of ideology, the individual is challenged by SIA through their agents. Interpellation is the call for the individual to become a subject, that is, for the individual to occupy a social position. Ideology "acts" in such a way that it "recruits" subjects among individuals (recruits them all), or "transforms" individuals into subjects by this very precise operation which we call interpellation. (Althusser, 1980, p. 99).

Let us consider as an example the situation in which a woman represses her son's bad behavior. The family SIA, through its agent, is the woman who occupies the social position of mother, challenges the boy, transforming him into a misbehaving son, another social position. In these positions, the mother has the right to reprimand and the child the duty to obey.

Discourse is always pronounced based on given discursive production conditions (Pêcheux, 1997). The set of statements in the mother's speech, subject to the family SIA, is conceived at this stage as a discursive machine.

The discourses explored in this phase were considered homogeneous and closed in themselves, being for Mussalim (2009) little controversial, that is, with little room for variation in meaning due to a greater silencing of the "other". This is one of the criticisms of DA-1, because when considering the primacy of the "same", otherness becomes absent.

Because there were no connections between the discursive machines, the approximation between the positions occupied by the same subject was impossible. For these reasons, discourse analysis had a more linguistic than social focus.

2.2.2 DA-2

The problem of the primacy of the "same" gave rise to this second phase, in which the concepts presented in DA-1 are reviewed and new concepts such as ideological formations, discursive formations and interdiscourse are introduced. The conditions of discursive productions introduced in DA-1 become a mechanism within discursive formations. Another notable event in DA-2 is that the "other" begins to be considered as different from the "same" and can influence what is or is not said by the subject.

Mussalim (2009) explains that, although the subject begins to play different roles according to the various positions he occupies in the interdiscursive space, he is not totally free, as he is oppressed by the Discursive Formation (DF), which is regulated by an Ideological Formation (IF), which he states.

The premise of IF is the DA-1 concept of ideology, which considers the existence of AIE, which challenges the individual and transforms him into an ideological subject. According to Brandão (2006, p. 46), this ideological interpellation consists of making each individual becoming aware of the impression that they are masters of their own words. That is, to be led to take position in one of the groups or classes of a given social consistency.

Ideological formation is the complex set of attitudes and representations that are neither "individual" nor "universal" but are more or less directly related to the positions of classes in conflict with each other. (Pêcheux and Fuchs, 1997, p. 166).

Attitudes and representations are generalized only in their field of activity, that is, they do not expand to the whole. To generalize, it is necessary to delimit. The relationships between subjects (or classes) can be one of antagonism, domination or coalition. The limit between the text and its context is the place of study of DA, where IF, the articulation between language and discourse, is translated into DF, what can and should be said.

The concept of discursive formation (DF) was conceived by the French philosopher Michel Foucault (1926-1984). For the author, DF makes it possible to move from dispersion to regularity, as it acts as a system of relationships between objects, enunciative types, concepts and strategies. DF was defined by his father as the general enunciative system to which a group of verbal performances obeys. A system that does not govern it alone, since it also obeys, and according to its other dimensions, the logical, linguistic and psychological systems (Foucault, 2008, p. 131).

Pêcheux borrowed the concept of DF, but not in its entirety. Foucauldian DF is conceived in terms of knowledge/powers while Peucheuxian DF is in terms of ideology and class struggles. DF is conceptualized by Pêcheux as:

That which, in a given situation, determined by the state of class struggle, determines what can and should be said, articulated in the form of a harangue, a sermon, a pamphlet, an exhibition, a program, etc. (Pêcheux, 1997, p. 160).

The DF is a set of statements, as demonstrated in Figure 4, marked by the same regularities or "formation rules" conceived as control mechanisms that determine its internal, what belongs, and its external, what does not belong. FI has as one of its components one or several interconnected DFs, as Pêcheux states:

The characteristic of every DF is to conceal, in the transparency of the meaning that is formed in it, the contradictory material objectivity of interdiscourse, which determines this DF as such, a material objectivity that resides in the fact that "something speaks" always "before, in another place and independently", that is, under the domination of the FI complex. (Pêcheux, 1995, p. 162).

In other words, the DF is a collection of possibilities for saying what is actually said, in a given social position, grouped by semantic fields (e.g. paternal, academic and positivist), where it is established what can and what cannot be said by such a subject, in a given situation, within an ideological formation. For Maingueneau (2015, p. 83), the interest of the notion of DF is to allow the constitution of heterogeneous corpora, freely bringing together statements originating from different types of topical units.

The notion of DF involves two functions, the paraphrase and the pre-constructed (Brandão, 2006, p. 48). The first is the resumption of statements and reformulation in a constant effort to close their borders in search of preserving their identity, being a procedure for discovering DA. The latter is a set of relationships and statements that exist before what is said, helping in the interpretation of undeclared sayings, in other words, it is a mechanism, at the level of discourse, to give meaning to what is said with a non-literary meaning., based on discursive memory.

Discursive memory is social and collective, and represents the stabilization of discourse, the possibility of repetition. It is the "already said", the previously constructed meanings, absorbing the event, that is, placing it within the already existing regularity, making it predictable and appropriate.

Discursive memory would be that which, in the face of a text that appears as an event to be read, comes to reestablish the "implicit" (that is, more technically, the pre-constructed elements, cited and reported elements, transversal discourses, etc.) of which its reading requires: the condition of the legible in relation to the legible itself. (Achard et al., 1999, p. 52).

A discursive event located at the meeting point of a current situation and a memory (Pêcheux, 2006, p. 17) is something unexpected, it is something that is born in an implosion, at a given moment and from this causes resignifications of the already existing discourse. Indursky's (2003) understanding of discursive event is:

Slips of meaning may occur, which are responsible for the reorganization of memory. These movements, which lead to the return of memory, also allow us to establish a rupture with the network of formulations to which the statement is related and to inaugurate a new network of formulations. In other words, this rupture indicates that the meaning has derived, becoming different.

When the event manages to cross/interrupt the discursive memory, it displaces and deregulates the implicit elements associated with the series of statements (discourses) that try to adapt it. Otherwise, the discursive event can destabilize the discursive memory, breaking the regularity of the series of utterances (Achard, 1999, p. 52), but it cannot produce the "forgetting" of the other-meaning that precedes it (Indursky, 2003).

Maingueneau (1984, p. 27), proposes that the discursive universe is the set of different DFs that interact in a given situation. This interaction occurs through interdiscourse, which is responsible for articulating the various discourses (of knowledge, of a certain social group) to produce a new meaning.

For Pêcheux (1995, p. 162), interdiscourse is a complex whole with dominant DFs and subject to the complex of FIs. Interdiscourse is located in the interstices of several DFs, as demonstrated in Figure 3. It is the effect of this interdiscursivity that the "other" begins to relate to the "same".

Figure 3. Location of interdiscourse

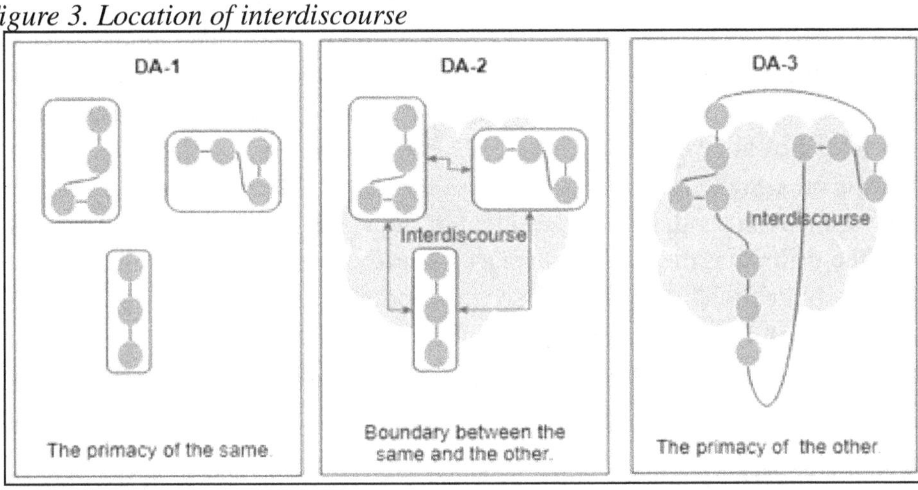

Pêcheux began to question the essentiality of DF as the basis of discourse and the validity of discourse identity as interdiscourse began to occupy more space in DA. The subject's speech will always depend on the "other". These reflections cause the implosion of the FD in the third phase.

2.2.3 AD-3

(..) every failed act is a successful speech (...)

Under the influence of Lacanian psychoanalysis, the subject becomes divided, cleaved, divided between the conscious and the unconscious, no longer being just a social position. The subject ceases to be his own master due to the loss of the centrality of the "I", where the "other" becomes part of his identity (Musalim, 2009).

At this stage, it is considered that interdiscursive relationships structure the identity of FDs. Gregolin (2005) understands that interdiscourse is the place where the objects taken by the enunciator in his speech (discourse dominated by a certain DF) are constituted, as well as the articulations between these objects, through which the enunciator will give coherence to his purpose within the intradiscourse, the discursive sequence that it enunciates.

Objects do not have the same meaning for different subjects; therefore, a statement does not have a universal meaning. Pêcheux (2006) defines this phenomenon as an oblique game of the declared opacity of the statement, where there are several meanings for a single statement.

Thus, because every statement is intrinsically susceptible to becoming "other", different from "itself", of discursively detaching itself from its meaning to derive to another meaning (Pêcheux, 2006, p. 53), it gives rise to the drift model, offering room for interpretation and perceiving a greater importance of the "other" over the "same". Using this model, it is possible to work with the possibilities of meanings of statements according to positions and their movements. Pêcheux presented this model in one of his works, Structure or Event, when interpreting the meaning of the phrase on a gagné, said by thousands of people when announcing the electoral victory of the then candidate for president of France, François Mitterand, in 1981. It is in the drifts that the subject displaces already established meanings, and the possibility of drifting causes discourses to shift, remake themselves, strengthen or disappear, given that the subject is not subject to conformity to the language and condemned to repeat discourses eternally (Souza, 2014).

3. APPROPRIATENESS OF USAGE OF DISCOURSE ANALYSIS

In elaborating on the appropriateness of adoption of discourse analysis as a qualitative strategy, it is worthy to establish the types of research questions that are addressed. The research questions are the cornerstone that suggest the usage of discourse analysis. The questions are such as (1) How is discourse structured in the text? (2) How are social actions as a cultural tool accomplished in the discourse? (3) How are identities, phenomenon or personalities negotiated in discourse? and (4) How are ideologies construed or misconstrued in the discourse. So, if the DA is interpretative perspective, it will allow the researcher to analyze the structures and patterns of the spoken languages, understand the significances therein and what it tends to communicate. And if the DA is on critical perspective, it will allow the researcher to understand how the discourse shapes social, economic and political practices, identities and power relations.

Therefore, it is interesting adopting discourse analysis in the studies that toe qualitative paradigms with critical perspectives analysing the power relations. Particularly having intent to critically investigate some assumptions, why it has been said and particularly on what contexts. Chouliaraki and Fairclough (1999) emphasizes critical discourse analysis focusing both relational-dialectical.

4. ACT OF DATA ANALYSIS WITH DISCOURSE ANALYSIS

The act of data analysis in discourse analysis hinges on duo possibilities of drawing conclusions. Firstly, upon a retrieved or constructed data which can be through what a language use reflects. Secondly, how it reinforces social structures and identities. In other words, how language creates and maintains social identities and how language creates and maintains power relations.

Table 1. Sketch of discursive approach

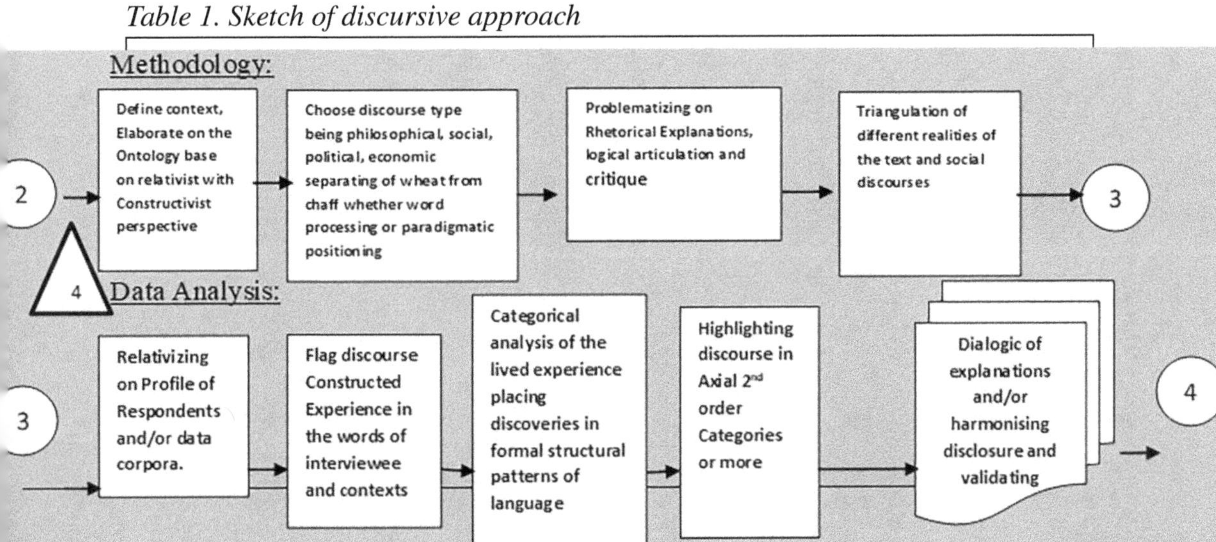

In Table 1 we demonstrate some elements that spurs discursivity with their interrelationship already mentioned. They are possibly not as sequential as shown in the figure because the processes could be recursive. However, the interaction of all these elements provides a roadmap to attaining the objective of the analysis.

5. CONCLUSION

Overall, discourse analysis provides insights into the dynamic relationship between language, society, and power, enabling researchers to explore how communication shapes and is shaped by social forces. It is a valuable tool for understanding both the explicit and hidden meanings in texts and conversations.

In summary there are key skills users need to possess in order to adopts discourse analysis, they are: the definition and types of discourse; theoretical foundations of discourse analysis, backing theoretical or the phenomenon under analysis (post-structuralism, critical theory, agency theory, etc.); how to formulate research questions and select relevant data; and the process of collecting and analyzing discourse data. Others are analytical tools and techniques, such as identifying themes, linguistic features, and discursive strategies; how to interpret discourse in relation to broader social and cultural contexts and lastly the ethical considerations, especially when using human subjects in research.

REFERENCES

Achard, P., Davallon, J., Durand, J.-L., Pêcheux, M., & Orlandi, E. P. (1999). *The Role of Memory*. Pontes Publishers.

Almeida, C. (2020). The Analytical Procedure Through the Discourse of the External Auditor. Available at https://teses.usp.br/teses/disponiveis/12/12136/tde-29062020-141256/pt-br.php

Althusser, L. (1980). *Ideology and Ideological State Apparatuses*. Editorial Presença.

Bakhtin, M. (1997). *Marxism and the Philosophy of Language: Fundamental Problems of the Sociological Method in the Science of Language*. Hucitec Publishing.

Chouliaraki, L., & Fairclough, N. (1999). *Discourse in late modernity: rethinking critical discourse analysis*. Edinburgh University Press.

Fiorin, J. L. (1990). Trends in Discourse Analysis. *Cadernos de Estudos Linguísticos*, *19*, 173–179.

Florêncio, A. M. G., Magalhães, B., Sobrinho, H. F. S., & Cavalcante, M. S. A. O. (2009). *Discourse Analysis: Foundations & Practice*. Federal University of Alagoas Publishing.

Foucault, M. (2008). *The Archaeology of Knowledge*. Forense Universitária Publishing.

Gregolin, M. R. (2005). Discursive Formation, Memory Networks, and Social Paths of Meaning: Media and Identity Production. Paper presented at the II Discourse Analysis Seminar (SEAD), at UFRGS, Porto Alegre, 2005. A more concise version was published on CD-ROM in the proceedings of the II SEAD.

Harris, Z. (1952). Discourse analysis. *Language*, *28*(1), 1–30. DOI: 10.2307/409987

Huan, C., & Guan, X. (2020). Sketching landscapes in discourse analysis (1978–2018): A bibliometric study. *Discourse Studies*, *22*(6), 697–719. DOI: 10.1177/1461445620928814

Indursky, F. (2001). From the Heterogeneity of Discourse to the Heterogeneity of the Text and Its Implications in the Reading Process. In: *Reading and Writing as Discursive Practices*, Org: Ernst-Pereira, A. & Funck, S. B., Ed. EDUCAT: Pelotas.

Indursky, F. (2003). Lula There: Structure and Event. *Journal of the Institute of Letters of UFRGS*, *17*(35), 101–121.

Maingueneau, D. (1984). *Geneses of Discourse*. Mardaga.

Maingueneau, D. (2015). *Discourse and Discourse Analysis*. Parábola Editorial.

Pêcheux, M. (1995). *Semantics and Discourse: A Critique of the Assertion of the Obvious*. UNICAMP Publishing.

Pêcheux, M. (1997a). Discourse Analysis: Three Periods (1983). In Gadet, F., & Hak, T. (Eds.), *Towards an Automatic Discourse Analysis: An Introduction to the Work of Michel Pêcheux* (pp. 311–319). UNICAMP Publishing.

Pêcheux, M. (1997b). Automatic Discourse Analysis (ADA-69). In Gadet, F., & Hak, T. (Eds.), *Towards an Automatic Discourse Analysis: An Introduction to the Work of Michel Pêcheux* (pp. 61–162). UNICAMP Publishing.

Pêcheux, M. (2006). *Discourse: Structure or Event*. Pontes Publishing.

Pêcheux, M., & Fuchs, C. (1997). On Automatic Discourse Analysis: Updates and Perspectives (1975). In Gadet, F., & Hak, T. (Eds.), *Towards an Automatic Discourse Analysis: An Introduction to the Work of Michel Pêcheux* (pp. 163–252). UNICAMP Publishing.

Pereira, A. D. (2011). *Auditing of Financial Statements: A Legal and Accounting Approach*. Atlas Publishing S.A.

Pereira, F. A. M. (2012). The Evolution of Institutional Theory in Organizational Studies: A Research Field to be Explored. *Journal of Organizations in Context*, *8*(16), 275–295.

Perrow, C. (1986). *Complex Organizations: A Critical Essay*. McGraw-Hill.

Pfeffer, J., & Salancik, G. R. (2003). *The External Control of Organizations: A Resource Dependence Perspective*. Stanford University Press.

Pontes, E. S., et al. (2017). National Academic Production in Accounting: Analysis of Theses and Dissertations Produced Between 2007 and 2016. *REAd | Porto Alegre*, 23(1), 239–258.

Porte, M., Saur-Amaral, I., & Pinho, C. (2018). Research in Auditing: Main Topics. *Journal of Accounting & Finance*, *29*(76), 41–59.

Possenti, S. (n.d.). Introduction to Discourse Analysis.

Robertson, J. C. (1996). *Auditing*. Irwin.

Rodgers, W., Mubako, G. N., & Hall, L. (2017). Knowledge Management: The Effect of Knowledge Transfer on Professional Skepticism in Audit Engagement Planning. *Computers in Human Behavior*, *70*, 564–574. DOI: 10.1016/j.chb.2016.12.069

Sampieri, R. H., Collado, C. F., & Lucio, P. B. (2006). *Research Methodology*. McGraw-Hill.

Scapens, R. W. (2006). Understanding Management Accounting Practices: A Personal Journey. *The British Accounting Review*, *38*(1), 1–30. DOI: 10.1016/j.bar.2005.10.002

Silverman, D. (2014). *Interpreting Qualitative Data*. SAGE Publications.

Smith, M. (2003). *Research Methods in Accounting*. SAGE Publications. DOI: 10.4135/9781849209809

Soeiro, T. M., & Wanderley, C. A. (2019). Institutional Theory in Accounting Research: A Review. *Organizações & Sociedade Journal*, *26*(89), 291–316.

Souza, S. A. F. (2006). *Understanding Discourse Analysis: Language, Society, and Ideology*. Valer Publishing.

Souza, S. A. F. (2014). *Discourse Analysis: Methodological Procedures*. Census Institute.

Spence, C., & Carter, C. (2014). An Exploration of the Professional Habitus in the Big4 Accounting Firms. *Work, Employment and Society*, *0*(0), 1–17.

Sweeney, J. T., Suh, I. S., Dalton, K. C., & Meljem, S. (2017). Are Workpaper Reviews Preparer-Specific? *The British Accounting Review*, *49*(6), 560–577. DOI: 10.1016/j.bar.2017.09.003

Thompson, S. K. (2012). *Sampling*. John Wiley & Sons. DOI: 10.1002/9781118162934

Tolbert, P. S., & Zucker, L. G. (1983). Institutional Sources of Change in the Formal Structure of Organizations: The Diffusion of Civil Service Reform, 1880-1935. *Administrative Science Quarterly*, *28*(1), 22–39. DOI: 10.2307/2392383

Tomas, C. W., & Henke, E. O. (1983). *Auditing: Theory and Practice*. Kent Publishing Company.

Tourangeau, R., Rips, L. J., & Rasinski, K. (2000). *The Psychology of Survey Response*. Cambridge University Press. DOI: 10.1017/CBO9780511819322

Trompeter, G., & Wright, A. (2010). The World Has Changed: Have Analytical Procedure Practices? *Contemporary Accounting Research*, *27*(2), 669–700. DOI: 10.1111/j.1911-3846.2010.01021.x

Turley, S., Humphrey, C., Samsonova-Taddei, A., Siddiqui, J., Woods, M., Baioudis, I., & Richard, C. (2016). *Skills, Competencies, and the Sustainability of Modern Audit*. The Institute of Chartered Accountants of Scotland and the Financial Reporting Council.

Varpio, L., Paradis, E., Uijtdehaage, S., & Young, M. (2020). The distinctions between theory, theoretical framework and conceptual framework. *Academic Medicine, 95*(7), 989–994. DOI: 10.1097/ACM.0000000000003075 PMID: 31725464

Vergara, S. C. (2013). *Research Projects and Reports in Administration.* Atlas Publishing.

Vinuto, J. (2014). Snowball Sampling in Qualitative Research: An Open Debate. *Temáticas, 22*(44), 203–220. DOI: 10.20396/tematicas.v22i44.10977

Westermann, K. D., Bedard, J. C., & Earley, C. E. (2015). Learning the Craft of Auditing: A Dynamic View of Auditor's On-the-Job Learning. *Contemporary Accounting Research, 32*(3), 864–896. DOI: 10.1111/1911-3846.12107

Westland, J. C. (2017). An Empirical Investigation of Analytical Procedures Using Mixture Distributions. *Intelligent Systems in Accounting, Finance & Management, 24*(4), 111–124. DOI: 10.1002/isaf.1405

Wood, L. A., & Kroger, R. O. (2000). *Doing Discourse Analysis: Methods for Studying Action in Talk and Text.* SAGE Publications. DOI: 10.4135/9781452233291

Zucker, L. G. (1987). Institutional Theories of Organization. *Annual Review of Sociology, 13*(1), 443–464. DOI: 10.1146/annurev.so.13.080187.002303

Chapter 7
Constraints and Challenges of Conducting Discourse Analysis in Qualitative Research

William Chakabwata
https://orcid.org/0000-0002-4224-5239
University of South Africa, South Africa

ABSTRACT

The chapter unpacks the evolution of discourse analysis and its use in qualitative research. The chapter explored the evolution of discourse from modernity to postmodernity. The challenges of pinning the definition of discourse was also reviewed. The chapter traces the shift from logocentrism to post -structural discourse analysis in qualitative research studies. The place of structuralism and its assumption in shaping communication was explored. The chapter utilized deconstruction, interpretivism and social constructionism to guide the discussion of the epistemology of discourse analysis. Finally the chapter submitted the constraints of conducting discourse analysis in qualitative studies and postulated ways in which these could be tackled.

INTRODUCTION

This chapter explored the challenges and constraints of discourse analysis in qualitative research. The chapter starts with a background that portrays the evolution of discourse analysis in qualitative research in the context of modernity and post-modernity. The epistemological basis of discourse analysis within the framework of deconstruction, interpretivism and constructionism are also explicated. The three

DOI: 10.4018/979-8-3373-4432-4.ch007

predominant strands of discourse analysis namely discourse theory, psychological discourse and Critical Discourse Analysis (CDA) are examined. The subsequent sections are discussed as discourse analysis in qualitative research, intertextuality and interdiscursivity, multimodal analysis among many other topics.

BACKGROUND

Wertz, et al., (2011) acknowledged five approaches to conducting qualitative research in the social sciences, which included case study, ethnography, phenomenology, grounded theory, discourse analysis, narrative research and intuitive inquiry. Other scholars have added basic qualitative, critical qualitative, participatory research and clinical research.Grochalska (2020) noted that the meaning of discourse has slippage and ambiguity. This observation has resulted in multiple ways of conceptualising and practising discourse analysis. In order to elucidate the meaning of the term discourse terms which includes medical discourse, political discourse, or scientific discourse have been used. The approach to discourse analysis has been categorised as modernism, structuralism, and postmodernism (Psychology Learners,2024; Sudartini, 2024). The discourse under modernism focalised on the assumption of continous human progress based on the natural and social laws of science which could be universalised and which generate theories that were founded on certainty and could make life predictable. Modernity monopolised discourse in the 19th and 20th century with its claims on objective reality and a single ideology influencing cultural formation. This confidence in continous and linear progress of mankind suffered a setback with the outbreak of World War 1, economic depressions and social challenges that emerged after World War 2. Discourse under modernity had assumed that talking and could serve the purpose of communicating. Modernity was marked by a belief in logocentrism, a faith in a metaphysical system that held the assumption in the existence of an all- powerful truth logos behind all phenomenon that occurs. In written texts the author was assumed to have authority over the interpretation they wanted to convey (Hendricks, 2016; James, 2023).

Rakita (2017) argued that the agenda of modernity discourse was to recreate and legitimise hegemonic narrative of capitalist means of reproduction and necolonialism for sustaining western imperialism after World War 2.The same author added that the concept of modernity reflects the social and economic formation of Anglo-European societies from the late 15th century to the late 20th century. Structural theorists who included Ferdinard de Saussure and Jacques Lacan postulated that human actions and social structures revolve around language and can be conceptualised as an interrelated system (Fernandes & Sargentini, 2022;Al Umma, 2015). Structuralism emerged in the 20th century with the offering of a course in lingustics by Ferdinard de Saussure and founded the structuralism. Sausure's theorising on language foregrounded the critical role of meaning and signification in shaping human life.

Postmodernity emerged as a reaction to the perceived limitation of the modern era (Yazdani et al., 2011;Campbell, 2018; Sudartini, 2024).Postmodernity is perceived as an attitude, a mental disposition, that came to the fore to eschew the absolutist claims to knowledge that marked modernity. The term postmodernity can be traced to the 1960s to indicate a perception that a new era that was marked by a steady collapse of ideologies and beliefs that had characterized modernity. The salient feature that distinguished modernity and postmodernity was the character of discourse. Postmodern theorists rebutted the modernist assertions that there can only be one theory that can be universalised. Postmodern scholars were more curious to identify differences between individuals and groups' experiences as opposed to commonalities. In addition postmodernity spotlighted the central tenets of life as change, uncertainty and diversity. More importantly postmodern scholars do not perceive language as a neutral tool that is free from values, but as an arena of political contestation which displays debate over meanings, values and ideology. Language is perceived as a social construct that emerged from specific contexts. Postmodernity reflected a dissatisfaction with the assumptions of structuralism and it surfaced as resistance to interrogate modernity and supersede its assumptions (Sudartini, 2024). Postmodernity is not a homogeneous paradigm, but reflects an assemblage of premises, worldviews and concepts that foregrounds "language and discourse; power and ideology; conflict and resistance" (Sudartini, 2024, p.5).

Postmodernism rebutted the logocentrism embedded in modernity arguing that the text does not present a metaphysical presence that is autonomous at the time of writing, but presence is always there at the time of writing. It is further averred that there is no text that can stand apart from other texts, as a text always exist in conjunction with other writings (Hendricks, 2016). It was due to the objection on the assumptions of logocentrism that deconstruction emerged as postulated by Derrida (The Editors of Encyclopaedia., 2024).

Wodak (2008) explicated the use of the term discourse had a broad application to social sciences, and prominent names that have been cited in relation to this subject includes Michael Foucault, Jurgen Harbermas, Chantal Mouffe and Ernesto Laclau. Wodak. The term discourse covers spectrum that ranges from objects that are associated with memory, a political, a political trajectory, narratives, text, talk and and language. In mundane parlance discourse may be identified as gendered, racialised, media discourse, and discourses of the past among a plethora of other categories. For example at the time of writing of this chapter the world was preoccupied with the Russian War on Ukraine and also Israel invasion of Gaza. Some media spotlighted the Russian invasion of Ukraine and the attendant death of women and children, while in Gaza South Africa described Israel invasion of Gaza as a genocide and some western media were more concerned with Hamas and the captured Israel citizens. The nature of discourse was so contested that the international court was entangled and had to define what a genocide is. The ascendancy of Donald Trump to power in the US, led to new discourse that completely ignored the two state solution in favour of an Israel takeover of Gaza.

Khafaga (2017) indicated that discourse analysis is constructed and is socially determined. It means that this approach forefronts not only the linguistic components a text, but the manner in which a text can be deciphered and comprehended and the social context in which it may be generated and used. The term discourse subsumes multiple approaches and subject matter and has been used in disciplines which includes anthropology, literary subjects, philosophers, psychologists, sociologists, education and law. Discourse analysis encompasses spoken discourse and that expressed through other medium which includes written, data from sign language and visual forms of data presentation (Fairclough, 2003; Khafaga, 2017; Dicks, 2019; Dahlborg, Boman et al., 2023). Although discourse analysis may occur at various levels, in this study, the focus is on discourse as used in qualitative studies such as in focus group interviews, ethnographic research, participant observation and interactions between the researcher and the participants. In discourse analysis activities such as recording, transcribing and pondering in meaning and significance. Some discourses are given prime significance while others are marginalised. The Hamas Israel conflict of 2024 highlights exactly how one discourse can be forefronted, while others may be suppressed.

Michael Focault (1926-1984) ushered discourse analysis to a new level of discussion (Foucault, 1972). Foucalt (1972) perceived a discourse as tool that furnishes the basis for conscious knowledge.

Researchers focalise on discourse analysis as a way of attaining some of the listed goals:

- Ways in which language is employed to generate or create social identities: Scholars can proble how linguistic choices can enable identity formation and expression and ways in which identity is used in diverse contexts;
- Ways in which language is employed to sustain disparate power dynamics: This also refers to ways in which power is utilized to challenge power structures or to sustain inequalities.

Zotzmann and O'Regan (2023) explicated the existence of multiple forms of Discourse Analysis (DA), which includes critical discourse analysis (CDA),critical discourse studies (CDS), and critical applied lingustics (CALx) (Catalano & Waugh, 2020). CDA expounded that CDA is an institutionalised qualitative research methods which has generated numerous approaches under its category. CDA functions by isolating and constrasting a variety of discourse strands and scrutinising hidden values and beliefs, which are normally taken for granted. CDA, CDS and CAKx commences with a social problem prior to spotlighting the role that discourse/ semiosis plays in that. The variety of perspectives span a spectrum from neo-Marxism, and Foucauldianism to a number of poststructural perspectives (Zotzmann & O'Regan, 2023). The analysis varies from those that provide a normative perspective and critical, they are also interdisciplimary spanning disciplines such as sociology, philosophy, psychology and anthropology. CDS represents a transition from CDA reflecting a wider theoretical engagement in counter-hegemonic way with issues that included issues such as sexuality, colonialism, decoloniality and discrmination. CDA has its etymology in the field of lingustics founded in the 1970s by proponents who included Hodge and Kress (1979) and Fowler et al.,(1979). These authors began to articulate how contextual variables constitute the spoken and written text, and how these discursive and practices and potrayal, in turn impact on how to apprehend a certain part of the natural and social world. In discourse analysis a text is delineated in a comprehensive and methodical way, focusing on the semiotic resources that are available and the meanings which this use appeal to and also taking into account what the text makes ambiguous or deliberately obscure.

EPISTEMOLOGICAL FOUNDATION OF DISCOURSE ANALYSIS

Discourse analysis holds that complete objectivity can never be attained, due to the explication that reality is socially constructed on assumptions that may be invisible to us. Discourse analysis helped to expand studies in linguistics. This connotes that at the level of epistemology, all knowledge is generated in social engagements and is capable of changing and there is no clear demarcation between the observer and the observed object or experience (Dahlborg et a., 2023). It is averred that language can therefore not be an objective way of framing reality or the carrier of reality. This observation played a central role in the linguistic turn. When viewed from a constructionist ontology, language serves a key role in the constituting of reality.

Dahlborg et a., 2023) categorises discourse studies in three major categories which includes discourse theory, discourse psychology and Critical Discourse Analysis (CDA). Discourse theory avers that all practices are circuitously constructed. The typical proponent for discourse theory is Laclau and Mouffe. Discourse psychology focalises on rhetorical approaches employed when writing speech, with a special emphasis on how discourse is used to portray social positioning and power relationships. Critical Discourse Analysis (CDA) integrates linguistics analysis with societal discourse. CDA reveals how language is utilised in day –to- day engagements in a manner that reflects causes and consequences in subtle way.

CRITICAL DISCOURSE ANALYSIS (CDA)

Van Dijk, (1993) averred that discourse analysis focalises on a study of the interaction between discourse, power, and the positionality of the analyst in that situation. Critical discourse analysis is a multidisciplinary field. Critical discourse analysis examines the place discourse plays in production or reproduction of disparities. Dominance connotes the practice of "elites, institutions, or groups that results in social inequality, including political, cultural, class, ethnic, racial and gender inequality" (Van Dijk,1993, p.250). In more precise terms discourse analysis assumes a clear socio-political stance. Some critical scholars collude with elites and legitimise elites who are middle class, heterosexual, white males. Dominance in day to day engagements may be reproduced in a subtle way through genres that appear natural and appropriate (Patel, 2020). Power and hegemony is related to the control over discourse and ultimately control over people minds.

EduBirdie (2024) noted that since the 1970s, Discourse Analysis (DA) has evolved into profound sub-categories the major one being Critical Discourse Analysis (CDA), which posits discourse as a social practice. DA does not have a homogeneous theory.

Jaeger (2001) the canon of discourse analysis as influenced by Michael Foucault displays the following attributes:

- The constitutive nature of knowledge at a given place and time;
- How this form of knowledge evolves;
- How this knowledge is transmitted;
- The function the knowledge serves in the constitution of subjects and in the shaping of society;
- The effect this knowledge has on the progress of society;

Knowledge is used to signify the nature of content that constitute consciousness and diverse forms of interpretations used by people over time to extract meaning and influence reality. People obtain knowledge from their specific context in which they are born and in which they are actively involved throughout their life. Discourse analysis encompasses the knowledge that is deemed valid at a specific time, and probes the immediate context of knowledge/power and subject it to interrogation (Mahmud, 2017). Discourse analysis ranges from a spectrum of knowledge that includes the knowledge available on media to day to day engagements within families and institutions to knowledge that is generated from various academic fields.

Blackledge (2012) averred that language is central in reproducing disparate relations of power. CDA reflects an array of political strands and methods that theorise on the relationship between power and language. The primary focas of CDA is social challenges and not merely language. CDA takes up cudgel against those who language in order legitimise their authority. CDA emerged as a reaction to the divide that existed between lingustics and social sciences as Sociology. CDA has a political trajectory which it doesn't even try to deny.

DISCOUSE ANALYSIS IN QUALITATIVE RESEARCH

Discourse analysis provides a lens to analyse research that seeks to address challenges in the communities.It coalesces diverse interdisciplinary and multiple approaches in analysing data which includes television interviews, online debates,or visual images or semiotic entity which reflects intertextuality and embeded in a given socio-political context (Wodak, 2008). Analysis in qualitative research has therefore to take into account the relationship among various texts and the contested meaning. Analysis has therefore to intertextual, interdiscursive between text and genres and the social variables, the history of institutions under research and the frames of references used by those organisations and the context. Intertextuality connotes the way in which one text is related to other multiple texts in contemporary times and in history. This analysis of qualitative data would entail recontextualisation, where reference is made to the same actors, transfer of meaning of a text from one document to another or refering to a topic that resonate with the one being researched on. The component being interpreted would then assume a different meaning because meanings are generated in use.

Turner (2020) adduced that focalises on what people say during data collection from interviews, diaries, social media, or documents. In addition qualitative researcher are more attentive to what people say and how they articulate it. Discourse analysis is the holistic system that people communicate and it is a broad application of language. It incorporates all forms of communication and the wider concepts that sustain the language and communication and how languages is transformed. It also reflects ways in which people articulate certain words in communication and the dialect that they use.

Researchers in qualitative data analysis have to function within a wide range of discourse analysis techniques which includes the use of lingustic analysis, critical discourse analysis and thematic analysis.

Wodak (2008) indicated that there was a trend to analyse open-ended questionnaires using quantitative methods and thereby neglecting the detailed arguments in such documents as well as the unfolding group dynamics in that text. The trend was challenged by developments called linguistic turn, in courses such as sociology, political science, anthropology, and history and text are now transcending quantitative analysis by using content analysis.The evolution of new forms of communication which includes online debates engagements, emails, and websites have led to what is termed a visual turn, which involves the analysis of visual information and is termed multimodality. Data analysis in qualitative research is expanding to encapsulate media, focus groups, political speeches among multiple other forms.

INTERTEXTUALITY AND INTERDISCURSIVITY

Intertextuality is in its widest meaning refers to the academic fascination with the situational context of texts and meaning in other texts. It was a reaction to the assumptions inherent in Aristotle claims about words and the world and explicated meanings of words, are not located in the text in isolation, but are located in the meaning of texts in relation to other texts (Bartesaghi, 2016). The origin of these concepts has been traced to other philosophical movements such as pragmatism and structuralism, although the concepts gained ascendency with the work of Julia Kristeva, who inagurated the use of the term intertextuality. The meaning of intertextuality is marked by slippage and ambiguity. There are different conceptions among scholars on what constitute a text and intertextuality functions and what intertextuality is able to do. The two terms intertextuality and interdiscursivity have been very productive in the fields of linguistics anthropology and sociolinguistics. A universal feature across disciplines has been to contextualise the location of meaning within a clause or isolated engagement, foregrounding instead the situational foundation of meaning, taking into account the impact of genre, discourse and ideology. Intertextuality has been used to analyse political encounters, mass media, advertising and practices in education using Critical Discourse Analysis (CDA).

Kuzmina (2015) opined that intertextuality was adopted from critical literary theory to linguistics in the 1980s, and was employed in diverse researches to reflect relations between texts and provided a mechanism to demonstrate genre change and transformation developing into interdiscursivity. The concept intertextuality was first used by Kristeva (1980). Nadu (2015) averred that Kristeva made a huge contribution to intertextuality. She formulated the term and underscored the potential power that is situated within a text. A text is dynamic and embeds multiple meanings which includes cultural, social and creative. A text does not have a linear interpretation. Kristeva (1980) developed her ideas basing them on the work of Bakhtin concepts of intertextuality. Bakhtin understood a text as possessing a dialogic nature, which means that text tends to be linked to other texts or genres. Bakhtin (1986) submitted that there is no single meaning to language that is perceived to be heteroglossic. The connection between genres and future genre is dialogic. Bakhtin explication adduced the relationship between texts and subsequent texts or future texts demonstrating how genres are interlinked. Dialogism reflects ways in which other people and values and norms influence a text. Furthermore Bakhtin displayed that multiple voices can be discerned in a text.

Interdiscursivity refers to how a text is influenced by other text conventions, genres, and discourses. Interdiscursivity refers to the entire language structure in a text.

MULTIMODAL ANALYSIS

The advert of technology has marked major shifts in ways in which qualitative research is conceptualised and the generation of multimodal genres and the combination of resources capable of generating many varied meanings.

Dicks (2019) asserted that multimodal analysis connotes the process of interpreting and making meaning of qualitative data in research that intergrates verbal and non-verbal forms of data in addition to video, audio, images and 3-D materials. Three predominant influences on the evolution of multimodal analysis are social semiotics, interactional sociology and sensory anthropology. Each category has developed into unique categories that have different ontologies and approach to analysis in terms of reflexivity, context, and essence. There are also pragmatic distinctions in terms of the way data is are categorised, transliterated and analysed and ways in which cameras and employed to capture the pictures and how the researcher perceives the field of research. Data analysis tends to be multifaceted because language is also complex encapsulating elements such as body posture, bodily movements, virtual forms, which reflects human and non-human forms of materials.

D'Angelo and Marino (2024) indicated that genre from a multimodal perspective may be explored from perspectives which included Systematic Functional Multimodal Discourse Analysis, Social Semiotic Multimodal Analysis, Multimodal Corpus Analysis. The analysis of Genre has its genesis among linguists from the 1980s. Genre analysis has over the years taken the form of researchers studying varied forms of texts for the purpose of identifying structure and organisation in order to appreciate how they related to the intended audience and whether it is erudite or not erudite (Deng et al.,, 2021; Kuzmina,2015). In the past the focus was on monomodal texts whether they were written or oral. The rise of multimodal genres as epitomised by YouTube, Facebook, websites, TV programmes, and PowerPoint presentations has led to new ways of analysing data that is termed multimodal analysis in the 1990s. However the way multimodality is operationalised in diverse disciplines tend to be different and there are no universal practices.

Modes have been identified as being either written or oral and those that are not related to language but are aural, visual, or based on gestures and images, soundtracks and software and hardware. They are culturally and socially determined due to their essence relies on the contexts in which they are situated. The nature and meaning of a mode or semiotic component relies on each discipline. Multimodality offers an opportunity for qualitative researchers to traverse their discipline to generate theory by the way we use different and multiple modes to create meanings.

SYSTEMATIC FUNCTIONAL MULTIMODAL DISCOURSE ANALYSIS (SF-MDA)

This is an interdisciplinary approach to studying multimodal texts that was advocated by Halliday (1978) and was further developed by O'Toole (2011). Halliday posited three functions of languages:

i) Ideational – which refers to the way language may be employed to classify human experiences and then transform them into multiple patterns.
ii) Interpersonal-which connotes the way social reality is mediated through language.
iii) How diverse elements are coalesced in multimodal text and to demonstate coherence.

SF –MDA is indispensable for discourse analysis.

SOCIAL SEMIOTIC MULTIMODAL ANALYSIS

Social semiotic which was introduced by Halliday and developed by Kress and Hodge (1985) and others averred that the traditional separation of language from the people who use it is not plausible. In addition, the social semiotic analysis postulate that meanings emerge in day to day interactions and engagements of people using semiotic devices. This lens therefore seeks top comprehend ways in which multimodal resources are produced and used in society.

MULTIMODAL CORPUS ANALYSIS (MCA)

This approach involved the processing of large corpus in order to isolate patterns. At times the corpus used may be restricted, because if it becomes to huge a reseacher may find it daunting. A corpus may constitute websites, newspapers, technical manuals a mong many other resources.

CONTENT ANALYSIS

Nicmanis (2024) averred that content analysis was at first used as a an approach to identify quantitative data in qualitative research has developed into a major approach in qualitative research. The dataset for qualitative research which is constituted of interview transcripts, and content from social media pose a huge challenge for researchers. The genealogy of content analysis within the framework of quantitative research methods used to analyse qualitative data has expanded into multiple variations and others which are termed content analysis and are contradictory. For instance, different approaches differ in the way the perceive content analysis and the allocation of data to clear distinct category in content analysis, compounding the complexity of the data analysis. Haradhan (2018) explicated that content analysis is a method that can be employed to analyse verbal, visual or written communication messages. Content analysis was at first utilised to probe hymns, newspapers, political speeches and magazine, articles and advertisement. Content analysis had a wider application in disciplines which includes journalism, psychology, communication and business. In the case of journalism, researchers would unpack the nuances in newspapers identifying the tone, if it was negative, positive or nuetral. The traditional approach used in content analysis entailed a researcher counting the frequency of certain words, phrase or in a dataset.

In DA analysis the meanings of concepts tends to be fluid and are determined by context. The absence of rigid meaning is impacted by the acknowledgement that meanings are socially constructed. Hopf (2004) argues that CD is not crafted to fit into a priori coding as what occurs in content analysis. DA acknowledges that discourses are situated in open social systems. DA is conceptualised as an approach to inquiry as well as a political theory. The meaning of any text is not rigid, but depends on the given text and the situated it is being used. In addition the meaning is not merely situated within the text, but in the relationship of that text to other text, which is termed intertextuality. Furthermore since, it is impracticable to demarcate in a definite way the point where a text ends it is therefore an open system and declarations about meanings require to be dealt with caution. It is for that reason that one requires to be cautious about the nature of conclusions that they can declare over qualitative findings. This is the essence of the rejection of the concept of logocentrism, in which the meanings are far from being ultimate and are socially configured in intersubjective space and are impacted by forces that may not be found in any other context or at a different time (Deng et al.,2021; Nganga et al., 2023; Osakinle, 2024). DA therefore holds that meanings are social and are not prescirbed and therefore contrasted in a dialogic way and not in comparison to some objective reality. Intertextuality means that meanings can be given by a researcher who holds assumptions of being objective, but that the meanings may not be assumed to be rigid over time and geography. DA offers no pretensions at addressing any complex problems but acknowledges the complexity and insurmountability of issues and admits that problems do not require another technique to solve them and that researcher need to be more humble in terms of the claims they make about knowledge. Hopf, (2004, p.31) succinctly opined that " epistemology and ontology trump methodology". This connotes that knowledge assertions about social phenomenon do not have need a lot of grandeur.

DA tends to differ with CA in terms of its epistemological predispositions, and doesn't nt make loft claims to trascend them, but aver moderate claims because of them. DA also takes into account two elements that CA leave unnoticed, issue of absences and irregularities. DA would be concerned with a lack of representation of certain marginalised groups in a study, whereas CA, would not theorise about absent data. DA would also be fascinated with irregular features such as for example the presence of a person from marginalised group in a high profile actor in a movie, which may be perceived as being counter hegemonic. In CA, anomalies may be neglected from the theorising as noise and mya be ignored. CA in terms of its inherent assumptions about the use of language, seem to be undergirded by the ideology of liberalism which holds the presupposition thatPeople chose words to explain exactly what they want to say and neglect to realise that people are in part constituted by language and the daily activities they engage in.

Agustian et al., (2020) averred that discourse works as reviewing social practices and social generated knowledge which are created in social contexts. Jaspal (2020) noted that content analysis, thematic analysis and discourse analysis display resemblances in the sense that they are all preoccupied with highlighting patterns in the dataset and a process that display a high degree of analysis. Content analysis, thematic analysis and discourse analysis also reveal subtle differences. The utilisation of each method are shaped by epistemological assumptions of each approach. Content analysis, thematic analysis and discourse analysis are able to reveal a nuanced and subtle analysis in varied ways. As in thematic analysis and content analysis, discouse analysis has a distinct focus and subsumes a methodological approach with multiple branches. Two major approaches in discourse analysis are discursive psychology approach and the approach influenced by Foucault. Discursive psychology focalises on what people do "with their talk" (Jaspal, 2020, p. 13). For instance a person can declare that I am not chauvinistic and then proceed to make a chauvinistic statement is to dispel the accusation before it is raised, and thereby presenting himself or herself as a non-chauvinistic person. Foucaldian discourse analysis forefronts the power structures in enabling us to talk and reflect as we do.

Discourse analysis has its roots in social constructionist epistemology. Discourse analysis spotlight ways in which language is employed to construct social reality.

THEMATIC ANALYSIS

Braun and Clarke (2012) noted that thematic was widely employed method of qualitative research, yet not adequately defined. The same authors observed that thematic analysis is increasing coming into prominence in qualitative research as other methods such as grounded theory and discourse analysis. Braun and Clarke (2012) conceptualized Thematic Analysis (TA) as a "

method of systematically identifying, organizing and offering insight into patterns of meanings (themes) across a dataset, TA allows a researcher to see and make sense of collective or shared meanings. Identifying unique and idiosyncratic meanings and experiences within a single dataset is not the focus of TA.

This means that TA is preoccupied with identifying commonalities within a dataset and to explore the meanings within the dataset. Naeem et al., (2023) acknowledged the use of TA as a pre-eminent practice among qualitative researchers particularly when probing thick data. TA is popular in inductive and deductive research and also in intepretive phenomenology, grounded theory, ethnographic approaches, constructivist and positivistic methodologies. Dovetail Editorial Team (2023) postulated that a researcher's personal experience plays a central role in identifying themes within the dataset.

Dovetail Editorial Team (2023) averred that there is a multiplicity of approaches to thematic analysis which includes:

- Inductive thematic analysis- involves the entails extracting meaning and themes from data with no preconceived framework. The data is probed with no predetermined expectation.
- Deductive thematic analysis- In deductive thematic analysis data is probed with a set of predetermined outcomes. Previous knowledge, research or theoretical lens underpins data analysis.
- Semantic thematic analysis- Semantic analysis entails the underlying interpretation of data is overlooked in favour of the stated meaning or facevalue of the data.
- Latent thematic analysis-This is based on a premise that is contrary to the semantic approach and involves the probing of meaning in the underlying data and questions the semantic data. The extracted meaning is then theorised and data is not embraced at face value.

Each approach will rely on the nature of the research design. Jaspal (2020) explicates discourse analysis in psychology as being epitomised by discursive psychology and Foulcaldian psychology analysis. Discursive psychology epitomises a well developed methodological intimate analysis of language, and it focalises on what "participants are doing with their talk" (Jaspal,2020,p13). For example,a person opine that I am not sexist and then proceed to make a sexist statement. This is a reflection of ways in which an individual can utilise their talk to dispel potential accusation, while proceeding to make the a controversial utterance. The Falcaudian approach spotlight discourse in a given social and political context due to the fact that it is this context that shapes the discourse that people subsequently draw on. In this manner, it is argued that idscourse shapes the subject. The discourse that emanate from the institutions of power determines the what can be uttered and by whom and at what time (Willig, 2008).

The advantages of using thematic analysis were noted as flexibility, appropriate for large data, makes possible to develop inductive codes from data, can address all research questions and a researcher can utilise experiential knowledge (Mark, 2024). Thematic analysis is able to address all research questions because of the epistemology of qualitative research that embraces the subjective nature of reality. In its philosophical underpinning thematic analysis embraces experiential knowledge.

Jaspal (2020) expounded that there is a unifying motif that that span across a dataset which is influenced by the probing what the data is revealing to us in vis -a -vis the research questions. The central role of the code is normally identified by referring to the research questions and resisting the temptation to determine the frequency a code appears. Thematic analysis is often embedded within a realist epistemology which hold the presupposition that language facilitate access to the participant cognition and drivers of behaviour and that through an urbane, careful and situated penetration of participant's accounts the researcher is able to determine how a participant thinks and feels ove a given phenonmenon. Briefly thematic analysis is utilised to extract themes across a dataset using either a realist or social constructionist perspective.

TA had been noted to have some advantages and demerits. Some of the demerits were enumerated as challenge of focusing, there is no restrictions on the use of language from the respondents, a tendency to abide by a theoretical framework in spite of the thick data that is available (Mark, 2024). The challenge of focusing occurs when working cumbersome datasets, which may pose a huge constriction for novice researchers. The language employed in qualitative data is not restrictive and participants may use mother tongue or English which poses a challenge to data analysis.

THEORETICAL UNDERPINNING

A theoretical basis for illuminating this discussion is interpretivism and deconstruction.

DECONSTRUCTION

This chapter acknowledges that discourse is always permeated by ideologies influences and it is therefore critical to identify a lens that is able to isolate these influences such as poststructural deconstruction theory. Defining the term deconstruction is not an easy task (Given, 2008; De Man, Bloom, & Miller, 1979; Sikirivwa, 2020). The challenge of advocating a definition of deconstruction is seen in the context that the proponent Derrida did not offer a definition. The concept deconstruct is seen to be linked to the term in French that is called "deconstuire" which in English denote taking something to pieces (Sikirivwa, 2020, p. 44). In the field of philosophy the term is associated with Martin Heidegger, (1889-1976) who utilized the term "deconstruire"derived from Germany. What is available in literature is more of an explication of what it is not as opposed to what it is. Sikirivwa (2020) opined that deconstruction is not a method that can be used by objects neither is it a a thing. Deconstruction forefronted the breaking down of structures as in structuralism, logocentrism and lingustic terms. Deconstruction is preoccupied with the arragement of systems.

One of the theoretical underpinning used in this chapter is deconstruction which owes its etymology to post structuralism as espoused by Jacques Derrida (1966). Post structuralism endeavors to spotlight and upend dominant narratives (Leavy, 2017). Post structuralism is committed to the process of dismantling dominant narratives in order to unmask what was conceived to be invisible or subtle. Leavy (2017) highlighted the way the media employed hegemonic terms such as riots in place of protests during the black lives matter outcry, in order to frame a dominant discourse that delegitimized the demands for an end to racial discrimination and police brutality. There are multiple political groups that challenge dominant discourse which includes feminists, indigenous studies, queer studies, and race focalise on racial and sexual hierarchy, patriarchal dominance by white males and social and political justice.

Deconstruction explicate structuralism as being ahistorical. In addition they contend that life and thought are historical, they transform different relations at different times and with varied elements (Alqudah, 2019). The method of deconstruction is linked to a term that Derrida termed metaphysics of absence. Deconstructionism rebut the claims of structuralism that assumes that there are standards for meanings which can be applied universally that were developed prior to language. Structures are perceived by deconstruction thinkers as lacking historical basis and are provisional and situational and subject to ongoing changes.

INTERPRETIVISM

Discourse analysis is founded on the assumptions of social constructivism in which spotlights sociocultural interactions as the key to attaining knowledge. Yan-Yan et al., (2011) adduced that one of the dominant paradigm that has been employed in qualitative research is interpretivism. At times interpretivism is used interchangebly with social constructivism, constructionist or interpretivist paradigm which is pited against the postivist paradigm. When perceived from a theoretical and methodological stance, there are multiple strands in interpretivism all of which hold a common assumption that human intepretation is the point of departure for generating knowledge about the social world. The common assumption is that it is vital to comprehend people's subjective reality is the basis for appropriating social reality (Phillips, 2023; Kessler, 2024). This premise runs contrary to the assumption of positivism, which perceived mankind as assuming the status of passivity, reactive and mechanical status.

Nickerson (2024) interpretivism explicate social reality as constituting understanding the worldview, motivations, and the way people process information in a social context as vital to decoding the essence of data that is gathered on a social phenomenon. In sociology interpretivism has multipe strands and the chief onces are hermeneutics, phenomenology and ethnomethodology. Interpretivism eschews the postulate that there is a single dominant approach to data analysis and therefore is flexible to employ both quantitative and qualitative approaches. Guba and Lincoln (2005), Morgan (2007)Kivunja and Kuyini (2017) explicated the attributes of an interpretive research as:

- acknowledges that the social world can not be comprehended from a position of a single individual;
- embrace the social construction and multiplicity of reality;
- Admission that knowledge construction inextricably bound with the role a researcher and their subjects;
- Acknowledges that context is important for knowledge construction and knowing;
- The importance of developing an understanding of indiividuals as opposed to generating universal laws;
- The explcit aknowledgement that causes and effects are intertwined;
- The foregrounding of the context in any methodological search for understanding.

Rooted in the conception that we can only comprehend reality discursively multiple proponents of postmodern and poststructural analysis uphold epsitemic relativism, which means that they perceive any claims to reality or truth to be ideological (Zotzmann & O'Regan, 2023).

SOCIAL CONSTRUCTIONISM

Social constructionism subsumes the belief that knowledge, concepts and theories are all capable of being socially constructed. Andrews (2012) explicated while social constructionism is intimately related to interpretivism, they are different. It has its origins the field of sociology and scholars such as Bacon, who questioned the capacity of obervation in depicting the actual reality. In addition social constructionism has been associated with postmodenity and it is opposed to realism in research. Chief individuals who have been associated with social constructionism are Durkheim, Mead, Marx and Schutz. The writing by Berger and Luckman (1991), the construction of social reality had a major effect on its development.When perceived from a standpoint of philosophy, social constructionism may be viewed as a collection of varied approaches that includes critical psychology, discursive psychology deconstruction and discourse analysis among many others. Social constructionism is not homogeneous.

Constructionism and interpretivism hold similar assumptions regarding the way in which meanings are created, worked out, strengthened and changed. The chief aim is to apprehend the world from the lens of the lived experiences of the people. Intepretivism distingushes between the objective world and the objective world and it therefore aims to generate ways of apprehending the world using the logical empricism. Constructionism is not preoccupied with the objective world, but is rather fascinated with the subject experiences of the participants.

CONSTRAINTS AND CHALLENGES

Using discourse analysis in qualitative research work is noted to be flaunt with challenges which includes constrictions of identifying discourses, preoccupation with the text itself, the intricate dynamics between the text, communication intent, linkages of researcher to participants and social positions have a potential to obscure the identification of discourses (Scispace, 2024). The constraints that is salient in using discourse analysis in qualitative research includes the use of too much brevity, use of numerous quotations,

Grochalska (2020) indicated that one of the primary challenge in using discourse analysis to analyse qualitative research resides in the use of the term discourse itself which is very ambiguous and slippery to define. The term subsumes numerous diverse practices that includes political discourse, public discourse, medical discourse among many other categories. This lack of a precise definition leads to multiple approaches to data analysis in research.

Three domains of discourse have been identified by Gąsior-Niemiec (2008) as text, discourse practice and social practice. This leads to three levels of analysis identified by (Grochalska,2020,p.100)such as

-Study of vocabularly structure and coherence of the text
-the process of production, distribution and consumption of the text
-emphasizing the political and idelogical meaning of the text

The nature of the social world is marked by conflict, uncertain and abitrary making it plausible to interpret all social phenomenon as discourse. Discourse analysis may assume take the following steps the choice of an area for study for indigenous knowledge system, and in the subsequent step an explanation of the theory of indigenous knowledge system and then applying different theories the researcher would seek to intergrate theory with observation. In discourse analysis three levels of analysis may be applied namely the Lingustic analysis, socio-cogntive analysis and unpacking of intersubjectivity, interdiscursivity and the context. In short the lingustic analysis focalise on analyses of language, pronouns and an analysis of power relations, the socio-cognitive analysis examines the way language is coded to potray social reality and the final stage is the probing of how the text is organised (Grochalska, 2020).

A variety of ways of analysing data was postulated by different authors such as Fairclough (1995) and van Dijk (2001) among many others. Fairclough (2001) postulated the use of the four steps in CDA namely

-focus on a distinct social issue such as gender based violence;
-isolation of the major discourses, genres and styles;
- taking into account the range of diversity within styles, genres and discourses;
-identification of the areas of resistance to colonisation processes by major discourses.

van Dijk (2001) postulated six steps to the data analysis process which included:

-probing of the themes focusing on structures of language;
-Scrutinising the local meanings allusions, doubts and presuppositions;
-Probing the nuanced structure;
-unpacking the global and local forms of discourses;
-scrutinising the specific use of language;
-Analysis of the content

The steps in the data analysis may vary, but when implemented correctly they may be able expose the nuanced power relations within power institutions. Discourse analysis collects data through the use of focus group interviews and in-depth interviews. The chief constriction that novice researchers have to confront when working with discourse analysis of qualitative data is the multiple discourse analysis, it is challenging to select an appropriate approach. The process of data collection methods needs to be comprehensive enough that a person is able to follow the textual analysis.

Innovative research practices as seen in discourse qualitative research face severe criticism from researchers who are immersed in positivistic thinking, who may find the diversity of methods disturbing. The novice researcher are therefore compelled to remain within the traditional research methodology framework rather than risk to face censure over methodology when using discourse analysis (Grochalska, 2020).

CDA mixed studies are devoid of stability in the meaning of terms that are used in research leading to a situation where all the phenomenon becomes subject to negotiation and interpretation. Approach research and data analysis from a CDA may upend the usage of established terms such as identity, feminism, gender and marginalised groups. Such disruption may be very disconcerting, although it is also fascinating. Each approach in CDA has been lambasted, for instance conversation analysis has been censured for being narrow, while Foulcadian analysis is critiqued for being to expansive. However, most of the critics acknowledge the value-laden nature of text and the challenge of them potraying reality in an undisputable way. However there are also strategies to ensure that these criticism are addressed which includes ensuring that there is a clear explanation of the steps used for data collection and analysis and to enable other researchers to trace the process and also the criticism of asserting a massive claims from very limited data can be addressed from the the way the evidence is examined (Pervin & Mokhtar, 2022). CDA is explanatory, interpretive and subjective. CDA has also been censored for forefronting context over text and in order to address this point, it is vital that CDA must begin with textual analysis before engaging in contextual analysis. CDA is driven by political agenda and not by lingustic needs and it is therefore it is imperative to employee an ecclectic approach to enhance the analysis intergrating the social, political and historical situation being researched and how discourse obfuscate the issues.

It is not possible to conduct research in CDA without functioning from value- laden perspective, and it is consequential that a researcher has to confess their positionality on issues they are researching on. It is submitted that CDA must gather data from the producers and consumers of discourse rather than merely depending on the view of the researcher. CDA research methods requires a humility and continuous commitment to learning new terms and concepts, given the bricolage technique that characterise the approach. Its an approach that demands highly honed lingustic skills. CDA is very appealing, however, using it for research can be extremely arduous. CDA is tedious and can take more time than intended.

DISCUSSION

In this chapter the way discourse evolved from modernity to postmodernity was examined. It was spotlighted that discourse analysis is very elusive to define. It was also noted that that the lack of a clear definition tends to add to the complexity of using DA in qualitative studies as a result of the multiple strands that are derived from it. Structuralism and its assumptions in the area of communication were also addressed. The chapter noted that structuralism is sought to advance total and unified explanations. Contrary to structuralism, post structural theory questions the assumptions of structuralism especially its totalizing claims to reality. It is also fundamental to foreground that post modernism is not a unitary position, but reflects a grouping of a number of political and ideological positions.

The chapter also critiqued the claims of structuralism leading to the emergence of deconstruction perspective. The interdisciplinary nature of discourse studies was also underscored. The chapter also highlighted the way discourse is viewed in the context of qualitative studies which includes ethnography, GT among other strands of qualitative research. The guises that discourses may assume were also highlighted. The emergence of DA in the 1970s as a political way of doing research with an agenda to spotlight hidden power disparities were also examined.

The challenges associated with using discourse analysis for novice researchers were unpacked. Some of the challenges that were noted were the challenge of defining the term discourse, the complex methodology used in discourse studies, the issue of researcher positionality and a tendency to compare qualitative discourse analysis unfavourably with traditional positivistic research and claims that discourse analysis is too preoccupied with the context rather than the text.

RECOMMENDATIONS

The chapter has discussed the constraints of discourse analysis. Some of the constraints that were noted were a tendency to use very short summaries in research data collection, which can make data analysis complex, the positionality of the research impinging on the ideological analysis of the issues in the research, the limitations related to identifying discourses among many other constrictions. Based on these observations then the following recommendations are offered:

i) It is significant that researchers who would like to use CDA or DA immerse themselves in the theory in order to be able to apply the concepts and theory of discourse analysis in their work;
ii) Researchers who use DA or CDA must develop very proficiency in the use of language in discourse analysis in order to execute the research methods properly;
iii) Researcher who employee discourse analysis must also get acquainted with the theory of CDA or DA so that they can incorporate the criticism that has been leveled against it in their research work;
iv) CDA functions from a political ideology and doesn't claim any assumptions related to positivistic objectivity, and researchers who use this approach must be able to demonstrate their ideological commitment in their work;
v) It is critical that researchers display reflexivity on their work so that there political position is clear to other researchers.

CONCLUSION

The chapter discussed the use of discourse analysis in qualitative research. The epistemological commitments of DA were also examined. The theoretical frameworks that informed this work were deconstruction, interpretivism and constructionist theories. The chapter also highlighted the constraints that are associated with using DA in qualitative research.

REFERENCES

Agustian, A., Malik, J., & Yuwono, U. (2020). Content analysis approach in critical discourse analysis of newspapers' news. [IJHI]. *International Journal of Humanities and Innovation, 3*(1), 1–6.

Al Umma, K. (2015). Ferdinand De Saussure: Structuralism and his role in modern lingustics. *Jurnal Lisanu ad-Dhad Vol. 02, No. 01*, 357-376.

Alqudah, M. (2019). Deconstruction: A cornucopia of Esoteric meanings . *International Journal of English Literature and Social Sciences (IJELS) Vol-4, Issue-2*, 500-504.

Andrews, T. (2012). What is social constructionism? [online.]. *The Grounded Theory Review, 11*(1), •••.

Bakhitin, M. (1986). *Speech genre and other late essays*. Texas Press. DOI: 10.7560/720466

Bartesaghi, N. C. (2016). Intertexuality and interdiscursivity. In *Oxford Bibliographies in Anthropology* (pp. 1–44). OUP.

Blackledge, A. (2012). Discourse and power. In Gee, J., & Handford, M. (Eds.), *The Routledge handbook of discourse analysis* (pp. 616–627). Routledge.

Braun, V., & Clarke, V. (2012). Thematic analysis. In H. Cooper, P. Camic, L. Long, D. A. Panther, D. Rindskopf, & K. J. Sher, *APA handbook of research methods in psychology vol 2: Research designs: Quantitative,Qualitative neuropsychological and biological* (pp. 57-71). Washington: American Psychological Association.

Campbell, M. (2018). Postmodernism and Educational Research. *Open Journal of Social Sciences, 6*(7), 67–73. DOI: 10.4236/jss.2018.67006

Catalano, T., & Waugh, L. (2020). *Critical Discourse Analysis, Critical Discourse Studies and Beyond*. Springer. DOI: 10.1007/978-3-030-49379-0

D'Angelo, L., & Marino, F. (2024). Multimodal analysis. In Kessler, M., & Polio, C. (Eds.), *Conducting genre-based research in applied linguistics: A methodological guide*. Routledge.

Dahlborg, E., Boman, A., Eriksson, H., & Tengelin, E. (2023). Encircling discourses—A guide to critical discourse analysis in caring science. *Scandinavian Journal of Caring Sciences, 00*, 1–8. PMID: 37421160

De Man, P., Bloom, H., & Miller, J. (1979). *Deconstruction and criticism.* Routledge and Kegan.

Deng, L., Laghari, T., & Gao, X. (2021). A genre-based exploration of intertextuality and interdiscursivity in advertorial discourse. *English for Specific Purposes, 62*, 30–42. DOI: 10.1016/j.esp.2020.11.003

Dicks, B. (2019). Multimodal analysis. In Atkinson, P. D. (Ed.), *Research Methods Foundations*. SAGE.

Dovetail Editorial Team. (2023, 02 06). *How to do thematic analysis.* Retrieved from Dovetail: https://dovetail.com/research/thematic-analysis/

EduBirdie. (2024, 09 7). *The evolution of discourse and development of discourse analysis*. Retrieved from EduBirdie: https://edubirdie.com/examples/the-evolution-and-development-of-discourse-analysis/

Fairclough, N. (2001). Critical discourse analysis as a method in social scientific research. In Wodak, R., & Meyer, M. (Eds.), *Methods of critical discourse analysis* (pp. 121–138). Sage.

Fairclough, N. (2003). *Analysing Discourse: Textual analysis for social research.* Routledge. DOI: 10.4324/9780203697078

Fernandes, C., & Sargentini, V. (2022). Saussure and Foucault; language and discourse. *Revista da Anpoll, Florianópolis, 53*(2), 53–67. DOI: 10.18309/ranpoll.v53i2.1692

Foucault, M. (1972). *The archaeology of knowledge and the discourse.* Phanteon Books.

Fowler, R., Hodge, B., Kress, G., & Trew, T. (. (1979). Language and control. London: Boston and Henley.

Gąsior-Niemiec, A. (2008). Teoria dyskursu Laclau i Mouffe jako narzędzie analizy socjo logicznej – przypadek dyskursu „osiedli grodzonych. In Horolets, A. (Ed.), *Analiza dyskursu w socjologii i dla socjologii* (pp. 291–316). Wydawnictwo Adam Marszałek.

Given, L. (2008). *The SAGE Encyclopedia of qualitative research methods.* Sage. DOI: 10.4135/9781412963909

Grochalska, M. (2020). Research practices in critical discourse analysis: Constraints and challenges. *Kultura i Edukacja, 2020*(2), 98–112. DOI: 10.15804/kie.2020.02.06

Guba, E., & Lincoln, Y. (2005). Paradigmatic controversies, contradictions & emerging confluences. In Denzin, N., & Lincoln, Y. (Eds.), *The sage handbook of qualitative research*. Sage.

Haradhan, M. (2018). Qualitative Research Methodology in Social Sciences and Related Subjects. *Munich Personal RePEc Archive*, *7*(01), 23–48.

Hendricks, G. P. (2016). Deconstruction the end of writing:'Everything is a text, there is nothing outside context. *Verbum et Ecclesia*, *37*(1), 1–9. DOI: 10.4102/ve.v37i1.1509

Hodge, B., & Kress, G. (1979). *Language as ideology*. Routledge.

Hopf, T. (2004). Discourse and content analysis: Some fundamental incompatibilities. *Qualitative Methods*, (Spring), 31–33.

Jaeger, S. (2001). Discourse and knowledge: Theoretical and methodological aspects of a critical discourse and dispositive analysis. In Wodak, R., & Meyer, M. (Eds.), *Methods of critical discourse analysis* (pp. 32–62). Sage.

James, I. (2023). Metaphor in Biosemiotics and Deconstruction. *Oxford Literary Review*, *45*(2), 229–250. DOI: 10.3366/olr.2023.0417

Jaspal, R. (2020). Content analysis, thematic analysis and discourse analysis. In Breakwell, G., Wright, D., & Barnett, J. (Eds.), *Research Methods in Pyschology* (pp. 1–52). Sage.

Kessler, M. (2024). References. In Kessler, M. (Ed.), *Digital Multimodal Composing: Connecting Theory, Research and Practice in Second Language Acquisition* (pp. 133–147). Blue Ridge Summit Multilingual Matters. DOI: 10.21832/9781800416680-011

Khafaga, A. F. (2017). Discourse Interpretation: A Deconstructive, Reader-oriented Approach to Critical Discourse Analysis. *International Journal of Applied Linguistics and English Literature*, *6*(2), 138–146. DOI: 10.7575/aiac.ijalel.v.6n.2p.138

Kivunja, C., & Kuyini, A. (2017). Understanding and applying research paradigms in educational contexts. *International Journal of Higher Education*, *6*(5), 26–41. DOI: 10.5430/ijhe.v6n5p26

Kristeva, J. (1980). *Desire in language: A semiotic approach to literature and art*. Columbia University Press.

Kuzmina, J. (2015). Intertextuality and interdiscursivity in information technologies organisational discourse. *Baltic Journal of English Language. Literature and Culture*, *5*, 75–89.

Kuzmina, J. (2015). Intertexuality and interdiscursivity in information technologies organisational discourse. *Baltic Journal of English Language. Literature and Culture, 5*, 75–89.

Kuzmina, J., Bartesaghi, M., & Noy, C. (2015). Interdiscursivity. In Karen, T., Ilie, C., & Sandel, T. (Eds.), *The International Encyclopedia of Language and Social Interaction* (pp. 1–6). JohnWiley & Sons.

Leavy, P. (2017). *Research design, quantitative, qualitative and mixed methods, arts based and community based participatory research approaches*. The Guilford Press.

Mahmud, M. (2017). Doing discourse analysis: An introduction. Kalangan: Phoenix Publisher.

Mark, L. (2024, 06 23). *What is thematic analysis advantages and disadvantages*. Retrieved from Acadmemic Papers: https://www.theacademicpapers.co.uk/blog/2021/10/02/what-is-thematic-analysis-advantages-and-disadvantages/

Morgan, D. (2007). Paradigm lost and pragmatism regained: Methodological implications of combining qualitative and quantitative methods. *Journal of Mixed Methods Research, 1*(1), 48–76. DOI: 10.1177/2345678906292462

Nadu, C. (2015). Text/Texts: Interrogating Julia Kristeva's Concept. *Research Journal of Humanities and Social Sciences, 3*, 77–80.

Naeem, M., Ozuem, W., Howell, K., & Ranfagni, S. (2023). A step-by-step process of thematic analysis to develop a conceptual model in qualitative research. *International Journal of Qualitative Methods, 22*, 1–18. DOI: 10.1177/16094069231205789

Nganga, C., Maroko, G., & Ong'onda, A. N. (2023). Intertextuality and interdiscursivity in postcolonial language policies in Kenya. *European Journal of Applied Linguistics Studies, 6*(2), 214–214. DOI: 10.46827/ejals.v6i2.480

Nickerson, C. (2024, 02 13). *Interpretivism and research philosophy*. Retrieved from Simply Psychology: https://www.simplypsychology.org/interpretivism-paradigm.html

Nicmanis, M. (2024). Reflexive content analysis:An approach to data analysis, reduction, and description. *International Journal of Qualitative Methods, 23*, 1–12. DOI: 10.1177/16094069241236603

Osakinle, P. (2024). Discourse of Intertextuality in the language of political advertisements in selected Nigerian newspapers. *Elsya: Journal of English Language Studiesvol, 6*(1), 1–19. DOI: 10.31849/elsya.v6i1.16121

Patel, D. (2020). Scope and usage of discourse analysis as a research method in English studies. *International Journal for Innovative Research in Multidisciplinary Field*, *6*(3), 234–238.

Pervin, N., & Mokhtar, M. (2022). The interpretivist research paradigm: A subjective notion of a social context. *International Journal of Academic Research in Progressive Education and Development*, *11*(2), 419–428. DOI: 10.6007/IJARPED/v11-i2/12938

Phillips, M. (2023). Towardsa social constructionist, criticalist Foucauldian informed qualitative research approach: Opportunties and challenges. *SN Soc Sci 3, 175*, 174-199. Retrieved 9 10, 2024, from DOI: 10.1007/s43545-023-00774-9

Poucher, Z., Tamminena, K., Carona, J., & Sweet, S. (2019). hinking through and designing qualitative research studies: a focused mapping review of 30 years of qualitative research in sport psychology. *national Review of Sport and Exercise Psychology Volume 13, Issue 1*, 164-186.

Psychology Learners. (2024, 09 07). *Discourse Analysis: Assumptions, theories and steps*. Retrieved from Psychology Learners: https://psylearners.psychotechservices.com/2015/09/solved-ignou-assignment-mpc005-q3.html

Rakita, M. (2017). Modernisation discourse and its contents. *Studia Ethnologica Croatica*, *29*, 103–148. DOI: 10.17234/SEC.29.4

Scispace. (2024, 10 23). *What is the weakness of discourse analysis in accurately identifying discourses?* Retrieved from Scispace: https://typeset.io/questions/what-is-the-weakness-of-discourse-analysis-in-accurately-3s3ehn0s41

Sikirivwa, M. (2020). Deconstruction theory and its background. *American Journal of Humanities and Social Sciences Research (AJHSSR) Volume-4, Issue-4*, 44-72.

Sudartini, S. (2024). Postmodernism paradigm and its use on the existence of language. *Rev. Gest. Soc. Ambient.*, *18*(7), 1–26.

The Editors of Encyclopaedia. (2024, 09 7). *Deconstruction*. Retrieved from Encyclopedia Britannica: https://www.britannica.com/topic/deconstruction

Turner, D. (2020, 09 29). *Analysis: Introducing discourse analysis for qualitative research*. Retrieved from Quirkos: https://www.quirkos.com/blog/post/discourse-analysis-qualitative-research/

van Dijk, T. (2001). Critical discourse analysis. In Scchiffrin, T. D., & Hamilton, H. (Eds.), *Handbook of discourse analysis* (pp. 352–371). Blackwell.

Van Dijk, T. A. (1993). Principles of critical discourse analysis. *Sage (Atlanta, Ga.)*, *4*(2), 249–283.

Wertz, F., Charmaz, K., McMullen, L., Josselson, R., Anderson, R., & McSpadden, E. (2011). *Five ways of doing qualitative analysis: phenomenological, grounded theory, discourse analysis, narrative research and intuitive inquiry*. The Guilford Press.

Willig, C. (2008). *Introducing qualitative research in psychology*. Open University Press.

Wodak, R. (2008). Introduction: discourse studies - important concepts and terms. In Wodak, R., & Krzyżanowski, M. (Eds.), *Qualitative discourse analysis in the Social Sciences* (pp. 1–29). Palgrave. DOI: 10.1007/978-1-137-04798-4_1

Yan-Yan, C., Shek, D., & Bu, F.-F. (2011). Applications ofinterpretive and constructionist research methods in adolescent research philosophy, principles and examples. *International Journal of Adolescent Medicine and Health, 23*(2), 129–139. PMID: 21870675

Yazdani, N., Murad, H., & Abbas, R. (2011). From modernity to postmodernity: A historical discourse on western civilization. *International Journal of Business and Social Science, 2*(11), 249–256.

Zotzmann, K., & O'Regan, J. (2023). Critical discourse analysis, critical discourse studies, and critical applied linguistics. In Zotzmann, K. (Ed.), *J. O'Regan, The Routledge of Applied Lingustics* (pp. 57–67). Routledge. DOI: 10.4324/9781003082637-7

KEY TERMS AND DEFINITIONS

Interpretivism: a perspective that holds that reality is shared by the context and the people who make it

Deconstruction: it refers to a method that involves breaking down of concepts and language in order to obtain a better understanding of them.

Constructionism: a method that social reality is created by people in social context at specific time

Intertextuality: Comparison of different texts in order to obtain meaning

Interdiscursivity: use of diverse discourses, social meanings or genres within a single text

Multimodal: made up of a variety of different modes or events

Positivism: a belief in objective reality that was associated with traditional sciences

Modernity: a philosophical view associated with the era of technological invention and industrial revolution, which created the idea of continuous progress

Discourse: a theoretical or political position on any issue that is contested

Chapter 8
Research Protocol in Studies With Semi-Structured Interviews

Deivisson Freire
Universidade Federal de Pernambuco, Brazil

Cláudio de Araújo Wanderley
UUniversidade Federal de Pernambuco, Brazil

Ana Lúcia Fontes
Universidade Federal de Pernambuco, Brazil

Umbelina Lagioia
Universidade Federal de Pernambuco, Brazil

ABSTRACT

Provide a comprehensive and detailed understanding of how to construct and implement research with semi-structured interviews in applied social sciences. By the end of this chapter, readers will have a clear, practical grasp of conducting research with semi-structured interviews and see how to apply the research protocol in practice. The chapter will consist of eight topics: introduction, semi-structured interviews, how to prepare a good semi-structured interview script, testing and refining the script, data collection, data analysis in semi-structured interviews, strategies to ensure the credibility (validity) and reliability of data, research ethics, and practical activity. This chapter does not aim to exhaust the topic or imply these are the only possible strategies.

DOI: 10.4018/979-8-3373-4432-4.ch008

1. INTRODUCTION

In-depth study is a qualitative approach focused on achieving a detailed understanding of a phenomenon, group, individual, or specific situation. Rather than broad generalizations, this research method thoroughly explores complex and subtle aspects of study subjects, capturing the richness of human experiences for a profound, contextualized understanding (Silva, Oliveira, & Neves, 2021).

Vieira (2004) observes that quality issues in qualitative research usually stem not from the chosen method but from its inadequate use. The author notes rigor and reliability as key features of strong qualitative research. A well-defined research protocol is essential for structured, rigorous, and ethical studies, providing a solid foundation for all research phases and ensuring results are valid, reliable, and significant. Additionally, it promotes transparency and accountability, supporting review and acceptance by the scientific community and other stakeholders.

2. SEMI-STRUCTURED INTERVIEWS

This qualitative data collection technique involves detailed conversations with participants to explore their perspectives, experiences, and the meanings they attribute to certain phenomena. It is valuable because it allows an in-depth exploration of participants' perceptions and experiences, adapting to the natural flow of conversation and welcoming emerging themes. This method fosters trust between interviewer and interviewee, encouraging honest and detailed responses. The interview should be conducted in a comfortable, private setting, with permission to record for data accuracy (Silva, Oliveira, & Neves, 2021). Transcription and thematic analysis are recommended to identify significant patterns and insights.

3. HOW TO DEVELOP A GOOD SEMI-STRUCTURED INTERVIEW GUIDE

Before developing the guide, it's essential to have a clear understanding of the research objectives and the topics that need exploration. Additionally, according to Leitão (2021), researchers should review relevant literature, familiarize themselves with similar studies, and identify gaps and contributions. It is also crucial to reflect on scientific paradigms related to the research problem and to consider time, human resources, and financial limitations.

According to Leitão (2021), questions should be open-ended to allow participants to share detailed experiences and opinions. Avoid closed or leading questions.

Types of Questions:

- **Open-Ended Questions**: Allow long and detailed responses.
  Example: "Can you describe your experience using the internal audit consulting service?"
- **Probing Questions**: Deepen understanding of a specific topic.
  Example: "What else can you tell me about this situation?"
- **Reflective Questions**: Encourage reflection on experiences or opinions.
  Example: "How do you think this experience influenced your view of audit services?"
- **Hypothetical Questions**: Encourage creativity and alternative thinking (Leitão, 2021).

Tips

Organize questions in a logical sequence, beginning with general topics and moving to more specific and complex ones (Leitão, 2021).

Structure the guide into thematic blocks that group related questions, which facilitates in-depth exploration and guides the interviewee in reflective and exploratory responses (Leitão, 2021).

While the guide should be structured, allow flexibility for the interviewer to explore emerging topics.

Use clear, accessible language, avoiding technical terms or jargon that might confuse participants (Leitão, 2021).

3.1 Recommended Interview Guide Structure

- **Introduction:** Presentation of the researcher, explanation of the interview's objectives, and discussion of ethical considerations. It is essential to clarify participant anonymity and how collected data will be disclosed, as these aspects are crucial for building trust and encouraging genuine collaboration during the interview. Additionally, clear and objective consent from the interviewee and permission to record should be obtained (Leitão, 2021).
- **Initial Topics:** Opening questions that help establish rapport with the participant, using introductory or "icebreaker" themes to introduce the topic smoothly (Leitão, 2021).
- **Central Topics:** Once a trusting relationship is established, you will be ready to ask questions that address the main research objectives.
- **Trigger Questions:** These are questions that complement and deepen the initial responses to the items on the outline, encouraging the production of further meanings around the topic. Example: "What does the expression 'x'

mean to you?"; "How so?"; "Why?"; "Could you give me an example?" (LEITÃO, 2021).

- o **Exploration Topics:** Questions that delve into specific themes that arise during the interview.
- o **Closing:** The interviewer should leave the end of the interview to address lighter topics or discuss something that contributes to the participant's well-being. At the conclusion, it is important to allow a moment for the interviewee to make spontaneous comments or ask questions, giving room for information that may not have been anticipated in the outline. It is worth mentioning that relevant comments often emerge after the formal conclusion of the interview, when the recording has already been turned off. In such cases, the researcher can inform the participant about the importance of the content and resume recording to capture it.

4. TESTING AND REFINING THE INTERVIEW GUIDE

Before using the guide in real interviews, it is recommended to test it in one or more pilot interviews (DUARTE, 2004; LEITÃO, 2021). This helps identify potential issues, such as confusing or redundant questions, and allows for refinements that will improve the quality of the interviews.

Tips

Conduct **pilot interviews** with individuals who share similar characteristics with the research participants. This will provide valuable feedback.

Revise the guide based on feedback from the pilot interviews and **critical reflection** on the process.

A well-developed in-depth interview guide is an essential tool for collecting rich and detailed data in qualitative research. By following the steps highlighted above, the researcher will be better prepared to conduct interviews that truly capture the nuances and complexities of the studied phenomenon.

Additionally, we emphasize that flexibility is a key aspect of in-depth interviews. The interviewer should be prepared to adapt the guide as the conversation unfolds, exploring new topics that may spontaneously arise without losing sight of the main research objectives. Thus, while the guide serves as a roadmap, the interviewer's skill in conducting the interview and responding to the dynamics of the interaction will be crucial to successful data collection.

During the interview, the researcher should develop attentive listening and treat the participant with respect, as suggested by Ludke and André (1986). They highlight that, besides respecting the culture and values of the interviewee, the interviewer needs to encourage a natural flow of information without steering the responses, creating a trusting environment where the participant feels comfortable expressing themselves freely. A trusting environment fosters detailed accounts, richer information, reflections, and even moments of openness (SILVA, OLIVEIRA, and NEVES, 2021).

5. DATA COLLECTION

It is important for data collection to be rigorous and systematic, meaning it should be carried out in a structured and consistent manner. In addition to the **interview guide** (Leitão, 2021), the researcher should use an **audio recorder** (SILVA, OLIVEIRA, and NEVES, 2021), an essential tool for accurately recording the interviews, and a **field notebook**, which allows the researcher to take notes during and after interviews, documenting observations, impressions, reflections, and additional insights that may arise.

Tips

To increase the internal validity of your research, **train interviewers** to ensure they all follow the same procedures and approaches, record all interactions in **detail**, and conduct interviews according to the established **protocol**.

6. DATA ANALYSIS IN SEMI-STRUCTURED INTERVIEWS

This is the process of examining, categorizing, and interpreting the collected data to identify patterns, themes, and meanings. Unlike quantitative research, where data is analyzed numerically, qualitative research focuses on understanding participants' narratives, contexts, and experiences. It seeks to capture the complex and subtle meanings that emerge from social interactions and individual experiences, taking into account the social, cultural, and historical context in which the data was collected, providing a rich and detailed interpretation. It allows for continuous adaptation and revision of analysis methods as new insights emerge from the data.

Although there are other ways to analyze qualitative data, this chapter will focus on content analysis, one of the most commonly used data analysis techniques (CHUEKE and LIMA, 2012).

6.1 Content Analysis

Content analysis is a qualitative technique that involves coding data into categories or themes that emerge from participants' statements, documents, or other textual materials. It can be used in a quantitative manner (counting the frequency of themes) or qualitatively (interpreting the meanings associated with themes). Before starting the analysis, the interview must be transcribed, preferably by the person who conducted it. Transcription is essential, as without it, it would not be possible to make inferences about how people speak, "their pauses, inflections, emphases, how they assimilate systems of coherence and construct their identities, how they deal with power relations, etc." (BASTOS; SANTOS, 2013, p. 30).

Once transcribed, the interview should undergo a reliability check, which involves listening to the recording again with the transcription in hand, following and verifying each sentence, change in tone, interjections, interruptions, etc. (DUARTE, 2004). Transcribing and reading each interview before moving on to the next stage helps to correct errors, avoid leading responses, and reassess the research directions (ALBERRTI, 1990).

According to Duarte (2004), interviews can and should be edited. Unless a discourse analysis is intended, excessively colloquial expressions, interjections, repetitions, incomplete sentences, language crutches, mannerisms, and grammatical errors should be adjusted in the edited transcription. However, it is crucial to keep both the original and the edited versions of all transcriptions. Duarte also adds that responses obtained through suggestive, ambiguous, biased, or leading questions that caused the interviewee to confirm or deny statements proposed by the researcher should also be excluded in the edited version.

One way of analyzing is to fragment the content and reorganize it based on new assumptions. This involves breaking down the participants' speech into units of meaning—essential excerpts for understanding—and interpreting each unit, relating them to each other to formulate explanatory hypotheses. The interpretation requires these units to be grouped into analysis categories, such as "school memories" or "initial training," to facilitate material organization and understanding. In this case, all text units related to, for instance, "initial training," should be grouped under this category (DUARTE, 2004).

According to Duarte (2004), the researcher can define categories beforehand, based on theoretical knowledge or knowledge of the empirical field, to analyze interviews. The interviews are organized into units of meaning and grouped into thematic categories. For ease, three or four major thematic axes can be chosen, within which specific subthemes can be created. This allows for the cross-referencing and articulation of responses to interpret recurrences, divergences, and patterns. The researcher should maintain a critical stance and "question" their interviewees, as suggested by Magnani (1986). Duarte (2004) also highlights that interviewees may start responding with what they believe the researcher wants to hear. For this reason, it is essential to review and cross-check the results with field observations and other sources, ensuring a comprehensive view and avoiding distortions. Throughout the analysis process, empirical material must be examined, interpreted, and revised considering the scientific literature that serves as the researcher's reference. Empirical material should be valued, as it can challenge or even refute the researcher's initial theories, requiring an openness to the unexpected in the field.

Content Analysis Steps:

1. Transcription and reliability check.
2. Identification and categorization of text segments (sentences, paragraphs) into categories or themes.
3. Segmenting participants' speech into units of meaning, extracting the minimum text necessary for the analyst's understanding (DUARTE, 2004).
4. Grouping units into analysis categories.
5. Interpreting each unit, relating them to one another to formulate explanatory hypotheses.
6. Interpretation requires these units to be grouped into analysis categories.
7. Analyzing the identified themes to understand their meanings and implications within the study context.
8. Generating insights and constructing explanatory narratives based on emerging themes.

Table 1. Advantages and Limitations of Content Analysis

Advantages	**Limitations**
Helps to systematize large volumes of textual data in a structured way.	Code interpretation can be subjective, depending on the researcher's perspective.
Can be applied to different types of textual data, such as interviews, documents, and media.	Focus on categorization may sometimes lead to the loss of important contextual nuances.

Source: the authors.

7. STRATEGIES TO ENSURE DATA CREDIBILITY (VALIDITY) AND RELIABILITY

Although the concepts of validity and reliability are often associated with quantitative research, in qualitative research, they assume meanings adapted to the exploratory and interpretive nature of the method. According to Maykut and Morehouse (1995), the term "internal validity" in quantitative research is equivalent to "credibility" in qualitative research. Below are the definitions of validity and reliability in the qualitative context and strategies to ensure that the data collected are both valid and reliable.

7.1 Credibility

For Guba and Lincoln (1989), credibility in qualitative research refers to the authenticity of the results, assessing the extent to which the researcher has built trust in them based on the study design, participants, and the investigated context. It determines whether the findings faithfully reflect the information obtained from the original data and accurately convey the participants' perspectives.

Table 2. Recommended Strategies to Ensure Credibility

Strategy	Description	Additional Comments
Triangulation (Chueke e Lima, 2012; Denzin e Lincoln,1994; GODOY, 1995)	Use multiple data sources, methods, theories, or researchers to corroborate findings and increase the depth and robustness of interpretations.	Combine in-depth interviews with participant observation and document analysis to gain a more comprehensive view of the studied phenomenon.
Participant Verification (Chueke e Lima, 2012; CRESWELL, 2007; Denzin e Lincoln,1994; Duarte, 2004; GODOY, 1995; LINCOLN e GUBA, 1985; Moreira, 2018)	Involve participants in validating findings and interpretations, allowing them to review and confirm the researcher's conclusions. According to Mertens (2005), this is one of the most important strategies for enhancing the credibility of qualitative studies, noting that it can be formal or informal.	At the end of an interview, the researcher can summarize what was discussed and confirm with the participant if the notes accurately reflect their opinion. Additionally, a draft report may be provided for participants to read and offer comments (MOREIRA, 2018).
Thick Description (Moreira, 2018)	Provide detailed and rich descriptions of the context, participants, and interactions, allowing other researchers or readers to understand and contextualize the findings.	Although generalization is not the objective of qualitative researchers, providing thick descriptions allows others to make comparisons and assess the transferability of the results.
Extended Fieldwork (Denzin e Lincoln,1994; GODOY, 1995; LINCOLN e GUBA, 1985; Mertens, 2005; Moreira, 2018)	Spend a significant amount of time in the field, interacting with participants to build trust and gain a deeper understanding of the phenomenon.	There is no set timeframe. According to Moreira (2018), when the researcher feels information is repeating, it is time to conclude data collection. This is what researchers (Bogdan and Biklen, 1994; Seidman, 2013; Blandford, 2013; among others) call data saturation, meaning interviews stop bringing new perspectives and content, leading to redundant information.

7.2 Reliability

Reliability in qualitative research refers to the consistency and repeatability of results—that is, whether other researchers, following the same process, would reach similar conclusions. While exact replication is challenging in qualitative research due to its interpretive nature, reliability focuses on ensuring that the research process is transparent, systematic, and well-documented.

Table 3. Recommended Strategies to Ensure Reliability

Strategy	Description
Richness in the description of methodological processes adopted (Chueke and Lima, 2012)	Maintain a detailed record of all research process steps, including field notes, methodological decisions, transcripts, and analyses, so others can audit the process. This should also include reflections on why certain interpretations were made.
Consistent Positioning (Chueke and Lima, 2012)	Develop and apply a consistent coding system to analyze data, ensuring that the same criteria are used throughout the study. Example: a detailed coding manual can be created, explaining how each code was defined and applied.
Peer Review (Denzin and Lincoln, 1994; Mertens, 2005; Moreira, 2018)	Peer review requires an external qualified researcher to examine the research processes and data interpretations. This allows for review and discussion of interpretations and conclusions to ensure analyses are robust and free of individual biases. According to Mertens (2005), in theses and dissertations, advisors can play this important role. Moreira (2018) suggests that peer review can be conducted in various ways. One option is to discuss emerging results with colleagues in the same field; another is to present and defend the method and conclusions at national and international conferences, seeking constructive feedback from other researchers. Yet another option is to present results and their implications to groups interested in the topic.
Researcher Bias (Moreira, 2018)	According to Creswell (1998), the researcher should clarify their position from the start of the study. This is important for readers to understand the researcher's perspective and any biases or assumptions that may influence the investigation (MOREIRA, 2018).
Reflexivity (Denzin and Lincoln, 1994; Duarte, 2004; Moreira, 2018; Romanelli, 1998)	Recognize and document the researcher's assumptions, biases, and influences on the research process to mitigate their effects on the results. Example: keeping a reflective journal where the researcher records personal thoughts, biases, and how these may have influenced interpretations.
Negative Case Sampling (Ary, Jacobs, and Sorensen, 2010; Creswell, 1998; Guba and Lincoln, 1989; Lincoln and Guba, 1985; Mertens, 2005; Miles and Huberman, 1994)	Negative case sampling involves identifying and analyzing cases that contradict the researcher's expectations. This process seeks and examines data elements that do not support, or even contradict, emerging patterns or explanations. This strategy serves to enhance the analysis by allowing the researcher to explain or justify most cases, ultimately strengthening and confirming identified patterns (MOREIRA, 2018).

Source: the authors.

Applying the strategies cited in this section to ensure data validity and reliability in in-depth qualitative research will help the researcher produce results that authentically reflect participants' experiences and perspectives while maintaining a transparent and replicable research process. These practices contribute to a significant impact and value in both academic and professional contexts.

8. ETHICS IN RESEARCH

Ethics in qualitative research is a central issue that permeates all stages of the investigative process (Silva, Oliveira, and Neves, 2021), especially in-depth studies where researcher-participant interactions are intensive and often involve sensitive topics. In this context, it is essential for the researcher to be aware of the ethical implications of their practices, ensuring that participants' rights, dignity, and well-being are respected, and that the research is conducted honestly and transparently (Rosa & Arnoldi, 2008). This approach preserves the credibility and validity of the findings. Therefore, submitting the research project to an Ethics Committee is an important procedure for conducting interviews (Silva, Oliveira, and Neves, 2021).

Ethical considerations, including **informed consent** and **data confidentiality**, are essential to ensuring that the study is conducted responsibly, respectfully, and fairly. These principles protect participants' rights and dignity and contribute to the integrity and credibility of the research. By rigorously adhering to ethical practices, the researcher ensures that the results obtained genuinely reflect the participants' experiences while minimizing risks and maximizing the benefits of the investigation.

8.1 Informed Consent

Informed consent is the process by which participants are fully informed about the purpose, methods, risks, and potential benefits of the research before agreeing to participate (Rosa & Arnoldi, 2008). This fundamental ethical principle ensures that participation is voluntary and based on a clear understanding of what the study entails. This communication can be achieved through a Free and Informed Consent Form (FICF). The table below presents essential aspects of the FICF.

Table 4. Important Aspects that Should Compose the FICF

Essential Aspects	Description
Complete Information	Participants must receive a detailed explanation of the purpose of the research, the procedures involved, the required time, potential risks, expected benefits, and how their data will be used. It should be clear that interviews will be recorded and transcribed for later analysis (Silva, Oliveira, and Neves, 2021).
Understanding	The researcher must ensure that participants fully understand the information provided, using clear and accessible language, and addressing any questions or concerns.
Voluntariness	Participation must be entirely voluntary, with no coercion or pressure. Participants should know that they can withdraw from the study at any time without having to provide a reason and without any consequences (Rosa & Arnoldi, 2008).
Documentation	Informed consent must be recorded in writing, usually through a consent form that participants sign before data collection begins (Silva, Oliveira, and Neves, 2021).

Note: Verbal consent may also be used in certain contexts with adequate justification.
Source: the authors.

To conduct the interview, the researcher should schedule it in advance, considering the participant's availability regarding date and time. It is important to inform the estimated duration of the interview so that the participant can allocate specific time without concern about other commitments. This planning helps the participant reflect more on the topic and organize their responses (Silva, Oliveira, and Neves, 2021).

In addition, the researcher should select an appropriate location for the interview. The environment should be private to ensure the participant's identity is preserved, quiet to avoid distracting noises, and welcoming to provide a comfortable space where the participant feels at ease to share impressions, experiences, and personal perspectives in the investigated educational context (Silva, Oliveira, and Neves, 2021).

8.2 Data Confidentiality

Data confidentiality refers to the researcher's commitment to protect participants' privacy, ensuring that the information they provide is not disclosed in a way that could identify them without their consent (Rosa & Arnoldi, 2008).

By ensuring confidentiality, the researcher fosters a trustworthy environment where participants feel secure in sharing personal and honest information. Maintaining confidentiality is an ethical responsibility that ensures the collected data is used responsibly and respectfully.

Table 5 . Strategies to Ensure Confidentiality

Strategy	Action
Anonymization	Remove names, addresses, and other information that could reveal the identity of participants, and replace them with pseudonyms or codes during the interview transcription process (Silva, Oliveira, and Neves, 2021).
Secure Storage	Ensure that all collected data is stored in secure locations, such as password-protected computers and physical files kept in locked places, accessible only to authorized researchers (Silva, Oliveira, and Neves, 2021).
Limited Disclosure	Share personal data only when absolutely necessary, limiting access to interview transcripts to a small number of researchers directly involved in the data analysis.
Data Destruction	Establish a timeline for the secure destruction of personal data after the study is completed, in accordance with the Ethics Committee's recommendations (Silva, Oliveira, and Neves, 2021).

Source: the authors.

Practical Activity

We suggest reading the article *"The Multiplicity of Performance Management Systems: Heterogeneity in Multinational Corporations and Management Sense-Making"* by David J. Cooper, Mahmoud Ezzamel, and Keith Robson, published in *Contemporary Accounting Research* (Vol. 36, No. 1, pp. 451–485, Spring 2019).

While reading Cooper, Ezzamel, and Robson's article (2019), which strategies did you observe that the authors employed?

Authors' Comments on Cooper, Ezzamel, and Robson's Research (2019)

Below is an analysis of the strategies the researchers employed and the measures they took to ensure a high-quality study. Here are some of the strengths we observed:

- **Methodology**. The interviews were semi-structured, and the exact content varied depending on each interviewee's experience and background. As themes emerged, they were further explored in depth.
- **Internal Validity**. The researchers provided verbatim transcripts of the segments that led them to specific conclusions. This approach helps reduce researcher bias. The study included precise and detailed information, such as comprehensive interview records, enhancing the data's credibility.
- **Research Context**. The researchers provided context for the study, allowing other researchers to assess the relevance and applicability of the findings in various scenarios.
- **Contribution to the Literature**. The study addresses an underexplored area in the literature, namely the interaction and use of multiple performance measurement systems.

REFERENCES

Alberti, V. (n.d.). *História oral: A experiência do CPDOC*. Fundação Getúlio Vargas.

Ary, D., Jacobs, L., & Sorensen, C. (2010). *Introduction to research in education* (8th ed.). Wadsworth Cengage Learning.

Bastos, L. C., & Santos, W. S. (2013). *The interview in qualitative research: Perspectives in narrative and interaction analysis*. Quartet Faperj.

Blandford, A. E. (2013). Semi-structured qualitative studies. In Soegard, M., & Dam, R. F. (Eds.), *The encyclopedia of human-computer interaction* (2nd ed.). The Interaction Design Foundation.

Bogdan, R., & Biklen, S. (1994). *Qualitative research in education: An introduction to theory and methods*. Porto Editora.

Chueke, G. V., & Lima, M. C. (2012). Qualitative research: Evolution and criteria. *Revista Espaço Acadêmico, 11*(128), 63–69.

Cooper, D. J., Ezzamel, M., & Robson, K. (2019). The multiplicity of performance management systems: Heterogeneity in multinational corporations and management sense-making. *Contemporary Accounting Research, 36*(1), 451–485. DOI: 10.1111/1911-3846.12416

Creswell, J. W. (1998). *Qualitative inquiry and research design: Choosing among five traditions*. Sage Publications.

Creswell, J. W. (2007). *Qualitative inquiry and research design: Choosing among five traditions* (2nd ed.). Sage.

Denzin, N., & Lincoln, Y. (Eds.). (1994). *Handbook of qualitative research*. Sage.

Duarte, R. (2004). (in Review). Interviews in qualitative research. *Educating*, (24), 213–225.

Godoy, A. S. (1995). Introduction to qualitative research and its possibilities. *RAE - Journal of Business Administration, 35*(2), 57.

Guba, E., & Lincoln, Y. (1989). Fourth generation evaluation. *Sage (Atlanta, Ga.)*.

Leitão, C. (2021). The interview as an instrument of scientific research: Planning, execution and analysis. In *Methodology of scientific research in informatics in education: Qualitative research approach* (Vol. 3).

Lincoln, Y., & Guba, E. (1985). Naturalistic inquiry. *Sage (Atlanta, Ga.)*.

Lüdke, M., & André, M. E. D. A. (1986). *Research in education: Qualitative approaches*. EPU.

Magnani, J. G. C. (1986). Discourse and representation, or how the Baloma of Kiriwina can reincarnate in current research. In Cardoso, R. (Ed.), *The anthropological adventure: Theory and research* (pp. 127–140). Paz e Terra.

Maykut, P., & Morehouse, R. (1995). *Beginning qualitative research: A philosophic and practical guide*. Falmer Press.

Mertens, D. (2005). *Research evaluation in education and psychology: Integrating diversity with quantitative, qualitative, and mixed methods* (2nd ed.). Sage.

Miles, M., & Huberman, A. M. (1994). *Qualitative data analysis: An expanded sourcebook* (2nd ed.). Sage.

Moreira, M. A. (2018). Criteria and strategies to ensure rigor in qualitative research. *Brazilian Journal of Science and Technology Teaching*, *11*(1), 405–411.

Romanelli, G. (1998). The anthropological interview: Exchange and alterity. *Journal of the Graduate Program in Psychology of the Faculty of Philosophy, Sciences and Letters of Ribeirão Preto*, 119–133.

Rosa, M. V. F. P. C., & Arnoldi, M. A. G. C. (2008). *The interview in qualitative research: Mechanisms for the validation of results*. Autêntica.

Seidman, I. (2013). *Interviewing as qualitative research: A guide for researchers in education and social sciences* (4th ed.). Teachers College Press.

Silva, L. S., de Oliveira, G. S., & Neves, E. H. C. (2021). Interview in research in education with a qualitative approach: Some theoretical and practical considerations. *Prisma Magazine*, *2*(1), 110–112.

Vieira, M. M. F. (2004). For good qualitative research. In Vieira, M. M. F., & Zouain, D. M. (Eds.), *Qualitative research in administration* (pp. 13–28). Editora FGV.

Chapter 9
Phenomenological Research in Education Studies:
Mertis and Pitfalls

William Chakabwata
https://orcid.org/0000-0002-4224-5239
University of South Africa, South Africa

ABSTRACT

In spite of the burgeoning of phenomenological research in recent times in multiple disciplines such as sociology, anthropology, economics, health sciences and journalism among many others, in education the methodology is yet to gain popularity. This chapter explores the reasons for the limited use of phenomenology in education at all levels, although it has much to contribute to professional growth of educational instructors as well in addressing learning challenges. Some of the challenges that were noted as limiting the use of phenomenology in education are the philosophical nuances and the fluid nature of the research process. Various philosophical strands of phenomenology are explored as well as sampling, data collection, analysis and validity and reliability. Social Constructivism and existentialism are used to underpin the chapter. The chapter also offer suggestions on ways to increase the use of phenomenological research in education.

DOI: 10.4018/979-8-3373-4432-4.ch009

INTRODUCTION

The chapter acknowledges the exponential growth of phenomenological research in fields such as social work, health sciences, sociology, economics, and journalism among many other fields, although in education the trend reflect constricted use of the method. The paucity of phenomenological studies in educational research may be accounted for by multiple factors which includes the complexity of apprehending the philosophical foundation of phenomenology and the lack of prescribed approach to the research methodology and dearth of resources on methodology to guide aspiring researchers in phenomenology. It is for that reason that the chapter examines diverse strands of phenomenology to guide researchers in the use of phenomenology. Various strands of phenomenological research are examined which includes descriptive, interpretive, hermeneutic, narrative and existential perspective. Husserl is credited with the development of phenomenological research, although later he lost many disciplines who questioned his key postulates of bracketing, free imaginative variation and eidetic reduction. Martin Heidegger (1889-1976) a student of Husserl rebutted some of the central tenets of Husserl phenomenology and proposed a new way of looking at objects that was distinct from Cartesian dualistic thinking. In order to help expand the discursive use of phenomenology in education key tenets used in practice such as sampling, data collection, data analysis, validity and reliability are also probed. The chapter is based on philosophical orientations of existentialism and social constructivism. The merits and pitfalls of using phenomenological research are explored. Finally, suggestions are offered to help expand the use of phenomenology in educational research, as it seems to be very appropriate in educational research.

BACKGROUND

Farrel (2020) noted that there has been a burgeoning of phenomenological studies in disciplines such as psychology, nursing, tourism and health sciences, while education seem to be left behind. This is an oddity particulary when viewed from the perspective of education as a discipline that is committed to experiential learning and can benefit massively from the use of phenomenological studies. The term phenomenology has its genesis from the Greek language from the word "phainomenon" which means "things appearing to view" (Farrel, 2020, p. 2).

Stolz (2022) noted a modest growth of research in phenomenology in education. This expansion of research that claims to be rooted on phenomenological espiteme has also led to a contestation on what may be regarded as phenomenology and especially in educational research. Finlay (2009) adduced that researchers concede that it is essential that we focus on embodied knowledge and experiential research in an effort to apprehend complex and rich experiences as they are demonstrated in concrete contexts. The contestations on how to engage in phenomenological research in education is daunting for novice researchers.

Finlay (2009) avers that multiple research approaches are undertaken under the banner of phenomenology. Phenomenological research, although it is diverse tends to display core tenets which includes the arduous descriptive nature of the research, use of reduction, probing of intentional relationship between people and contexts, and revelation of essences or structures of meaning, which suffuse human experiences through utilising imagination in different situations. Phenomenological research methods are encapsulated in three intersecting ways, namely phenomenological reduction,description and search for essence.

Internet Encylopedia of Philosophy (IEP, 2024) explicate phenomenological reduction as the experience where we find ourselves in the world with no experience or preconceived theories. When we know a phenomenon from experience with no a priori assumptions, it is distinct from approaching it with prior experience and theories. The phenomenology of experience is both a descriptive and prescriptive approach that makes a researcher to deliberately activate the power of astonishment so that they are able to know from astonishment when exploring mundane experiences. Hursel emerged with this perspective when he realised that the basis on which scientific research was premised on, was corrupted by the framework of science and the assumptions that a scientist brings to work. It was for that reason that a research had to engage in phenomenological reduction, which is a process of emptying one's mind in a regimen similar to meditation, in order to experience a new transformation that can help one to apprehend reality.

Cohen et al.,(2018) noted that phenomenologists differ in terms of the structure of consciousness and reflection which leads to an acquisition of knowledge, and this led to the emergence of a number of perspectives which included transcendental phenomenology of Husserl and existential phenomenology typified by Schutz. Husserl who was preocuppided with opening up new avenues of knowing that were distinct from positivism, urged researchers to use bracketing in order to liberate ourselves from the theories and misconceptions we obtained from the world. Schutz focalised on comprehending "the meaning of structure of a stream of consciousness that has no inherent meaning in themselves" (Cohen et al.,(2018,p 23). Schutz averred that meaning can be derived through a process of reflexivity.

Phenomenology is located within the category of qualitative research approaches (Pathak, 2017;Johnson & Christensen, 2017; Leavy, 2017). It is not a unitary methods to research, but a diverse approach with constestations and reflexivity, and dynamic in its nature, and it will continue to evolve in future. Phenomenology is interdisciplinary and used in all contexts.

It is an approach that is widely becoming ubiquitous in doctoral research. Immanuel Kant utilised the term phenomenology in a work titled Critique of pure reason, and he distingushed between mental representation of objects as the existence of the object in itself without interaction of experience (a priori knowledge automous from experience) and objects comprehended on the basis of experience which he termed empirical knowledge or a posteriori. Phenomenological knowledge marked a departure from the philosophical standpoint of Kant and is based on existential philosophy, as will be explored more fully later in the chapter.

Woodruff (2018) averred that phenomenology refers to the anatomy of the consciousness as experienced directly by an individual. Phenomenology is separate discipline, but shares some affinity with other key disciplines in philosophy such as ethics, epistemology and logic. The discipline has been practiced for many years, but was largely spotlighted by Husserl, Heidegger, Satre and Merleau-Ponty among many others. Phenomenology when taken from the essence of the term denote phenomena or the way we perceive things in our experiences. Phenomenology area of focus includes the importance of objects, self, time, and occurances as they are experienced by people (Saunders et al., 2019).

Odongo and Ntara (2024) explicated phenomenology as being preoccupied with completeness, and perceiving situations from varied angles, until an intergrated picture of the phenomenon or experience emerges. A phenomenological approach uses experience to secure holistic descriptions that facilitate reflection on the structure of a phenomenon. Moreno (2002) expounded that transcendent phenomenology has its origin in the work of Edmund Hussel at the beginning of this century as reflecting a science of human experience.Edmund Hurssel postulated that as human being we become conscious of the "external world" that is inhabited not merely by animals and objects, but also by us and other people.

McClelland (1997, p.108) declared that phenomenological research is the "study of everyday, lived experiences and meanings that people construct from them". Johnson and Christensen (2017) echoes the same goal for phenomenology as seeking to access the Lebenswelt (the lived experiences) of the research participants and appropriate the meanings they give to that experience. This means penetrating a participant inner space of consciousness and experience. Phenomenology holds as a tenet that there is a commonality to human experience which is a universal element, called invariant experience, which they seek to apprehend through research.

As in narrative inquiry, phenomenological research scrutinises intimately an individual interpretation of his or her experiences. Phenomenologists also seek to comprehend the experience from the frame of reference of the participant (Frechette et al.,2020). Phenomenologists do not always seek to generate a chronological account of events. In addition this form of research selects an experience and then probe it intensely and its attendant psychological meaning for the person. Cohen et al., (2018) indicated that phenomenology subscribes to the notion that our knowledge of the world is restricted to our current experiences and the mandate of the researcher is to explicate, understand, and interpret and account for these phenonemon or experiences. This form of research intends to unravel, describe and interpret an occurance by spotlighting the participants in it, both at individual and group level.

Because of the multiplicity of experiences, it follows that phenomenological research generates multiple perspectives. A researcher has to suspend judgement and foreground the experiences of the participants, particularly their emotions, experiences, meanings, values, attitudes and beliefs. The researcher is called upon to be faithful in recording the experiences of the participants in rich data and then proceed to categorise it and then handle it at an abstract level where theory may be generated.

Leavy (2017) underscored the observation that phenomenology hold the tenet that social life is constituted and reconstituted as a result of the interactions that transpires daily among individuals and the meanings that they attach to these engagements.

APPROACHES TO PHENOMENOLOGICAL RESEARCH

The types of phenomenological research may be enumerated as

i) Descriptive phenomenology
ii) Interpretive phenomenology
iii) Hermeneutic phenomenology
iv) Narrative phenomenology
v) Existential phenomenology

There are multiple types of phenomenological research, however the more universal ones are transcendent and hermeneutic researches.

Descriptive Phenomenology (DP)

Descriptive phenomenology is associated with Edmund Husserl (Husserl, 1913.2013). Holloway and Galvin (2017) declared that descriptive phenomenology focuses on the lived experiences of individuals. It is one of the approaches that is predominant in fields such as nursing and social work. The researcher seeks to find answers related to experiences and how these experiences are perceived and interpreted by participants. The experience may assume the guise of emotion and memory (Sinfield et al.,2023). Descriptive phenomenology is positioned within the intepretive research paradigm. Ayton (2024) expounded that descriptive phenomenology uses an approach that endorses bracketing, which is the act of stripping the phenomenon from its context for the purpose of scrutinising it. Interpretive phenomenology endorses the interlocking nature of individual experience with their social,cultural political environments. The researcher social markers are considered as a valuable source of knowledge and they only have to be acknowledged as a component of the research process.

Interpretive Phenomenology Analysis (IPA)

Interpretive phenomenology was postulated by Martin Heidegger. Sinfield et al.,(2023) explained that interpretive phenomenology entails a researcher extracting meanings from experiential accounts, drawing from each participant's account regarding an event or occurrence in their lives, without being influenced by a priori theory. Smith and Osborn (2015) added that IPA acknowledges that people are able to make sense of their experiences and the role of the researcher is to seek to comprehend the meaning that the people are attaching to their experiences. IPA prioritises the experience of an individual before navigating to a more general explanation of all participants. Heiddeger uses the concept "Daisen" a term that has Germany roots to explicate our position in the world. Britannica, T, Editors of Encyclopaedia (2023) adduced that Martin Heidegger (1889-1976), postulated that a human being should be thought of as "being there" and in the process of seeking to comprehending entitities as well as themselves. This was a departure from the view of Rene Descarte (1596-1650) view of consciousness as being symbolic of internal state and also the mind.

In the cartesian view, mankind was perceived as being a mind only (a mind spirit or soul) and is completely different from the physical world and therefore divorced from it cognitively (Patton & Broward, 2023; Floyd, 2020; Agbo, 2018). Heiddeger postulation rebutted Husserl concepts of a mental state, being indispensable for mediating the relationship between man and the physical world. In Dasein objects reveal themselves as they are, and are not unveiled. This capacity that was perceived as being mental, Heidegger perceived it as a feature of being human. What is fascinating about Heidegger concept of Dasein is that it viewed man as a complete being as opposed to being a mind or a combination of mind and body.

Hermeneutical Phenomenology

Hermeneutical phenomenology focalise on the interpretive structures of experiences and how we interacts with things around us human world and including ourselves and other people. The term hemeneutical phenomenology has its origins from the term "hermeneuin" which means to "translate or interpret" (Farrel,2020, p

3). The term is traced to the Greek mythology of a God called Hermes who had a duty to relay messages from the Zeus and other Gods to ordinary people. While Husserl was preoccupied with the issue of epistemology, Heidegger spotlighted more on ontology, the meaning of being in the world.

Hermeneutical phenomenology embraces the work of Heidegger particularly chaims on revealing and hiding truth and the chief proponent is Hans- Georg Gadamar. It is a theory that forefronts the need for understanding. Suddick et al.,(2020,p) defined the hememeutical cycle as the " ongoing, attentive,circular movement between part and whole". Heidegger perceived the hermeneutical cycles as constituting of foreunderstanding or forestructure as the demonstration of Daisen. Foreunderstanding means that the experience that a person brings to the situation and also the thinking that that will help to understand the issue.

Narrative Phenomenology

Narrative phenomenology foregrounds participants lived reality. In addition, it focuses on how society shape and influences individual perception of their experiences. Narrative inqiury looks at stories that people tell. Narrative inquiry endevours to appreciate the human experience and reflects stories that emanate from people's experience (Ellett, 2011). James Cook University (2023) expounded that narrative research is a part of qualitative research that forefronts studying people's experiences and stories in order to construct meaning from them. Narrative inquiry places focus on groups that are placed on the periphery of society either as groups or individuals to give them a voice. The form of data that is gathered includes display of emotions, artefacts, biography and story telling. The researcher and the participants engage in the process of generating meaning. Data analysis process in narrative inquiry assumes the form of isolating themes, patterns and stories.

The next section discusses existential phenomenology.

Existential Phenomenology

In this section existential phenomenology is discussed in this section. Existential phenomenology is associated with Martin Heidegger, Maurice Merleau-Ponty and Jean –Paul Sartre. The proponents of hermeneutical phenomenology postulated that it brings together nomothetic and idiographic components of human experience (Pandin & Yanto, 2023). The existential research methods are appropriate when one person want to comprehend different types of experiences, for example, emotions such as anger,humiliation, embarrassment and happiness. The research primary focus is on self- perception and what motivate certain emotions.

Seatle University (2024) averred that existential phenomenology has its roots in humanistic psychology. It doesn't subscribe to psychological approaches that treats people in a positivistic ways that are reductionist. Existential phenomenology can help address the issues of love, growth, spirituality and finding meaning. Zieske (2020) explicated existential phenomenology as an approach to pyschology in its diverse branches, especially in clinical and counselling psychology. Existential phenomenology and phenomenological phenomenology, although they share some common tenets, are not synonmous. The school of existentialism is basically a philosophy and theory that sought to foreground human issues such as existence, the need to collaborate and live with others, and ways in which people deal with issues of their mortality. Martin Heidegger is noted for intergrating the concepts of existentialism and phenomenology in a single study. Though they are intergrated they also exist as separate disciplines.

The next section discusses transcendant phenomenology.

Transcendental Phenomenology

Pathak (2017) noted that trascendental phenomenology focuses on the work of Edmund Husserl (1859-1938) who emphasised the idea of reduction. Given (2008) identify the key terms that are associated with transcendental phenomenology as eidetic reduction, intentionality and constitution of meaning. Husserl postulated that a philosopher is able to shift from a concept to another dimension of imagination where they can see the unchanging nature of the object.

Farrel (2020) explicated that Husserl was preoccupied with the need to establish the fundamental nature of things of essence of objects or human knowledge. It is a preoccupation that has dominated western philosophy from the time of Kant, Hegel, Brentano and Descartes (1650) among many others. Husserl (1998) admitted that the influence of Descartes had a major impact on the contemporary conception of phenomenology, particularly the dualism of man constituting the inner world of the mind/soul/ and the external world that we have to investigate. This dualism where a person experiences the external world of physical objects through the sensory data and subsequent mental impressions is foundational not only to Husserl thoughts, but western epistemology in general. Husserl held the assumption that consciousness is the foundation of all knowledge. Husserl averred that phenomenology could be a vehicle through which human consciousness could be studied. In addition Hegel held that everything that appears to human consciousness could be studied. Husserl opined that consciousness could be realised through a process he termed bracketing. Another chief feature of Husserl phenomenology is termed free imaginative variation, which implied that it is vital to descibe an object and separate the basic features which can not change in our minds. These key concepts are discussed briefly in the subsequent sections.

EIDETIC REDUCTION

Eidetic reduction refers to the way a philosopher migrates from the consciousness of a person and objects to elevated level that is above the senses and therefore attains a eidos (shape of a thing) (eidetic reduction, 2024). Phenomenology is a sciences that cetralises the basic structure of things which is called the eidos (the necessary structure of a thing). This structure excludes everything that that is not related to the thing to its principal nature or basic structure. The eidetic process starts with a philosopher studying a concrete item and then proceed to examine it imaginatively from a variety of angles. The philosopher will navigate from the perceptual dimension to evidence in the imaginative dimension, they can attain understanding of the nature of a thing at a level where the object can not change.

Intentionality

Alhazmi and Kaufmann (2022) expounded that intentionality is a central tenet of phenomenology. Leung (2022) noted that intentionality serves a pivotal place in Husserl's phenomenology. In Husserl philosophy uses intentionality of consciousness and the related concept of intentional consciousness. To claim that a thought is intentional it has to be focussed on any object or thing. To mention the intentional content of thought connotes that we are speaking of a mode or thought in which a thought is about any object. Given the way intentionality serves an immanent place in consciousness and experience in Husserl thought, it was evolved into systematic theory of consciousness, epistemology,theory of logic among other branches of philosophy (Internet Encylopedia of Philosophy,2024).

Franz Brentano (1838-1917) provided the basis for Husserl augumentation on intentionality. Brentano was in turn inspired by a desire to comprehend the nature of psychology. Other influences that inspired the study on intentionality is the work of Aristotle psychology, Descartes and among many others. Every mental thought has an object. Recognising intentionality as the function of the mental makes it possible to study the mind in relationship to objects and the different forms that this relatedness may assume such as imagination, perceiving and psychotic expressions. Intentional thought and intentional object are perceived as being separate.

Constitution of Meaning

The first text of constitutive phenomenology is based on Husserl's first publication. Constitutive phenomenology focalise on the constitution of things. To analyse the constitution of an object is not merely to identify its material make –up, but describe the build- up of intentive processes in conscious life, to which it relates as an intentional object (Embree, 2024). Cadena (2019) Husserl described consciousness as a position of knowledge revealing the constitution of the world. For Huserl the word constitution means unveiling or revealing. He postulated that knowledge doesn't impose anything on the world. To constitute is to assign meaning.

BRACKETING

Bracketing represents an efforts to encapsulate the views of the participants in a way that is not imbued with bias of the researcher (Sinfield,et al., 2023; Tufford & Newman, 2010; Habibullah et al.,2023). Bracketing has been assumed to be similar to taking a role as an actor, where a person seeks an exact display of the new character that they assume. Dörfler & Stierand (2020) averred that bracketing entails efforts by the researcher to hold their knowledge about a phenomenon in suspension in order to realise experiences and make sense of them. Bracketing is chiefly a theoretical and philosophical construct and it is vital to expound on its application to phenomenology.

Bracketing subsumes values, beliefs, feelings, attitudes and hypotheses. In addition this also includes the historical premises a researcher comes with to a phenomenological research, cultural assumptions and experiential knowledge that a person comes with to a research or philosophical inquiry. Furthermore academic premises are also bracketed. There is no agreement on what should be left out from bracketing. Habibullah et al.,(2023) explicate that internal prejudice may differ with each assumption and that researchers may be more conscious of their feelings than their cognitivities proclivities. The challenging process of bracketing will enable a researcher to experience varying states of consciousness. Suspending preconceptions is also fundamental to the process of self –discovery, as it enables the surfacing of hidden emotions and experiential knowledge (Thomas & Sohn, 2023). The process of self-discovery was opined as a task that can realised by collaborating with a partner. Husserl opined that if a researcher sustains a state of self-awareness as an ongoing activity, and act as an instrument in data gathering,data analysis, and interpretation could profit from bracketing. Thomas and Sohn (2023) questioned the possibility of people attaining a state of consciousness called bracketing. It is also admitted that in later works Husserl displayed less confidence in the possibility of bracketing in research (Emiliussen et al., 2021). Many phenomenologists have grappled with the effort to articulate accurately the essence of the term bracketing. Later Husserl proclaimed that it was not possible to completely practice bracketing, because even complex concepts would require an appreciation of fundamental terms. The more pointed criticism emerged from Heidegger when he postulated the concept termed "Daisen" which means that people always find themselves in a context and impacted by the situation in a way that affects social engagements and scientific research (Emiliussen,et al., 2021, p. 2). Husserl seem to have shifted in his postulation on bracketing multiple times making it difficult to declare with confidence his final position.

In the practice of bracketing there is no consensus among phenomenologists in terms of the possibility of using bracketing during an interview (Farrel, 2020). Some scholars argue bracketing should be employed during the process of data analysis and not during data gathering, while others aver that bracketing during data collection and subsequent stages can culminate in a good research. Husserl lost many of his adherents towards the end of his life, who had become skeptical of the possibility of bracketing completely the immanent influence of culture and history. The next sections explore data sampling in phenomenology.

POPULATION AND SAMPLING

Cohen et al., (2018) explicated one of the major challenge that confronts a researcher is the decision on the size of a sample size. There is no prescribed sample size in research because it depends largely on the type of research and also the type of population taking part in the study. However, it is possible to theorise on how a sample size may emerge in a research. Phenomenological research like other qualitative studies employees non-probability sampling. This means that in qualitative studies a researcher may focus on a group being fully aware that it doesn't typify the population, but represents itself.

In descriptive phenomenology a sample may vary from 10-15 participants. Palinkas, et al., (2015,p2) declared that purposive sampling is used in "qualitative research for the purpose of identification and selection of information- rich cases". This sampling procedures entails the selection and use of cases that are highly informed or knowledgeable about the issue being researched on. The sample size for research is fundamental for the purpose of meeting the empirical demands of a research. Sampling methods for qualitative research are intended to realise a broad appreciation of an issue. Phenomenological research sample may vary from 3-6 participants who are interviewed multiple times (Sarfo et al.,2021). There are financial and time implications for using too many people in sample size or very few people. The debate in qualitative study on sampling,has also revolved on whether a sample size can be determined a priori or not.

A central tenet of qualitative research is the concept of saturation, which means that there is no new information that is being obtained from the participants. Failure to attain data saturation has implications in terms of the trusthworthiness of the research and quality of the research. Data saturation tends to differ from form of qualitative research to another. The availability of participants in a population, has an effect on data saturation. Data saturation is not determined by the number of the participants, but by the quality of the data (Sarfo et al.,2021).

Phenomenological research focalise on the lived experiences of many people. A phenomenological research can utilise 6-10 people and other authors have opined that educational research may use between 8-31 participants (Sarfo et al.,2021;). It is notable that in phenomenological research sample size may be small and that it is fundamental that data collection continues until saturation (Hennink & Kaiser, 2022; Subedi, 2021). The subsequent section looks at ways data is collected in phenomenological studies.

DATA COLLECTION METHODS

The data collection process in phenomenology is shaped by the philosophy that holds that experience is diverse and composite, consequently researchers gather enormous amount of data from the participants over an extended period (Cohen et al., 2018; Lodico eta al,2010; Saunders et al.,2019). Some of the ways data is collected includes use of open unstructured interviews,video recording, diaries, document analysis and focus group interviews.

Document Analysis and Video Recording

The primary method that has been used to collect data in phenomenological research is the use of in-depth interviews, although documents analysis and video recording has also been used (Given, 2008). Phenomenology foregrounds human experience in data collection what is termed a second order experience.

Interviews

Data collection in phenomenological studies has been largely dominated by the use of interviews, although there are instance of data collection using document analysis and video-recording (Given,2008). Different forms of interviews are employed in qualitative research in order to secure information from the participants.

Elhami and Khoshnevisan (2022) asserted that in order to secure rich data, it is vital that the interview is conduted in an appropriate way. When an interview is held in real time, it enables data to be collected with the participants establishing a bond with the participant and being also able to probe the interviewee. Qualitative research such as phenomenology may employee open- ended and semi-structured interviews. Qualitative interviews demands adequate preparation in order to collect feelings, opinions and ideas (Doringer, 2020; Dursun, 2023; Cohen et al.,2018).

The focus of an interview in phenomenological research is to reveal the experiences of individuals and to decipher how they make sense of such experiences. The interview process is perceived an an engaging process in which a researcher gathers data. In the interview process the researcher explores the issues from the angle of the participant. Interview questions may be structured, semi-structured or unstructured (Given, 2008). Structured interviews assume a standardised form.Semi-structured interviews are more flexible than standardised interviews. Finally, in unstructured interviews, an interview can generate questions according to the flow of the interview. In phenomenological research the primary concern is on capture human experience, what is termed the second- order perspective. This perspective is normally perceived in the nature of interview questions which are aligned to the research questions.

Use of Diaries

Olorunfemi (2023) explicated diaries as a methodological tool for conducting qualitative research. A diary offers views of a reseacher generated from a participant being studied. Mishra and Gupta (2024) indicated that the use of diaries can help a researcher in qualitative studies to manage emotions behaviours, tendetious behaviour and provide catharsis during the interview process. Phenomenological researchers find the diary a very useful device when working with extremely emotionally exothermic topics.

A diary has capacity to enable a researcher to engage in reflections and to polish up interview questions. A diary enables a researcher to address the layers in a research and also to identify the nuances on an issue being researched on (Cudjoe, 2022). A diary facilitates the entry of events on a regular interval together with specific dates and time. A researcher can also work with the participants by encouraging them to keep a record of their activities that are aligned to research questions in a diary. Participants may be guided in the process of completing the diary and this can help provide a very rich source of data for the researcher in phenomenology in education.

Focus Group Interviews

There is no consensus on the use of focus group interviews in phenomenological studies. Some academics argue that phenomenological studies spotlight on individual experiences and for that reason focus- group can not be used for data collection (Bradbury-Jones,et al., 2009; Smith,2004; Beer, Undated). On the other hand there is also a postulation that focus group are congruent with phenomenological research data collection, because it is possible to retain an individual experience in a group. It is further underscored that the use of focus group interviews has potential to open up more possibilities to apprehend the issue being researched on in phenomenological studies.

The use of focus group in phenomenological research in education is still in its infancy. Gundumogula (2021) conceptualised a focus group as an extended interview that utilises an in-depth way of collecting data. The aim of a focus group is to offer in-depth information in a group. Different scholars have postulated different sizes for a focus group ranging from four participants to those who declare a maximum of 12 (Muijeen et al.,2020; Manzano, 2024; Hennink & Kaiser, 2022). Others that that there is no prescribed number for a focus group (Gundumogula, 2021).

DATA ANALYSIS

Descriptive phenomenology data analysis involves identifying common themes in a phenomenon. The primary focus of phenomenological study is the different ways in which participants experience a phenomenon. Given (2008) indicated that analysis proceeds through multiple stages, in which reading of various interviews will lead to the development of different areas of focus. A researcher generates a provisional level of data analysis during the time of data collection. Hossain et al., (2024) opined that phenomenological approach offers two ways of analysing data. One of the method of data analysis postulated by Giorgi and Giorgi, (2003) is inductive. A phenomenological reduction approach is used to comprehend the participants accounts using bracketing. A careful reading of the accounts leads to the creation of units that help to identify the themes in a study that forefront participants experience in the study.

Validity And Reliability in Phenomenology

Phenomenology approaches validity and reliability from the perspective of qualitative research as opposed to logical positivistic approach. Ahmed (2024) explicated that validity and reliability in phenomenological research is conceptualised within the framework of qualitative research which upholds credibility,transferability,dependability, and conformability. There are multiple frameworks for assessing the quality of phenomenological research. The central concern in qualitative research depends on the degree to which a research reflects the reality of the participants.

Lincoln and Guba (1985) postulated that the seminal framework for building trustworthiness in qualitative research. Researchers engaging with participants for an extended period may help to enhance trustworthiness in a phenomenological research (Stahl & King, 2020). Transferability and trustworthiness is foregrounded by Guba and Lincoln (1985) as central ways of ensuring that a researcher get as close as practicable to the phenomenological research context. In phenomenological research the goal is different from logical positivism, which seeks to ensure that knowledge is applied to all contexts. In phenomenological research, the patterns and descriptions that are identified in one research context may be applicable to another research context that has similar features. Credibility of phenomenological research may also be assured by dependability. This refers to the use of peer debriefing and receiving feedback from peers. This input from peers is critical for ensuring dependability within the context of trustworthiness. Confirmability is a vital component of trustworthiness and this refers to getting as close reality as possible.

THEORETICAL UNDERPINNING

The discussion of this chapter is guided by the philosophies of phenomenology and existentialism and the social constructivism.

Phenomenology and Ontology

Alhazmi and Kaufmann, (2022) foregrounded that phenomenology emerged as a reaction to positivistic sciences, especially in the presuppositions that traditional sciences made regarding the nature of reality, that is, if reality is external to us or internal. Traditional sciences postulated that reality is external to us, is independent of us and is objective. In its debut phenomenology was not preoccupied with ontological reasoning, but devoted energy to designing a different epistemology response to traditional postivistic augumentation.

Phenomenology sought to forefront the link between the knower and the known, which is widely an epsitemological perspective as opposed to being an ontological issue (Dörfler & Stierand, 2020). The trajectory of phenomenological inquiry was to acsertain if reality existed independently of human consciousness, whether we reflect or not on it. The fundamental assumption that is embeded in phenomenology is that we are inclined towards a version of reality that is provided by our consciousness. It is therefore postulated that knowledge is an outcome of a deliberate action towards an object or phenomenon. Phenomenology avers that truth can be located internally within the individuals and is mediated by culture, beliefs, history and semiotic system. Heidegger contributed significantly to the ontology of phenomenology by postulating on the philosophy of existentialism.

Phenomenology and Existentialism

In this part of the chapter, the relationship between phenomenology and existentialism is explored. Phenomenological research owes its philosophical origins to the philosophy of existentialism and phenomenology. Existentialism and phenomenology are both perceived as lens that function from the inside going outside. This means that the two lenses work by probing the deep psychological processes in order to spotlight the external behaviours and actions.

Edmund Husserl founded phenomenology (Panovski, 2024). The same author postulated that there is something that is definite, and it is our consciousness or our conscious thought activity. The central canon of phenomenology is that we live as subjects in a world that is replete with objects. Building on the canon of phenomenology on consciousness, Heidegger postulated the philosophy of existentialism, upon which Satre also added new insights. Existentialism evolved to become one of the major philosophy of the twentieth century. Husserl opined that if we want to construct knowledge on a firm foundation, we need to build on our conscious thought actions.

Heidegger averred that our consciousness can not be developed as a free - floating attribute, but in relation to something.The concept of awareness is a relational concept which can not be dissected in isolation. Based on that line of reasoning, it follows that it becomes very complex to isolate a state of consciousness and an object of consciousness. The state of consciousness and the object of consciousness become blurred in our day to day life although they are able to be divided at a conceptual level. Hursel averred that the object of our consciousness subsist in spite of the status of their existence that they may or may not have.As this objects are being probed, we become aware of the world and this is a state of consciousness. The rudimental idea is that consciousness is always directed towards something, in order to comprehend any aspect of it. Because human beings have this unique capacity to direct attention to objects they have intentionality.In essence phenomenology focuses on the objects of our consciousness, which provide a basis for building of knowledge.

Heidegger explicated that human beings can not be perceived as subjects who are completely removed from the reality that is waiting to be known.Existentialism averred that humanity is among the objects in the cosmos and dealing with that reality. We are beings in a world that is replete with other beings and we are coping.

Hiedegger does not eshew abstraction for its own sake nor the conscious reflection on the objects in the world, but he felt that it was secondary. Jean- Paul Satre started off as a phenomenologist, but after an encounter with Heidegger, he transformed to an existential philosopher. Satre is credited with the full development of the concepts of existentialism, and he explicitly rejected the dualism as espoused by Hursel, that imbued the thinking of Kant, which was based on the assumption of phenomenon and noumenon (the way things look like and how they are actually constituted). Satre averred that dualism opined by Kant is an illusion and that the representations are the only reality. Furthermore, Satre postulated the two concepts the first is being in itself and being for itself. Being in itself meant is in a fixed state and is not aware of itself and being for itself is aware of its consciousness and is unfinished and not fullfilled. Satre asserted that it is this element of being undefined which is undetermined that makes a human being a human.

SOCIAL CONSTRUCTIVIST APPROACH

Phenomenological research holds the assumptions that reality is socially constructed and is diverse. Creswell and Creswell (2018) asserted that constructivism is perceived as a typical approach to qualitative research. A number of influences have shaped constructivism which includes Mannheim, Berger and Luckman (1967) titled *The social construction of reality*, and many other publication on the same theme. Social constructivism is built on the premises that people desire to comprehed the world that they live in and on that basis they generate subjective meaning of their experiences in relation to objects or things (Erciyes,2020).

The meanings generated from social constructivism are multiple and diverse which means that the priority is not on creating single version of reality, but multiple versions. The method used in data collection includes open- ended interviews that are designed to facilitate the extractions of the broad views of the participants. The meanings that are created are mediated by history and cultural contexts. They are created in social engagements and that is why the perspective is termed social constructivism. Researchers also acknowledge ways in which their own cosmology impact on the research process.

Phillips (2023) expounds that some of the social makers that are capable of impacting on qualitative research includes sex, gender, enthnicity, socio-economic status, among many others.Qualitative research is based on participant's lived experiences and for that reason it tends assume an interpretive stance. Qualitative research acknowledges the intricacy of social life and eschews a belief in a rigid social world. The perspective embraces use of language, cultural symbols, concepts of a social world and reality. In qualitative researcher, meanings is created through reflection as opposed to the use of numbers (Flick, 2022).

MERITS AND PITFALLS OF PHENOMENOLOGY

Lewis-Hickman, (2015) expounded that there are merits and pitfalls that are associated with the use of phenomenology in education. In this section, the merits and pitfalls are explored. Phenomenological research design has a number of strength which includes the capacity of the researcher to function from intrinsic motivation to conduct a research. The way data is collected in phenomenology also serves as a merit of this research method. The researcher is able to secure in-depth information by using open –ended questions and also to change the focus during the data collection process in response to the trends in the data collection process.

Descriptive phenomenology holds the chief strength of being able to help people to comprehend peoples' personal experiences and is able to interrogate assumptions that people may hold regarding phenomenon and creates opportunities for generating innovative solutions. The approach holds numerous promises of generating new theories or effect new policy trajectory.

Brodowicz (2024) adduced that using phenomenology in education has a number of merits which include the provision of rich description of phenomenon in education, which facilitates a deep comprehension of an issue. Phenomenological research also foregrounds discriptive phenomenon which means that probing is used in addition to the use of observation. Each phenomenon is perceived as a way of representation which is designed to apprehend self- intepretation which is rooted in individual experience. In the educational context, the actors may work with the researchers in order to comprehend something meaningful to them.

Alsaigh and Coyne (2021) expounded that hermeneutical phenomenology lacks a rigid structure or steps when it is used in research. It is for that reason that researchers or students who may want to apply hermeneutical phenomenology to their work may face challenges to implement it, particularly in data analysis. Some of these impediments arises due to a lack of appreciation of the layered and nuanced approach to phenomenology. This challenge is also exacerbated by a lack of resources that provide a clear guidence on how to conduct a phenomenological research using a phenomenological method. The paucity of materials leaves a novice researcher with no adequate tools to conduct a qualitative research which uses phenomenology as a method. It is for that reason that many students or researcher would opt for the traditional accepted research approaches, which have prescribed steps and are perceived as being rigorous.

Phenomenology has merits of enabling a mode of inquiry that can facilitate understanding of themselves and facilitating transformation and advance comprehension of self- awareness. Phenomenology allows for reflexivity that is able to lead to personal transformation and also understanding of the issue being researched.

In spite of all these merits that are associated with phenomenology in education, it is faced with multiple pitfalls. Neubauer et al.,(2019) although the use of phenomenology holds tremendous potential in educational research, it can be intimidating for novice researchers (Farrel, 2020). Inexperienced researchers may find the process of isolating each philosophical position daunting, for instance distingushing between transcendental and hermeneutical research. The use of people in data collection serves as both a merit and a demerit in data collection in the research process. It is a merit in the sense that a researcher is able to collect in-depth data on the phenomenon of interests and it is also a demerit in the sense that the element of bias is ever- present in the research process at all stages of data collection, data analysis and also reporting of findings.

Descriptive phenomenology may not be appropriate in all disciplines for instance when gathering data in a health settings. For instance when conducting an evaluation of a research facility open- ended questions may not produce the needed responses, especially if the focus is the participant lived experience. In that scenario, it may be appropriate to gather data using highly structured questions. There are also impediments that researchers may encounter when utilising descriptive phenomenology research which involve language barriers in a situation where the participants may not be adequately articulate, due to multiple factors which include age or even language. The process of data gathering tends to be arduous. The challenge of descriptive data intepretation is to generate data that is not value- laden. The findings of descriptive phenomenology many not be generalised to all contexts (Cohen et al.,2018). Collecting data using phenomenology method can be time- consuming and arduous. Data analysis can involve working through enormous amounts of data, which is a demerit. Due to the use of non-probability sampling in phenomenological research, the data that is collected may not be universalised (Creswell & Creswell, 2018). Critics have forefronted the challenges of ensuring reliability and credibility in phenomenological research, which implies that a researcher has to demonstrate to other researchers measures undertaken to ensure credibility and reliability of the study.

Given (2008) indicated that phenomenology has been censored for not taking into account the research context. In addition it has also been argued that when context is ignored the researcher and the participants may not be talking of the same context. This critique has severe impact when perceived from the angle of validity of phenomenological research. In order to address this criticism it is important for the researcher to ensure that they are sharing the same frame of reference with the participants by refering to the context of the study. Phenomenological research findings with their emphasis on the lived experience of the participants are very fundamental in all fields especially in education, where it is vital to comprehend lived experiences from the simple to the more complex. For instance in education phenomenology may focus on how students face challenges to master certain knowledge or skills at school.

DISCUSSION

There is a growing body of research that employees phenomenological approach in the field of education. In this section, a number of these researchers that used phenomenological approach will be examined. In spite of the growth of phenomenological studies in other disciplines, it is notable that in education there has been very little use of this methodology to conduct research in education. Some of the reason that that has been advanced is the complexity of apprehending the philosophical underpinnings of the phenomenological research, which many practitioners in education may find intimidating.

Education is built on the presupposition that values experiential knowledge in the learning process, a feature that can make phenomenology very relevant to education in investigate issues such as studying the lived experiences of students who are from marginalised groups in school or the experience of learners who use English in school which is not their mother- tongue in learning certain subjects such as geography and mathematics. The use of phenomenology in educational research holds tremendous potential for enhancing learning and resolving multiple challenges facing the education system at all levels of learning from preschool to higher education.

Fuster (2019) indicated that hermeneutic phenomenology holds potential to help address numerous pedagogical issues we face regarding learning at all levels. Pedagogy is multidisciplinary and is situated in fields such as philosophy,psychology, economics, sociology and anthropology and forefronts the need to address challenges identified in the teaching and learning situation. Education has similar implication as pedagogy as it involves socialisation of a learner so that they grow from one level to another. The growth also transpires in the one facilitating the learning of values, culture and customs. The use of phenomenology in education is not merely an attempt to address pedagogical issues, but is a vehicle for taking stock in a reflexive way of the pedagogical elements an instructor must be involved in, for the purpose of identifying weakness and optimising learning.

In studying the essence of things phenomenology in education, explores realities that are very subtle to communicate, but imperative to apprehend the psychic life of an individual. When using phenomenology it is important to acknowledges the biases of both parties the researcher and the participants, in order to have a holistic picture of the phenomenon. (Guillen, 2019)

A researcher positionality in educational studies will also need to address the issue of positionality, by undertaking a number of questions which includes identification of personal values, religious inclination, culture among many other influences that taint the research process.

RECOMMENDATIONS

The study noted that very few researches are conducted using phenomenology methodology in education when compared with other disciplines. It is for that reason that measures have to be intriduced to ensure that educational practitioners are able to address learning and didactic issues using phenomenological research. The following recommendations are therefore postulated to ensure that practitioners are able to utilise more phenomenological research:

i) Practitioners in educational research must be trained to apprehend the philosophical dimensions of phenomenology and its various strands;
ii) Practitioners in phenomenological research also need training on data gathering approaches in phenomenology and the skills to effect data analysis and presentation;
iii) Some of the philosophical components of phenomenology require to be simplified to enable many people to access and utilise them in research;
iv) Experienced researchers in the use of phenomenology in education must be willing to offer support to novice researchers on how to conduct phenomenology at doctoral level and also on didactic aspects in the educational sector.
v) More resources can be produced and availed to practitioners on research in education to conduct research.

CONCLUSION

In this chapter it was noted that phenomenology has multiple strands. The multiplicity of the strands and their theoretical complexity makes phenomenology intimidating to novice researchers in education. The chapter has presented information on conducting sampling, data collection, data analysis and reliability and validity in phenomenological research. The pitfalls that are associated with conducting phenomenological research were explored and suggestions to address to address them were also presented. Finally, phenomenology because of its emphasis on the lived experiences of the participants is very relevant to addressing didactic challenges in education and also to ensure professional growth of the educators.

REFERENCES

Agbo, J. (2018). Heidegger through Husserl to phenomenological ontology. *An African Journal Philosophy VOL, 18*(2), 1–13.

Ahmed, S. (2024). The pillars of trustworthiness in qualitative research. *Journal of Medicine, Surgery, and Public Health, 2*, 1–4. DOI: 10.1016/j.glmedi.2024.100051

Alhazmi, A., & Kaufmann, A. (2022). Phenomenological qualitative methods applied to the analysis of cross-cultural Experience in novel educational social contexts. *Frontiers in Psychology, 13*, 1–12. DOI: 10.3389/fpsyg.2022.785134 PMID: 35548502

Alsaigh, R., & Coyne, I. (2021). Doing a hermeneutic phenomenology Research Underpinned by Gadamer'sPhilosophy: A framework to facilitate data analysis. *International Journal of Qualitative Methods, 20*, 1–10. DOI: 10.1177/16094069211047820

Ayton, D. (2024, 10 25). *Chapter 6 Phenomenology*. Retrieved from Open Educational Resources Collective: https://oercollective.caul.edu.au/qualitative-research/chapter/__unknown__-6/

Beer, S. (Undated). *Humanising Phenomenological analysis: using focus groups, food, and drink to collect data for Descriptive and Interpretative Phenomenological Analysis (IPA)*. Bournemouth University: Department of Tourism and Hospitality.

Bradbury-Jones, C., Sambrook, S., & Irvine, F. (2009). The phenomenological focus group: An oxymoron. *Journal of Advanced Nursing, 65*(3), 663–671. DOI: 10.1111/j.1365-2648.2008.04922.x PMID: 19222664

Britannica, T. Editors of Encyclopaedia. (2023). Dasein. Encyclopedia Britannica.

Brodowicz, M. (2024, 07 2). *Strength and weaknesses of phenomenology*. Retrieved from Aithor: https://aithor.com/essay-examples/strengths-and-weaknesses-of-phenomenology

Cadena, D. L. (2019). The constitution of objectivities in consciousness in ideas I and ideas II. *Rev. Filos., Aurora, Curitiba, v. 31, n. 53*, 503-519.

Cohen, L., Manion, L., & Morrison, K. (2018). *Research methods in education*. Routledge.

Creswell, J., & Creswell, J. D. (2018). *Research design qualitative, quantitative and mixed research*. Sage.

Cudjoe, E. (2022). Using diaries with interpretative phenomenological analysis: Guidelines from a study of children whose parents have mental illness. *International Journal of Qualitative Methods*, *21*, 1–9. DOI: 10.1177/16094069221084435

Dörfler, V., & Stierand, M. (2020). Bracketing: A phenomenological theory applied through transpersonal reflexivity. *Journal of Organizational Change Management*, •••, 1–22.

Doringer, S. (2020). "The problem-centred expert interview" Combining qualitative interviewing approaches for investigating implicit expert knowledge. *International Journal of Social Research Methodology vOL 24 Issue 3*, 265-278.

Dursun, B. (2023). A qualitative research technique: Interview. *The Journal of Interdsciplinary Qualitative Research*, *1*(13), 100–113.

Elhami, A., & Khoshnevisan, B. (2022). Conducting an interview in qualitative research:The modus operandi. *MEXTESOL Journal*, *46*(1), 1–7. DOI: 10.61871/mj.v46n1-3

Ellett, J. (2011). Narrative and Phenomenology as methodology for understanding persistence in art teachers: A Reflective Journey. *Marilyn Zurmuehlen Working Papers in Art Education 2011.*

Embree, L. (2024, 10 28). *Routledge Encylopedia of Philosophy*. Retrieved from Overview phenomenological movement: https://www.rep.routledge.com/articles/overview/phenomenological-movement/v-1/sections/constitutive-phenomenology

Emiliussen, J., Engelsen, S., Christianse, R., & Klausen, S. (2021). We are all in it!: Phenomenological qualitative research and embeddedness. *International Journal of Qualitative Methods*, *20*, 1–6. DOI: 10.1177/1609406921995304

Erciyes, E. (2020). Reflections of a social constructivist on teaching methods. *European Journal of Educational Sciences*, *7*(4), 16–26. DOI: 10.19044/ejes.v7no4a2

Farrel, E. (2020). Researching lived experience in education: Misunderstood or missed opportunity? *International Journal of Qualitative Methods*, *19*, 1–8. DOI: 10.1177/1609406920942066

Finlay, L. (2009). Debating phenomenological research methods. *Phenomenology & Practice*, *3*(1), 6–25. DOI: 10.29173/pandpr19818

Flick, U. (2022). *An introduction to qualitative research*. Sage.

Floyd, G. (2020). he Sense of Phenomenology: A response to Tarek Dika. *The INPR Journal. Disputatio*, *I*, 163–168.

Frechette, J., Bitzas, V., Aubry, M., Kilpatrick, K., & Lavole-Tremblay, L. (2020). Capturing lived experience: Methodological considerations for interpretive phenomenological inquiry. *International Journal of Qualitative Methods, 19*, 1–12. DOI: 10.1177/1609406920907254

Fuster, D. (2019). Qualitative Research Hermeneutical Phenomenological Method. *Propósitos y Representaciones*, *7*(1), 201–229.

Giorgi, A., & Giorgi, B. (2003). *Phenomenology*. Sage.

Given, L. (2008). *The sage encylopedia of qualitative research methods* (Vol. 1&2). Sage. DOI: 10.4135/9781412963909

Guillen, D. (2019). Qualitative Research Hermeneutical Phenomenological Method. *Propósitos y Representaciones*, *7*(1), 201–229.

Gundumogula, M. (2021). Importance of focus groups in qualitative research. *The International Journal of Humanities & Social Studies*, *8*(11), 299–302.

Habibullah, K., Mohammed, B. U., & Hamza, M. (2023). Conceptual framework in reflexive bracketing techniques in qualitative methodology. *Research Review Journal*, *4*(2), 849–857.

Hennink, M., & Kaiser, B. (2022). Sample size for saturation in qualitative research: A systematic review of empirical literature. *Social Science and Medicine vol vol 292*, 1-10.

Holloway, I., & Galvin, K. (2017). *Qualitative research in nursing and healthcare*. John Wiley., Retrieved from https://deakin.libguides.com/qualitative-study-designs/phenomenology

Hossain, S., Alam, K., & Ali, S. (2024). Phenomenological Approach in the Qualitative Study: Data Collection and Saturation. *ICRRD Journal*, *5*(2), 148–172. DOI: 10.53272/icrrd.v5i2.4

Husserl, E. (1913.2013). *General introduction to pure phenomenology*. London: Routledge.

Internet Encylopedia of Philosophy. (2024, 10 26). *Edmund Husserl: Intentionality and Intentional content*. Retrieved from Internet Encylopedia of Philosophy: https://iep.utm.edu/huss-int/#:~:text=For%20Husserl%2C%20intentionality%20includes%20a,just%20to%20name%20a%20few

Internet Encylopedia of Philosophy (IEP). (2024, 10 24). *The Phenomenological reduction*. Retrieved from Internet Encylopedia of Philosophy (IEP): https://iep.utm.edu/phen-red/

James Cook University. (2023). *Qualitative research methodologies*. Retrieved from James Cook university: https://jcu.pressbooks.pub/intro-res-methods-health/chapter/4-3-qualitative-research-methodologies/

Johnson, R., & Christensen, L. (2017). *Educational research,quantitative, qualitative and mixed approaches*. Sage.

Klinke, M., & Fernandez, A. (2023). Taking phenomenology beyond the frst-person perspective: Conceptual grounding in the collectionand analysis of observational evidence. *Phenomenology and the Cognitive Sciences*, *22*(1), 171–191. DOI: 10.1007/s11097-021-09796-1

Leavy, P. (2017). *Quantitative,Qualitative, Mixed Methods Arts-Based, and Community-Based Participatory Research Approaches*. The Guildford Press.

Leung, K.-H. (2022). Sartre and Marion on Intentionality and Phenomenality. *Theory, Culture & Society*, *39*(1), 41–60. DOI: 10.1177/02632764211024349

Lewis-Hickman, C. (2015). Strength and limitations of research. Walden: Walden University.

Lincoln, Y., & Guba, E. (1985). *Naturalistic inquiry*. Sage Publications. DOI: 10.1016/0147-1767(85)90062-8

Lodico, M., Spaulding, D., & Voegtle, K. (2010). *Methods in educational research* . USA: A Wiley Imprint Jossy-Bassey.

Manzano, A. (2024, 10 28). *Policy evaluation and approaches: Focus group*. Retrieved from esbc: https://scienceetbiencommun.pressbooks.pub/pubpolevaluation/chapter/focus-groups/

McClelland, J. (1997). Knowing and being known: Parents' experiences with rural schools. *Journal of Research in Rural Education*, *13*(2), 108–116.

Mishra, P., & Gupta, A. (2024). Exploring the Use of Diary Entries for Qualitative Researchers: Mitigating Challenges When Investigating Sensitive Topics on Indian Women. *The Qualitative Report*, *29*(0), 5, 1279–1299. DOI: 10.46743/2160-3715/2024.6330

Moreno, V. (2002). Validity Issues in phenomenological research: Bridging theory and practice in a study of IT driven radical organisational change. *Americas Conference on Information Systems (AMCIS)* (pp. 1-8). USA: AMCIS 2002.

Muijeen, K., Kongvattananon, P., & Somprasert, C. (2020). The key success factors in focus group discussions with the elderly for novice researchers: A review. *Warasan Wichai Witthayasat Kanphaet*, *34*(4), 359–371. DOI: 10.1108/JHR-05-2019-0114

Neubauer, B., Witkop, C. T., & Varpio, L. (2019). How phenomenology can help us learn from the experience of others. *Perspect Med Educ. 5;8(2)*, 90–97.

Odongo, I., & Ntara, C. (2024). Role of phenomenology in management research. *Int. J. Management Concepts and Philosophy*, *17*(4), 427–441. DOI: 10.1504/IJMCP.2024.142324

Olorunfemi, D. (2023). Diary studies in research: More than a research method. *International Journal of Market Research*, *66*(4), 1–7.

Palinkas, L., Horwitz, S. M., Green, C. A., Wisdom, J. P., Duan, N., & Hoagwood, K. (2015). Purposeful sampling for qualitative data collection and analysis in mixed method implementation research. *Administration and Policy in Mental Health*, *42*(5), 533–544. DOI: 10.1007/s10488-013-0528-y PMID: 24193818

Pandin, M., & Yanto, E. (2023). The what and how of existential phenomenological research. *The Qualitative Report*, *28*(3), 816–827. DOI: 10.46743/2160-3715/2023.6268

Panovski, A. (2024, 04 24). *5 Phases of phenomenology and the rise of existentialism*. Retrieved from The Collector: https://www.thecollector.com/phenomenology-rise-existentialism/

Pathak, V. (2017). Phenomenological Research: A Study of Lived experiences. *IJARIIE*, *3*(1), 1719–1722.

Patton, C., & Broward, J. (2023). The Giants and Forerunners of Phenomenology: Husserl, Heidegger, and their Predecessors. *American Journal of Qualitative Research*, *7*(4), 79–94. DOI: 10.29333/ajqr/13600

Phillips, M. (2023). Towards a social constructionist, criticalist, Foucaldian-informed qualitative research approaches:Opportunities and challenges. *SN Social Sciences*, *3*(175), •••.

Sarfo, J., Debrah, T., Gbordzoe, N., Afful, W., & Obeng, P. (2021). Qualitative research designs, sample size and saturation: Is Enough Always Enough? *Journal of Advocacy. Research in Education*, *8*(3), 60–67.

Saunders, M., Lewis, P., & Thornhill, A. (2019). *Research methods for business students*. Pearson Education Ltd.

Seatle University. (2024, 10 31). *Existential Phenomenological Psychology*. Retrieved from Seatle University: https://www.seattleu.edu/academics/all-programs/psychology-ma/epp/

Sheehan, S. (2014). A conceptual framework for understanding trancsendental phenomenology through the lived experiences of biblical leaders. *Emerging Leadership Journeys*, *7*(1), 1–6.

Sinfield, G., Goldspink, S., & Wilson, C. (2023). Waiting in the Wings: The Enactment of a Descriptive Phenomenology Study 22. *International Journal of Qualitative Methods*, *22*, 1–11. DOI: 10.1177/16094069231207012

Smith, J. A. (2004). Reflecting on the development of interpretative phenomenological. *Qualitative Research in Psychology*, *1*, 39–54.

Smith, J. A., & Osborn, M. (2015). Interpretative phenomenological analysis as a useful methodology for research on the lived experience of pain. *British Journal of Pain*, *9*(1), 41–42. DOI: 10.1177/2049463714541642 PMID: 26516556

Stahl, N., & King, J. (2020). Expanding approaches for research: Understanding and Using Trustworthiness in Qualitative Research. *Journal of Developmental Education*, *44*(1), 26–28.

Stolz, S. (2022). The practice of phenomenology in educational research. *Educational Philosophy and Theory*, *55*(2), 1–35.

Subedi, K. (2021). Determining sample size in qualitative research. *Scholars'. Journal*, *4*, 1–13.

Suddick, K., Cross, V., Stew, G., & Stew, G. (2020). The work of hermeneutical phenomenology. *International Journal of Qualitative Methods*, **•••**, 1–14.

Thomas, S., & Sohn, B. (2023). From uncomfortable squirm to self-discovery: A phenomenological analysis of the bracketing experience. *International Journal of Qualitative Methods*, *22*, 1–12. DOI: 10.1177/16094069231191635

Tufford, L., & Newman, P. (2010). Bracketing in qualitative research. *Qualitative Social Work: Research and Practice*, *11*(1), 80–96. DOI: 10.1177/1473325010368316

Woodruff, D. (2018). Phenomenology. *The Stanford Encyclopedia of Philosophy*. Retrieved 09 4, 2024, from https://plato.stanford.edu/archives/sum2018/entries/phenomenology/>

Zieske, C. (2020). A brief history and overview of existential-phenomenological psychology VOL 17 Issue 2. *American Journal of Undergraduate Research*, *17*(2), 45–58. DOI: 10.33697/ajur.2020.020

KEY TERMS AND DEFINITIONS

Phenomenology: A description of a person's lived experience.

Eidetic: Clear images or memories.

Daisen: The state of finding yourself in this world as in existential thought.

Bracketing: An attempt to rid yourself of all preconceptions or prejudice in order to record accurate record.

Data Saturation: The point at which a researcher starts getting the same data from participants.

Ontology: The nature of being.

Existentialism: A philosophy that places stress on individual free will.

Social Constructivism: A belief that knowledge is created by people in social and cultural settings at any given historical moment.

Chapter 10

Mapping Complexity as a Qualitative Approach to Operationalizing Actor–Network Theory:
The Brazilian Clinical Trial Regulation Case

Ricardo de Abreu de Abreu Barbosa
https://orcid.org/0000-0003-2022-6368
Universidade Presbiteriana Mackenzie, Brazil

Walter Bataglia
https://orcid.org/0000-0003-4233-9988
Universidade Presbiteriana Mackenzie, Brazil

Harry Sminia
Strathclyde University, UK

ABSTRACT

This paper explores the Cartography of Controversies as a methodological extension of Actor-Network Theory (ANT) to investigate complex socio-technical systems empirically. Through a case study on regulating clinical trials in Brazil, focusing on legislative debates over PLS 200/2015 and PL 7082/2017, the study demonstrates how controversy mapping can visualize the interactions of heterogeneous networks that shape and are shaped by regulatory processes. The findings show that regulation functions as both an actant and an outcome of interwoven relationships, highlighting socio-technical assemblages' dynamic and performative nature. This

DOI: 10.4018/979-8-3373-4432-4.ch010

approach underscores the utility of cartography for operationalizing ANT, enabling a richer understanding of fluid and evolving phenomena in qualitative research.

1. INTRODUCTION

In addressing the persistent challenge of translating Actor-Network Theory (ANT) from abstract principles into empirical practice, this chapter explores how the Cartography of Controversies offers a transformative methodological bridge. Conceived by Bruno Latour and Tommaso Venturini, this approach extends ANT beyond its theoretical origins, providing researchers with a structured, replicable framework for examining complex socio-technical landscapes. This method renders dynamic associations among diverse actors—human and non-human—visible, enabling a more nuanced exploration of the shifting forces shaping organizational and policy landscapes. Particularly suited to contexts where regulatory processes and social structures evolve through fluid and contested interactions, the Cartography of Controversies equips researchers with essential tools to trace the often invisible currents that define social reality.

Developed initially within Science, Technology, and Society (STS) studies by Bruno Latour, Michel Callon, and John Law, ANT offers an epistemological framework to describe realities as fluid, associational processes emerging from the agency of human and non-human actors. ANT proposes that organizations, groups, and institutions lack intrinsic, stable natures; instead, they are best understood through the relationships that connect them (Venturini, 2018). In ANT, actor networks are not sources of action but outcomes of associations among diverse entities (Latour, 1996, 2007, 2012). These networks form and transform through controversies — rhizomatic associations or relationships that constitute and reconstitute social realities, a process referred to as "translation" (Law, 1992; Callon, 2018).

ANT presents a promising approach for exploring complex, post-human issues (Walsham, 1997). However, its conceptual complexity has been criticized for challenging empirical research (Mitev, 2009). Latour (1996) acknowledges this challenge, explaining that ANT integrates three dimensions: a semiotic framework for constructing reality, an ontological affirmation of actors as networked, and a methodology for capturing the heterogeneity of this construction. This blend of theory and method poses operational challenges for researchers.

Despite Latour's description of ANT as a straightforward way to learn from actors without imposing predefined interpretations (Latour, 1999) and his provision of seven general principles (Latour, 1987), ANT has been critiqued for lacking clear methodological guidelines. Scholars argue that ANT proponents do not sufficiently address practical research techniques (Bloomfield & Vurdubakis, 1999; Kanger,

2017). The guiding instruction to "follow the actor" remains ambiguous, offering little specific guidance on empirical procedures (Blok et al., 2020).

This ambiguity extends to data collection and analysis. "*Case studies using ANT rarely clarify their research procedures or explain how interpretive leaps from specific instances to broader categories were achieved, leaving the impression of 'scholarly magic' rather than disciplined interpretation*" (Kanger, 2017, p. 24). This gap has raised questions about what differentiates ANT from other inductive approaches, particularly in justifying its practical utility in tracing and reconstructing associations or mobilizing controversies (Kanger, 2017).

Several techniques have been applied in ANT research, including socio-technical graphs (Latour et al., 1992), semiotics (Latour, 1984), and conversation analysis (Simpson, 2007). However, these studies have primarily focused on ANT's inter-disciplinary dialogue with fields like semiotics and linguistics rather than providing robust solutions to its methodological limitations. Additionally, citation databases reveal that these techniques have seen limited replication in ANT studies, indicating that the methodological challenges persist.

We propose that the Cartography of Controversies could address these issues. This set of techniques, designed for visualizing and exploring complex issues, has been adapted to guide students and researchers in investigating contemporary socio-technical debates. We suggest it can be applied in ANT-based empirical research to map heterogeneous elements and their evolving associations as they construct and reconstruct social assemblages.

Initially inspired by Deleuze and Guattari's method for studying subjectivity (Kastrup, 2007), cartography emphasizes processuality, capturing complexity without reducing it to simplistic schemes. In ANT, Latour introduced cartography, collaborating with Venturini and other STS researchers to create MACOSPOL — a European collaborative project for Mapping Controversies in Science and Technology for Politics. MACOSPOL aimed to promote cartography as a technique that enables scientists, professionals, and citizens to explore controversies through a web-based platform.

Since its inception, Latour and Venturini have endorsed the Cartography of Controversies as an effective means of examining the construction of social life (Latour, 2011; Venturini et al., 2016). Further developments have been made by scholars such as Venturini (2010a, 2010b, 2018), Hussenot (2008, 2014), Clegg and Tureta (2021, 2024), who recently developed a protocol for applying controversy mapping to anti-history research methods.

According to Venturini (2010a), based on the MACOSPOL project experience, cartography offers a method to describe reality flexibly without commitment to any single theoretical framework. It allows multiple perspectives by prioritizing the actors' voices rather than the researcher's interpretations.

Using the case of clinical trial regulation in Brazil, this study illustrates how controversy mapping supports ANT's application in qualitative inquiry. The Cartography of Controversies reveals the distributed agency among regulatory bodies, medical professionals, and ethics committees by documenting the regulatory debates and alliances that shape legislative reform. This mapping operationalizes ANT's theoretical premises—emphasizing socio-material entanglements and distributed agency—to generate detailed insights and descriptions of socio-technical assemblages.

This study contends that Cartography of Controversies bridges ANT's theoretical and empirical divide and broadens its utility in contemporary research, enabling qualitative scholars to engage more effectively with complex, evolving phenomena. This methodological framework holds broader implications, offering a scalable, systematic approach for exploring diverse socio-technical controversies and providing a replicable model for research that demands nuanced engagement with heterogeneous actor networks.

The following sections outline ANT's theoretical foundations and methodological approach, explaining the challenges posed by its abstract nature and advocating for the Cartography of Controversies as a solution for operationalizing ANT in empirical research. We then demonstrate the application of controversy mapping in the context of Brazilian clinical trial regulation. We discuss how cartography enables an expanded view of assembling complex artifacts, as illustrated through narratives of symmetrically distributed agency. Final considerations are done in the final.

2. ACTOR-NETWORK THEORY

Theoretical foundations and fundamental concepts

The Actor-Network Theory originated from the sociotechnical studies of the "Science, Technology and Society" (STS) school. It gained great prominence from the works published by Bruno Latour, Michel Callon, and John Law, seminal authors who started an alternative movement to try to understand reality from a procedural and heterogeneous view (CAMILLIS; ANTONELLO, 2016).

It represents an alternative lens to understanding reality that relies on four approaches that embody it. First, the influence of a lack of interest in an anticipated definition of the object stands out from the *relational sociology* (EMIRBAYER, 1997), which defends a relegation of an "a priori" labeling and an emphasis on tracking the propagation and reiteration of relational configurations (LAW, 1992). Second, the rejection that collective entities are conceived as stable and singular stands out from the *process approach* and the defense that they result from continuously gathered elements (LAW, 1994; LANGLEY AND TSOUKAS, 2016).

Third, the *performativity approach*, originated by Austin (1962) and developed by Callon (1998) in his analysis of economic models whose theorizations would not only describe pre-existing conditions for the occurrence of economic phenomena but would have the effect of interfering with their conditions. Fourth, the *socio-material approaches* touch on the symmetry principle or the recognition of an ontological status between humanity and objectivity (TONELLI, 2016).

ANT works with specific semantics whose operational definitions are worth considering.

Generalized symmetry

The notion of symmetry, which originated with David Bloor (1976), started with the critique that elements of nature and society were not being symmetrically considered to explain the difference between true and false sciences. Bloor (1976) defended the impartiality between true and false, between rational and irrational, and between success and failure, in addition to the need to transfer the right of explanations to the then-neglected pole of society.

Callon (1999) developed the notion of generalized symmetry, in which an accurate symmetric description could not start from any of the bars (nature or society). Callon and Latour's concept advanced to refer to symmetry, not only as an equal treatment between truth and falsehood in the history of science but also to establish an ontological symmetry between elements of nature and society and between humans and non-humans for the generation of negotiations and performativities (TONELLI, 2016). Thus, symmetry has become a heuristic tool used to understand the contingent role played by actants within an actor-network (CALLON, 1999) based on the assumption of complete planification between the elements of nature and society (LATOUR, 1994).

From the symmetry principle, any actor can participate in a controversy involving human beings, groups, and non-humans in an integrating "heterogeneous arrangement" (LAW, 1999). ANT does not claim an object has a volitional capacity like a human. Still, human action does not result exclusively from the will dissociated from its relationships with things. Every volition is manifested from symmetrical relationships in which any human or non-human actant is present, making them protagonists of controversies that function as a "hybrid forum" (CALLON; RIP, 1992) or a space of conflicts, debates, and negotiation in search of stabilization.

Controversies

Controversies are central to ANT (CALLON 1989), as associations between heterogeneous elements occur under friction. Moreover, the notion that collective life is made up of alliances and their dissolutions also emerges here, thus emphasizing the role of controversies in the constitution of society (VENTURINI, 2015).

Controversies are also considered one of the most complex phenomena observed in collective life (VENTURINI, 2010a), although the authors of ANT have little explored them. They involve anything that challenges *the status* quo among actors (HUSSENOT AND MISSIONER, 2010).

To develop controversy studies under ANT, Latour created MACOSPOL (Mapping Controversies on Science for Politics), a project funded by the European Union to establish a collaborative platform for mapping controversies under the actor-network approach. According to MACOSPOL documents (2007, p. 6), controversy is every piece of science and technology that needs to be stabilized or closed. This definition presents controversy as a state of shared uncertainty about facts of a scientific nature.

Controversy resists reduction (VENTURINI, 2010a), meaning that everything gives rise to discord as if there were an infinite regression that makes their closure impossible. Volcanic magma, which alternates between its liquid and solid states, is an excellent metaphor to illustrate the dynamics of controversies, which ceaselessly settle down or intensify (VENTURINI, 2010a).

As criteria for choosing a controversy, Venturini (2010a, 2010b) and Hussenot (2014) suggest that it is advisable to investigate where collective life becomes more complex, where the largest and most diverse variety of actants are involved, where alliances and oppositions are transformed more irrationally and inconsistently, where nothing is as simple as it seems, where everyone is screaming and fighting and where conflicts become more severe.

The closure of a dispute constitutes what ANT calls the "*black box*," a cybernetic term that means the concealment of the internal systems of a system or the result of an attitude of making opaque the functional processes that transform one data into another (LATOUR, 1999). A black box is formed, or a statement becomes solid as a fact, whenever it is introduced in new formulations as an unquestionable premise, to the point of deducing that the solidity of reality, or its stabilization, will always depend on all those who keep it moving, in a work of legions of allied and interconnected networks (LATOUR, 1999). Whenever a fact encounters opposition or obstacles in circulation, it will be pushed towards an assertion of "falsehood" in an environment of controversy in which black boxes will be reopened. When a fact becomes a black box, it creates autonomy and independence. When the black box opens, it means the reinstatement of controversy.

The emergence and development of controversies co-occur with the production of collective aggregates (or groups) that oppose each other. These are accompanied by elements enlisted to support their positions and arguments that delineate these group boundaries, enabling the construction of analysis maps of the established relationships. The controversies will be perceived as a result of these different senses of belonging of the groups and the translations formed from this work of the actors-networks.

Translation

Supposing controversies represent disagreement between heterogeneous actors engaged in action (VENTURINI, 2010a), the translations will seek to establish their contradictory interests (LATOUR, 1999), reshaping network actors and causing them to appear, change, or be excluded from reality (HUSSENOT, 2014).

Michel Serres, one of Latour's great influencers, created the notion of translation (1996). Serres also defended the practical impossibility of dualism between the sciences and the outside world and developed the idea of multi-temporality by stating that the phenomena of reality refer simultaneously to the past, present, and future (TONELLI, 2016).

These ideas outlined what translation consists of in Michel Serres's view (1996): processes of building connections, passages, or establishing communications, an act of invention or *bricolage* that emerges from the mixture of heterogeneous multitemporal elements (TONELLI, 2016).

This notion of translation did not only have linguistic or symbolic connotations but involved acts of displacement or substitution of knowledge, people, or things, and fell not only on elements that it translated but also on what was translated in a clear allusion to the performativity of relations (CZARNIAWSKA, 2005).

The seminal authors of ANT incorporated the notion of translation, which they defined as an "interpretation given by fact-makers to their interests and those of the people they enlist" (LATOUR, 1998). Later, it was conceived as a geometric notion of displacement and transformation, which offers new interpretations of interests and channels people into different positions (LATOUR, 1999, 2007).

Callon (1986, p. 203) defined translation as a "process during which the identity of the actors, the possibility of interaction, and the margins for maneuver are negotiated and delimited." This negotiation involves four movements: problematization, interessement, enrolment, and mobilization. In the problematization movement, the main actors establish their identities and goals to create a crossing point everyone must accept to achieve their interests. It is as if the rules for developing controversies were established (CLEGG, 1989). The interessement involves the imposition of other actors of the movements defined in the problematization, preventing the identities represented in the previous phase from being lost. The enrolment means defining the roles to be played by the actors and how the others will relate within the networks. Finally, the mobilization phase indicates a coalition in which the main actors borrow the strength of their allies, passive agents, and become their representatives or spokespersons. At this stage, dissenting voices are silenced, and the controversy ends with the commitment to negotiation, the result of which becomes legitimized.

From these concepts, it becomes evident that controversies drive the sustained dynamics of actor-network interactions. This process exemplifies distributed and symmetrical agency between human and non-human entities, which, through relational and performative acts, facilitate continuous translation. Such translations manifest in the assembly and reconfiguration of social groupings, underscoring ANT's perspective that the social is best understood as a constellation of associations—entities in perpetual motion and redefinition, embodying a fundamentally procedural and fluid ontology.

ANT as a set of methodological guidelines

Building on the foundational aspects of ANT, as outlined by Latour (1996), which emphasize its semiotic and procedural ontology and its role as a method for describing reality through the assemblage of associations, it is essential to note that ANT serves as an umbrella term encompassing various interpretations. For instance, it embodies at least seven distinct versions, reflecting its inherent flexibility (Kanger, 2017). Gad and Jensen (2010) suggest that ANT should be seen not as a rigid entity but as a 'multiple' framework, adapting and evolving through its diverse applications.

With this theoretical foundation established, we explore ANT as conceived by Latour, emphasizing its role as a methodological approach. This examination foregrounds ANT's structured yet adaptable principles necessary for operational clarity. Such an analysis prepares the ground for discussing the cartography of controversies as an essential tool for systematically applying ANT, enabling researchers to map and understand the intricate nature of socio-technical debates effectively.

Initially, ANT's methodological rules and principles were crafted to investigate scientific practices and processes. However, their scope expanded over time, enabling their application to comprehend any reality through ANT's lens. This extension highlights ANT's flexibility and potential to be applied broadly, making it relevant for exploring various social, political, and organizational phenomena.

Latour (1999) advocates for an approach to science that sees it as a dynamic process rather than a finished product. This perspective entails studying science at its formative stages, before concepts and technologies are "black-boxed," or reopening these settled "black boxes" to reveal the underlying disputes and developments. This approach showcases the inductive nature of ANT, insisting on descriptions that capture the roles of both human and non-human actors in shaping controversies. Latour's directive to "simply describe the facts at hand" underscores the necessity of refraining from theoretical preconceptions that may distort observations.

The objectivity of claims within ANT is redefined through the lens of transformations shaped by collective actions and disputes. Scientific knowledge is portrayed as inherently collaborative, molded by external scrutiny and interactions. Latour stresses that the rigor of scientific insights grows with the range of perspectives and conflicts involved, which contrasts sharply with the perceived rational calm of everyday life. Thus, progress emerges from negotiation and dissent rather than mere agreement.

Latour posits that disputes and resolutions build representations of nature and consolidate social structures. He argues that using these constructs to explain controversies would be fallacious, implying circular reasoning. Instead, human and non-human agents must be acknowledged as critical to stabilizing agreements. Artifacts and machines illustrate how non-human entities achieve and maintain these consensuses.

Additionally, the impartiality Latour emphasizes involves recognizing the interplay of internal scientific dynamics and external socio-political influences. Pure research is rarely insulated from broader societal factors, including media, legal, and political interactions. Thus, the robustness of scientific inquiry is bolstered by its contextual entanglements.

Latour suggests reexamining the network structure and observer perspectives rather than adhering to strict logical norms in response to critiques of irrationality. Scientific facts are products of extensive actor networks that gain strength by expanding through alliances. This relational aspect explains the uneven distribution of technoscientific power: while a minority wields significant influence, broader society interacts with outcomes shaped by these robust networks. The principle of relativity ensures that different beliefs and viewpoints can coexist, contributing to resolving controversies without dismissing them as irrational.

Finally, Latour warns against attributing superior cognitive abilities to scientists without scrutinizing their data collection and correlation methods. Psychological and cognitive explanations should be secondary to thorough network analysis. This methodological stance emphasizes how scientific representations are co-produced through human and non-human translations, underscoring the complex social fabric that generates accepted knowledge.

3. CRITIQUE OF LATOURIAN METHODOLOGY AND THE RESPONSE OF CARTOGRAPHY OF CONTROVERSIES

While Latour's methodological rules significantly broaden ANT's scope and application, it is essential to acknowledge that ANT does not provide a detailed, prescriptive research technique. This absence of explicit procedural guidance often demands researchers adopt an exploratory and adaptive approach. In response to these challenges, the cartography of controversies has emerged as a practical method that addresses these gaps and enables the systematic application of ANT in empirical research.

Cartography facilitates the mapping and analysis of complex interactions among human and non-human actors, capturing the dynamic translation and negotiation processes frequently overlooked by more conventional approaches. This tool allows scholars to visualize heterogeneous networks and the underlying tensions shaping social and technological phenomena, fostering a more nuanced and comprehensive analysis.

Despite the coherence of Latour's methodological rules, they often lack the clarity and precision needed for consistent empirical guidance. As Annemarie Mol (2010) aptly noted, "No one will hold your hand while you walk" (p. 261), highlighting the difficulties researchers face when operationalizing ANT in practice. Similarly, Gad and Jensen (2010) point out that ANT-based studies can sometimes appear to lack transparency in data collection and analysis, creating an impression of "academic magic" rather than systematic interpretation (p. 73).

Grounded theory, by contrast, offers a highly detailed, step-by-step procedural framework, emphasizing clear research pathways. This comparison underscores the challenges of showcasing ANT's unique strengths and making them evident in practice (Kanger, 2017). Proponents of ANT often assume that its methodological advantages—such as tracing and reconstructing associations or engaging with controversies—are self-evident. However, these strengths may be difficult for those new to ANT to discern due to the high level of abstraction inherent in its principles (Kanger, 2017).

The cartography of controversies effectively bridges this gap. Developed by Latour (2007) and expanded through the EU-funded MACOSPOL project, cartography offers a concrete method for conducting empirical analyses of complex and contentious situations. It provides visual and analytical representations that make these complexities more comprehensible and accessible.

As an educational extension of ANT, controversy mapping allows researchers to "live, know, and practice in the complexities of tension" (Law & Hassard, 1999, p. 12). This method preserves ANT's foundational principles while removing theoretical obstacles that might impede practical use (Venturini, 2010a). However, Venturini

(2010a) emphasizes that cartography has challenges; it must faithfully represent controversies while adhering to ANT's inductive emphasis on "just observing" collective life without resorting to prescriptive theories.

Achieving neutrality in cartography can be complex, but the practice should strive to integrate all relevant perspectives. Latour (2007) refers to this as pursuing "second-degree objectivity," where including diverse viewpoints enhances impartiality. The more cartography incorporates these varied perspectives, the more robust and objective its representation becomes.

Tureta and Clegg (2021) have proposed frameworks to operationalize the cartography of controversies, drawing on the work of Venturini (2010a; 2010b) and Hussenot (2014). Their model, structured around five relational and non-linear focal points, aligns with Latour's methodological principles and helps address ANT's practical shortcomings.

Table 1. Cartography of controversies process

Step	Function	Description
Sampling	How to identify controversies	**What types of controversies are at stake, and what are their connections?**
Scanning	How to map actors-networks	**Follow the network of actors formed around controversies, looking for those whose participation affected the actions of others. This will enable you to map who and what was involved in the controversies.**
Tracing	How to design the translation process	**Seek how network actors relate, connect, disconnect, and align, generating translations.**
Labeling	How to analyze the policies of the actors-networks	**Understand the relations and practices of power involved in the controversies, looking for what moves the actors to mobilize other actors and what represses or silences them.**
Describing	**How to describe multiple realities and power relations**	**Identify excluded actors, bring up hidden events, and bring out the different possibilities of reality.**

Source: Tureta; Clegg, 2021, adapted by the author

By applying Tureta and Clegg's model, researchers can operationalize ANT and effectively map the intricacies of socio-technical issues, preparing the groundwork for further analysis of real-world applications.

4. PANORAMA OF CLINICAL RESEARCH AND ITS REGULATION IN BRAZIL

A clinical trial represents a critical phase in the pharmaceutical research and development (R&D) process, focusing on the controlled administration of new products to human participants to establish their safety, efficacy, optimal dosage, and potential adverse effects (Gomes et al., 2012). The capacity to develop new drugs is deeply linked to the ability to conduct such trials, which historically led to the vertical integration of R&D within the pharmaceutical industry throughout the 20th century. Increased global competitiveness has introduced financial pressures and made regulatory compliance a pivotal factor in pharmaceutical innovation. This shift has extended timelines for approvals and amplified oversight, raising ethical concerns about research practices.

In Brazil, clinical trial regulation is divided into ethical and health oversight. Ethical regulation, necessary for research involving human participants, emphasizes their rights, safety, and well-being. Health regulation, conversely, pertains to the broader supervision of research and the procedures leading to product registration for eventual market access.

A key player in Brazil's ethical oversight is the National Health Council (CNS), a permanent deliberative body within the Unified Health System (SUS). The CNS establishes the protocols for ethical evaluations of clinical research through two primary entities: the National Research Ethics Commission (CONEP) and the network of Research Ethics Committees (CEPs). Collectively, these bodies form the CEP-CONEP System, tasked with critically reviewing research protocols. Any clinical trial involving human participants must first be submitted to a CEP, which evaluates its ethical feasibility.

The regulation of clinical trials often stirs significant controversy. It balances the interests of patients seeking innovative treatments, research participants, sponsors, investigators, research centers, and state regulatory agents. Central to these debates is the challenge of aligning efficient trial processes with the rigorous control measures needed to protect participants' rights and safety.

Efforts to modernize and strengthen Brazil's clinical trial sector have been notable in recent years. A significant legislative milestone was the introduction of Bill PLS 200/2015 in the Federal Senate, later reworked as Bill 7082/2017 and debated in the Chamber of Deputies. This legislative push aimed to create a new regulatory framework, streamlining the processes of the CEP-CONEP System to enhance the sector's efficiency and competitiveness by addressing delays.

These legislative efforts culminated in the enactment of Law No. 13,874/2024, which marks a pivotal change by establishing a comprehensive regulatory framework intended to overcome procedural inefficiencies and support the growth of clinical trials in Brazil. However, the new law has sparked debate. Critics question whether it effectively upholds its participant protection and ethical oversight goals, leading to ongoing discussions about its broader implications for research practices and moral standards in the country.

5. HOW THE CARTOGRAPHY WAS BUILT, AND THE NARRATIVES OBTAINED

The cartography

The methodological process was structured into three interconnected phases: data collection, analysis, and mapping. Although presented sequentially, these phases were iterative and did not occur linearly.

In the data collection phase, we gathered documents from various online sources directly or indirectly related to the clinical trial sector. Using targeted search terms such as "clinical research," "clinical trial," "regulation of clinical research," and specific legislative references like "PLS 200/2015" and "PL 7082/2017," we identified approximately 100 relevant sources. These included news articles, opinion pieces, interviews, and transcribed lectures, which provided initial insights into the controversies surrounding clinical trial regulation in Brazil.

The analysis phase involved a comprehensive examination of legislative records from the Federal Senate and the Chamber of Deputies that document the proceedings related to PLS 200/2015 and PL 7082/2017. This phase included reviewing legislative reports, amendments, proposals, and shorthand notes from committee meetings. We analyzed more than 200 documents—totaling around 1,500 pages—including legislative session transcripts, public hearings, and relevant normative acts produced by CONEP and ANVISA.

In the mapping phase, we applied the cartography of controversies to visualize and document the reassembly of the regulatory framework for clinical trials. This approach adheres to Venturini's (2010a) principles, positioning the process within a socio-technical context marked by regulatory, ethical, and political complexity. The cartography method was chosen for its ability to map interactions among various human and non-human actants, illustrating the intricate web of influences in regulatory debates.

Following Tureta and Clegg's (2021) model, we identified controversies (sampling) through systematic reading and organization of the collected documents into hierarchical trees or "Popperian trees." These visual structures depicted the complexity of the debates, aligning with Venturini's (2010b) assertion that controversies encompass many perspectives and cannot be reduced to binary oppositions.

Each controversy was detailed at four levels: issues, positions (convergences of actants around possible solutions), arguments (points raised to support each position), and the actants (human and non-human) involved (Ceretto & Domenico, 2016). This mapping was facilitated by Debategraph, a collaborative tool for visualizing complex debates.

Cross-coding and categorization methods (Corbin & Strauss, 1990) were used, allowing for a priori categories like "controversies" and "actants" and a posteriori categories such as "events," "decisions," and "interests." This coding supported the construction of hierarchic trees that disaggregated controversies into (a) themes, (b) positions, (c) supporting arguments, and (d) counterarguments.

Identifying controversies at these levels enabled us to extract both explicit and implicit interests involved. During the scanning phase, we mapped the actor networks present in each controversy, distinguishing between human actants, non-human objects, quasi-objects, and institutional macro-actors (Latour, 2007). Events linked to the legislative process of PLS 200/2015 and PL 7082/2017 were documented chronologically, with notes on the actants participating at each stage.

Cross-linking allowed us to connect controversies with specific events, associated actants, and expressed interests, mapping the influential actors and their impact on the broader network. This facilitated tracing how translations occurred, marking pivotal legislative decisions and reactions by involved entities. Legislative deliberations, such as the approval of amendments and shifts in Project versions, were detailed, alongside normative changes by the CEP-CONEP system and related responses.

The labeling phase employed abductive reasoning to analyze power relations and interactions, focusing on moments when these relationships influenced the legislative process. This phase emphasized how coalitions formed and acted as unified entities—an instance of actualization within ANT terminology.

Finally, the describing phase involved reconstructing a narrative of the clinical research regulation process, showcasing the analysis consistent with ANT's principles of procedural ontology, observer agnosticism, generalized symmetry, and the freedom of association.

The freedom to follow the paths of collecting and analyzing the empirical material of the research, opposing the strictness of traditional methodologies, allowed us flexibility in the process, as Czarniaswska (2006) recommended in works based on ANT.

The narratives

From the multiple ramifications and different traceable paths, we structured the reassembly of the controversy based on three narrative themes that helped us to understand the clinical trial regulatory dynamics: (a) clinical research, (b) the CEP/CONEP system, and (c) the Bill of Law (PLS 200/2015 – PL 7082/2017).

Figure 1. Narrative themes of the case

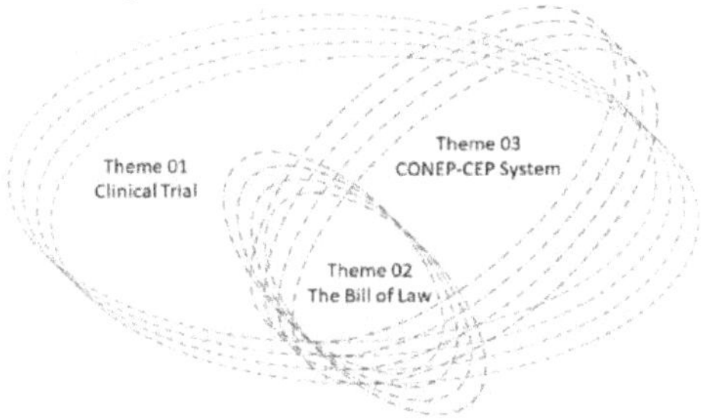

Source: the authors

The narratives were built from the search to understand the relations that described the dynamics regarding (a) the meaning attributed to clinical research and the regulation involved in it, (b) the controversies surrounding the ethical regulatory function (CEP-CONEP system), and (c) the legislative process that proposes a rearrangement of this regulatory dynamic but also suffers performative effects in its construction.

In a brief description of what was found in the work of actants, the cancer pill, the law that regulated it, and the decision of the Federal Supreme Court that prohibited it was who gave the initial contours to the first controversies.

Clinical trial regulation, as a socio-technical artifact, was assembled as a starting point by its relation with slowness and bureaucracy. The propagating effect of disease, death generator, and injustice promoter scares away talent and reduces wealth. Furthermore, due to its delay and bureaucracy, the current regulation of clinical trials has also been accused of hindering participants' access to the benefits of experimental treatments.

When the 'trial' element, as an actor, was introduced in the counterattack, the 'regulation' was resignified in a way that experimentation and any possibility of not finding what one is looking for, besides unanticipated and uncontrolled adverse effects of the research, must be avoided, or at least mitigated, especially when research participants are human beings.

'Regulation' and 'clinical trial' had enlisted new allies, expressing the artifact's ambivalence: on the one hand, regulation "saves lives"; on the other hand, clinical trials " produce deaths" as well.

A new debate arose related to the level of the consent of clinical trial participants, and the conflict between "freedom" and "safeguard" shifted to an apparent dilemma between "protecting the trial participants" or "taking care of the patients undergoing treatment."

Controversies arose over what regulation dosage one wants to exert on the clinical trial and who will make this decision.

The CEP/CONEP system reassembled the ethical regulation instance. The search for security, predictability, and stability had intensified precisely because CONEP and CEP were accused of not promoting such results.

The attack on the CEP-CONEP system was due to accusations of inefficiency and discredit. Counterattacks invoked the system's international credibility, history, and evolution. Ultimately, the controversy over the model of social control exercised by CONEP gained notoriety. It was the struggle over who or what would ultimately control regulation.

Finally, some controversies were described around the dismantling of CONEP, the reassembling of the CEPs, the reconfiguration of the level of instance decisions in the system, and the composition criteria of the new system.

The meta-controversy between agility and safety in clinical research activities resulted in sub-controversies organized in Popperian trees (VENTURINI, 2010b), which emerged according to the unfolding and complexity of the networks were developed. Thus, in a structure prioritized by themes, positions, and arguments, a summary of the controversies can be seen in Figure 3 below.

Figure 2. Sub-controversies listed

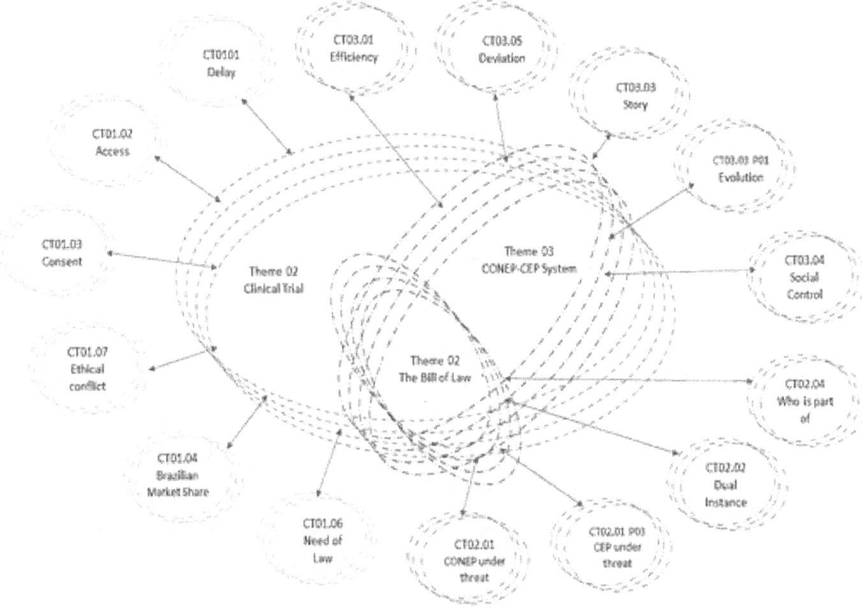

Source: The Author

The actants were not seen as substantive realities but as actors-networks or relationships arising from the association between human and non-human actors, whose participation is planned, horizontal, diffuse, and dispersed, in a word, rhizomatic.

Among the different types of participating actors that mediated the regulatory deconstruction and reconstruction process, the prodigality of macro-actors or institutional actors, which are collective aggregates constituted by punctual relations and that, for this reason, externalize reified relational patterns (LATOUR, 2007), was observed.

Although such macro-actors appeared in the narratives as supposedly ready-made or given entities, they were collective aggregates in formation whose supportive networks had their translation and performative dynamics.

The macro-actors took different positions, guided by the relational arrangements that supported them. The participation of macro-actors occurred through the speech of humans, whose vocalization was also the result of their associations with other actors, able to produce an effect in "reassembling" the regulation.

The dynamics of macro-actors were always produced by relational interactions with effects on the process of regulatory deconstruction and reconstruction, as the following examples: the effort of the CONEP network to demonstrate the reduction in deadlines for analyzing clinical research projects; the search for endorsement, both by CONEP and by other macro-actors, aiming at lending credibility to their actions. When one actant stated that "several instances have already moved on" during the billing process, it was recognized that the participative nature of CONEP's performativity became a mediator of the vocalized agency itself, regardless of human voluntarism. It was seen how CONEP tried to mobilize FIOCRUZ and the SENATE to support the preservation of the regulatory status that PLS 200/2015 sought to restore.

The agency, in addition to the participation of humans and institutional actors, also involved a set of non-human elements that we classified as (a) objects, (b) quasi-objects, and (c) artifacts.

The presence of these elements in the reconstruction of regulation could reflect their influence throughout the conflicts. Such features are necessary for the development of reality to be different in the relational course.

One of the first critical manifestations of the regulatory artifact proposed throughout the legislative process was the mediation of the well-known "cancer pill," whose agency was decisive not only in opening the black box that culminated in PLS 200/2015 but also in acting at times when elements such as "hurry," "agility," "efficiency," and "fight for life" were manifested in the network works.

Likewise, the texts, which are documents in the form of reports and legal opinions, were actors that resonated with the bill's decisions and were important in the struggle of disputed interests around the reassembly of the regulatory artifact.

The texts can be the result of conversations or be in written form. In the present case, from a process conducted by essentially formal legislative activity, we emphasize the agency of documents that gained a punctuated character as they began to be repeated and used in other moments of conversational interaction in the form of stylization, such as happened with the report on the experience of Indian regulatory review and with the legal opinion of the Minister of the Federal Supreme Court, Luiz Barroso.

These documents supported the arguments of the actors-networks and were part of the action throughout the process.

It was also interesting to observe the de-contextualization promoted by the documents, as their actions were translated and renegotiated in different spatiotemporal situations in which they were presented.

This was the case, for example, of the opinion of 'Luiz Barroso-minister' when he was still 'Luiz Barroso-lawyer' after being made a 'minister'; his decisions at the Federal Supreme Court conflicted with the legal position issued when he defended the need for a law to provide legal certainty to the regulatory artifact.

Facts also played a role in the relational process. These are episodes, events, phenomena, or situations that cannot be understood ontologically as persons or objects, although they can be taken as subjectified or objectified entities in ANT.

Cancer, for example, is a chronic disease that inevitably takes on a subjective character: "Cancer is ending my life"; "I live with cancer." On the other hand, "Blue November" is an objective program promoted by SUS to alert the male population to early diagnosis of prostate cancer. The disease and the educational event entered the network to join the other actants in defense of the bill, generating the performative effect of changing the regulatory artifact in its infra-legal instance. The fire of the "Joelma Building" was a tragedy that occurred on February 1st, 1974, in the central area of the city of São Paulo, Brazil, which caused the death of 187 people, left more than 300 injured, and is remembered by Brazilians as a synonym for tragedy. This situation was carried over to the death of patients with cancer who did not benefit due to the delay in clinical research. The rescue of this episode enlisted an actant that relativized the search for "the safety" of clinical research participants in favor of PLS 200/2015.

It also identified the participation of concepts in the composition of regulation. Thanks to the idea that it would be "bureaucratic," it was given a new meaning. On the other hand, the notion of "consent" influenced the extension and direction of the power that would derive from the regulatory relational arrangement under construction.

Several artifacts participated in remaking the regulation, which is also considered a hybrid of socio-technical relations. Shorthand notes recorded the presence of indicators, videos, slides, spreadsheets, and indexes, but here, we highlight the indicators and norms that acted performatively in the process. Thanks to hands underscoring the relevance of Brazilian clinical research globally, the credibility of the CEP-CONEP system was restored. Changes advocated by PLS 200/2015 (PL 7082/2017) were avoided when international standards reminded the networks that were references to the construction of the infra-legal system intended to be deconstructed. The resolutions of CNS (National Health Council), created after the beginning of the billing process, were the best example of the performative effect generated in the intra-relations of the regulatory artifact. Additionally, the Constitution of the Republic revived CONEP, bringing it back to the system, although reconfigured.

Some of the controversies mapped will be presented in detail, demonstrating how cartography that complies with ANT principles offers us an analytical view of the work of the actors-networks in remodeling the regulatory artifact of Brazilian clinical research.

6. SOME CONTROVERSIES MAPPED IN DETAIL

In 2015, significant mobilization around clinical trial research activities was re-ignited in Brazil, marked by assertions linking its regulatory framework to delays in approving potentially life-saving drugs. The debate gained initial momentum with the case of the phosphoethanolamine pill, regulated by Law 13269/2016 and subsequently prohibited by the Federal Supreme Court, which highlighted regulatory shortcomings as potential contributors to the deaths of cancer patients.

To illustrate the reconfiguration of the "regulation of clinical research" artifact and its associated relations, we mapped specific controversies that we believe were pivotal to this process. These controversies emerged during the legislative activity related to PLS 200/2015-PL 7082/2017, encompassing the rhetorical strategies employed by parliamentarians in legislative sessions and the inputs of various actants. These included representatives of sponsors, scientists, trial participants, ethics committees, and other regulatory bodies that played critical roles in public hearings and debates.

Additionally, we examined the agency of non-human actants, whose influence is reflected in the speeches of human actors as they embody and articulate the interests of the heterogeneous networks that mobilize them (Latour, 2007).

Controversy CT01.01: The Delay in Clinical Trial

The debate on regulating clinical research gained momentum within the Constitution and Justice Committee of the Federal Senate, where senators pushed for expediting medication approvals, with cancer as the most pressing concern.

"If you had cancer, what would you do? I would even eat dirt. (...) The right to life is in the Constitution (...), and we must do something for life. We have to shorten the time and make it easier" (Senator Ivo Cassol, CT0101 P01 b4, 11/10/2015).

The perceived delays disproportionately affected those with fewer resources.

"My grandson needed medication that was not authorized. We imported four ampoules. And what about the boy born at Climério?" (Senator Walter Pinheiro, CT0101 P01 b6, 11/10/2015).

"Even the president had cancer and was cured with a drug that had not been authorized at the time" (Senator Ivo Cassol, CT0101 P01 b5, 11/10/2015).

Concerns extended to researchers leaving the country due to bureaucratic hindrances.

"Cláudio Juazeiro, a top researcher at Novartis, left Salvador at 18. We have been trying to bring him back since 2007, but we cannot. He says: 'What I am going to do in Germany, I cannot do in Salvador'" (Senator Walter Pinheiro, CT0101 P01 c, 11/10/2015).

Delays also stifled industrial property development.

"An anticancer plant was discovered at the Federal University of Ceará, but bureaucracy meant the research had to be conducted in the USA" (Senator Simone Tebet, CT0101 P01 to ob002, 08/05/2015).

Critics of regulatory delays delivered a restrained but firm rebuttal. Networks opposing PLS 200/2015, led by CONEP, echoed Latour's maxim of minimizing controversy:

"Any effort to reduce deadlines will have our support" (Jorge Venâncio, CT0101 P01 f, 11/10/2015).

Their stance embodied the notion that "we all want the same things" (Latour, 1998, p. 312) and positioned them as aligned against delays. However, their assertions were countered by specific data:

"Response time at CONEP dropped from 10–11 months to 48 days. The claim that projects still take a year or more is completely unfounded now" (Jorge Venâncio, CT0101 P01 d, 11/10/2015).

"We recently reviewed a phase 3 proposal for the dengue vaccine from the Butantã Institute and approved it in 38 days: 17 days in the first phase, 21 in the second" (Jorge Venâncio, CT0101 P01 d3, 11/10/2015).

Nevertheless, numbers were contested:

"We acknowledge CONEP's efforts to reduce initial review times, but note that these come with numerous pending issues. Sometimes there are over 40 pending points" (Fábio André Franke, ALIANÇA BRASIL, CT0101P01 d2, 11/10/2015).

If CONEP's faster response times did not lead to actual advancements due to persistent bureaucracy, then "speed" became synonymous with inefficiency.

Cancer, the threat of death, injustice for the poor, talent flight, and intellectual capital loss were seen as effects exacerbated by a "bureaucracy" that hindered clinical research and amplified societal harm.

"Researching in Brazil is almost prohibitive" (CT0101 P02 b). The system was described as "inefficient, anachronistic, and marked by distortions" (CT0101 P02 a), "scaring industry away" (CT0101 P02 c), and "harming scientists" (CT0101 P02 d).

"There is no shortage of professionals or researchers; there is an excess of bureaucracy! It is time to end this suffering and encourage our researchers abroad to return. This country has potential, and this Commission is setting a positive example" (Senator Eduardo Amorim, CT0101 P01 h, 03/15/2016).

As a socio-technical artifact, regulation was seen as intertwined with slowness, bureaucracy, and cascading adverse effects—death, injustice, talent flight, and economic loss. The urgency that initially justified revising regulations shifted towards speeding up the legislative process.

New actants, such as the internal regulations of the Federal Senate and Chamber of Deputies, emerged to delay PLS 200/2015. Requests for legislative "views" suspended the bill's progress, leading to renewed calls for urgency:

"Since Senator Randolfe Rodrigues requested a view, I stress the urgency of this matter. Cancer patients cannot wait; they have exhausted chemotherapy, which harms their health. We must act" (Senator Ivo Cassol, CT0101 P01 g, 03/01/2016).

"A person with cancer cannot wait! People have died waiting for clinical research that could give them hope. Delaying this bill even a week is unacceptable" (Senator Ana Amélia, CT0101 P01 I, 11/23/2016).

Despite these appeals, legislative delays persisted, accompanied by defenses from political and technical networks.

"I reject any suggestion that my actions hinder this bill's approval. I speak as a senator, health professional, pharmacist, and citizen aware of Brazil's research needs. This project is highly controversial" (Senator Vanessa Grazziotin, CT0101 P01 j, 11/23/2016).

With repeated delays, PLS 200/2015, now PL 7082/2017, stalled without resolution.

"In the civilized world, research is registered in 60 days. This law has been pending in Congress for over two years" (Senator Ana Amélia, CT0101 P01 l, 02/13/2017).

While legislative urgency waned, CONEP's infra-legal changes reduced processing times, surpassing U.S. standards:

"When we started, reviews took 330 days; now it is under 30 days. We often approve research faster than the U.S." (Jorge Venâncio, CT0101 P01 m, 12/06/2018).

Proponents of PLS 200/2015 recognized that despite obstacles, regulation had advanced:

"Institutions began to move, as Dr. Paulo Hoff confirmed" (Senator Ana Amélia, CT0206 P01 b, 11/10/2015).

Controversy CT01.03: The Ethical Conflicts in Clinical Trials

Why hinder their advancement if clinical trials favor the population, the country, and scientific progress?

Moreover, 'trial' or 'research' itself, as an actor, is the key to understanding that all these benefits take their toll, as research involves discoveries, experimentation, and the possibility of not finding what one is looking for. There are also unanticipated and uncontrolled adverse effects that must be avoided or mitigated, especially when research participants are human beings.

The social-technical networks of regulation and clinical trial have enlisted new allies, expressing the ambivalence of the artifact: on the one hand, regulation, which promoted death, did not contain cancer, and was bureaucratic, now saves lives. On the other hand, clinical trials, which until then generated jobs, supported the SUS, and were responsible for the progress of science, now produce deaths as well.

If both clinical research and regulation can save and kill, the benchmark to find a balance in this antagonism becomes the consent of research participants, a new actant, which must be free and clear.

An essential ally of regulation was the Indian case, which followed the same steps (of deregulation) recommended in PLS 200/2015, thus making the rules for controlling clinical research more flexible. However, the experiment's results were catastrophic: 370 deaths in clinical studies in 2 years. Between 2005 and 2012, 2600 patients died. The Supreme Court of India then decided to suspend research until regulation was more adequate (Jorge Venâncio, CT0105, on 11/10/2015).

The newspaper "*The Times of India*" is enlisted to endorse CONEP by reporting clinical research performed with young girls, without parental consent, involving adverse effects of HPV vaccination tests. Follows a text (actant) reproduced by Jorge Venâncio at the public hearing of the Federal Senate on 11/10/2015:

The girls' parents were unaware that their children were part of a research study involving an experimental vaccine. Independent Ethics Committees may have believed that the Phase 4 trial posed a minimal risk, justifying consent from the inn's owner instead of contacting the potentially illiterate parents who might not comprehend the study. Alternatively, they may have deemed it culturally inappropriate to seek parental consent for the HPV vaccine, given the taboo surrounding adolescent sexual activity in India, which could lead to parents refusing consent (excerpt from the *Times of India*, as cited by Jorge Venâncio, CONEP, CT0103 P01 b, 11/10/2015).

At the same time that Indian deregulation is allied *a contrario sensu* to the defense of Brazilian regulation, an anti-history is rescued to recall the origins of ethical legislation and its function to protect participants' freedom of consent in clinical trials:

This need came from certain abuses during World War II. There was a whole international debate about these terrible experiences, and several norms emerged whose primary concern is protecting research subjects, especially regarding their autonomy and the need to obtain the subject's free and informed consent to participate in the research. (Fernanda Sobral, SBPC, CT0103 - P01, on 11/10/2015).

As the difference between drug and poison lies in the dose that is given, the discussion here arises once about whether the consent of clinical research is an act of the participant, who expresses himself freely, or whether the vulnerable and unable-to-discern participant has their freedom compromised or suppressed.

The clinical trial of the "cure of diseases" and "progress" combined with the artifact of "regulation that kills" will remain an individual act with the participant's consent.

The clinical trials regarding the "Indian experience" in the "Second World War," allied to the "regulation that saves lives," will remain with mitigated consent, where not even the participant can decide alone. It is the dispute between "freedom" and "security."

From the view of 'freedom,' regulation cannot be conceived to protect participants beyond what they wish to be protected. The final word belongs to the one who is protected and not otherwise.

It's as if we watched our house so much, putting up a security system, lookouts, and doorkeepers, that not even the house owner can get inside it. It is so guarded, so surrounded, that we deny our Brazilian patients access to treatments that could mean better life expectancy and a dignified treatment for their disease. Fabio Franke, ALIANÇA BRASIL, CT0107 P02 a, on 11/10/2015.

And, by the way, whoever is in the middle of a fire (or has a terminal illness) will do anything to survive.

"Joelma building" was on fire. Those people didn't throw themselves out of the building because they wanted to but because it was the only alternative they had to save their lives. If I'm going to burn to death, I might think, "I'm going to throw myself off the twentieth floor, and maybe something will happen, and I'll end up surviving this tragedy." (Senator Omar Aziz, CT01.07 P01 a, on 11/10/2015)

If a person has cancer or a terminal illness, that is a tragedy for their life. Any Brazilian citizen would choose to take tests to try to survive. I think that one should never deny someone an opportunity to survive (Senator Omar Aziz, CT0107 P01 b, on 11/10/2015)

However, from the point of view of 'security,' it is precisely despair that suppresses free and informed consent from people, requiring them to be protected even from themselves.

The primary function of an ethical regulation system in Brazil and globally is to protect research participants, who represent the vulnerable side of research. Participants, often in desperate situations such as a recent cancer diagnosis, focus solely on survival and may not consider whether their rights are fully respected. This inherent imbalance is why ethical regulation systems were established worldwide (Jorge Venâncio, CONEP, CT 0107 P01 c, 12/06/2018).

It is not just about mitigating the right to free consent of clinical research participants but also reconciling conflicting interests of those who want a cure and those who wish to profit from the cure despite the lives lost:

"Sometimes, eager to solve the problem, we have the fox taking care of the hen house. If we ignore details, especially those regarding private autonomy and the manifestation of the will of all sectors, we do run the risk of having the fox take care of the hen house. (Claudio Roberto Camperlingo Araujo, UFBA, CT0103 P01 c, on 11/10/2015).

The conflict between freedom and security shifts to an apparent dilemma between "protecting the participants" or "caring for the patients." It's as if we were faced with "Sophie's choice." Would it be acceptable to sacrifice participants to save patients? Or, to protect participants, will we live as much as necessary with the deaths of patients? (CT01.07 P02).

The dilemma is rejected:

It is perfectly possible to carry out research respecting the rights of the participants and having clear humanitarian criteria so that things are kept to a good standard, as we evolved after Resolution No. 96 that created Conep in 1996 (Jorge Venancio, CONEP, CT0107 P02 b, on 11/10/2015).

But "hurry" reminds regulation that the reasonableness of time for clinical research is also an ethical precept to be protected, including the very regulation.

Ethical control is necessary, and it takes time. The time required to evaluate each protocol ethically cannot counterbalance all the steps guaranteed by the CEP/CONEP System. (Rodrigo Stabelli, FIOCRUZ, CT0107 P03 a, on 11/10/2015).

The ambivalence of artifacts fighting for a regulation definition seems to recalcitrate. Time and ethics need to agree.

Controversy CT01.06: The Need for a Law to Regulate Clinical Trial

Parallel to the unbundling of the relations that support clinical trials and their regulation, the controversy that most interferes with the dispute that appears in the works of PLS 200/2015 (and PL 7082/2017) raises the issue regarding whether a law is essential for clinic research to improve in the country.

In defense of the law, rhetoric took on a characteristic of 'stylization' (which is a protection against the future through the repetition of the proposition to the point it detaches from its author): there is a "legislative gap" in clinical trial" (CT0106 P01), there is a lack of "legal validity" (CT0106 P01 a) to its regulation; "the system is fragile" (CT0106 P01 a); "we have laws for animals but not for people (CT0106 P01 b); "our rights must be enshrined by law and not by resolutions" (CT0106 P01 c); "without law we do not know how long the research will last" (CT0106 P01 d); "CONEP resolutions are authoritarian" (CT0106 P01 e).

The networks against the creation of a law enlisted their arguments: the regulation of clinical research is too dynamic to be regulated by law (CT0106 P01 g); existing gaps in regulation will not be resolved in a law (CT0106 P01 g1); if law is thorough, it will stifle research (CT0106 P01 h); we already have in CONEP resolutions (CT0106 P01 h1) what FIOCRUZ proposes in PLS 200/2015.

Two actants emerge within the discussion of whether a law is essential. The first one regards international standards for the regulation of clinical trials:

"International regulatory instruments for clinical research, such as the Nuremberg Code and the Declaration of Helsinki, were established to protect research participants, who are the most vulnerable in such settings. These regulations emphasize that scientific progress must be balanced with ethical values and human rights, aiming to prevent abuses similar to those committed during the Nazi experiments on Jews and other groups (Senator Otto Alencar, CT 0106 P01 j, 29/06/2016).

These norms would lend credibility to the resolutions created by CONEP once the whole set of infra-legal norms was supposedly issued inspired by international regulation.

"(...) it is essential to highlight that the current system that regulates clinical research with human beings is primarily based on international standards, such as the Nuremberg Code; the Declaration of Human Rights; the Declaration of Helsinki; the United Nations – UN International Agreement on Civil and Political Rights; the International Ethical Guidelines for Biomedical Research Involving Human Beings of the World Health Organization – WHO; and the International Guidelines for the Ethical Review of Epidemiological Studies, among others (Deputy Hiran Gonçalves, CT0106 P01 m, on 10/29/2018).

The second actant is an opinion written by the lawyer Luiz Roberto Barroso, now Minister of the Federal Supreme Court.

In this regard, we reproduce the understanding expressed by jurist Luiz Roberto Barroso, current Minister of the Federal Supreme Court, whose clear speech follows:

The exercise of regulatory power—at least in the terms it is traditionally understood—presupposes a previous formal law that disciplines the matter to be regulated.

In this case, no formal law in Brazil disciplines the ethical evaluation of clinical research with human beings: the only existing normative acts on the matter are resolutions of the National Health Council. In this context, the validity of these resolutions cannot be assessed based on material parameters contained in the law itself since they do not exist.

....

[...] Given the current normative system in the country, CNS resolutions are not competent to create and impose a system of ethical control of clinical research. Such a finding does not disqualify the Board, its importance, and the legitimate motivations that inspire its performance. However, its legislative nature is not aligned with the institutional arrangement in force in the country and with the democratic principle itself. (Senator Otto Alencar, CT 0106 P01 k, on 06/29/2016)

If the first actant (international norms) had the effect of rescuing the dignity of the set of infra-legal norms that developed over the last few decades, the second actant (in Minister Barroso's opinion) held his position that a law is necessary since only a law produced by parliamentarians or representatives of the people (and not by technicians) would ensure the democratic principle in the regulatory choices that should govern clinical research activity, besides providing the propagated "legal security."

Minister Barroso's opinion covered the entire legislative process, almost to the point of 'framing' (a technique described by Latour (1999) that promotes the exposure of the argument to as many people as possible to reduce the intensity of the controversy). Still, a mobilization, in the *Callonian* style (CALLON, 1986), was proposed by the rapporteur of the Science and Technology Committee (CCT), 'Afonso Mota+deputy,' already in the course of PL 7082/2017:

"However, before considering the explanation of the changes proposed by this rapporteur, I emphasize the importance of regulating the issue through a law approved by this Parliament. With the approval of this project, I believe that greater legal certainty will be ensured for those involved in clinical research on human beings. In addition to establishing a series of rights and obligations for those involved, the PL, authored by Senator Ana Amélia, also has the merit of defining the national system of ethics in clinical research with human beings, clearly distributing the attributions of the different instances, thus avoiding overlapping of functions and in a last analysis, making the process of ethical analysis more rational.

While I recognize the importance of establishing rules through law, I also understand that law has limits. The legislator can't glimpse all the issues that permeate the subject, not to mention glimpsing all its vicissitudes. This statement is even more accurate than complex and dynamic topics such as the one under analysis. Therefore, I understand that it is up to the legislator to establish general rules detailing only essential aspects. Through its bodies, the Executive Power will be responsible for detailing via administrative acts what is lawfully established by the Parliament. Therefore, the bill did not include many demands sent to this rapporteur. (Deputy Afonso Mota, CT 0106 P01 l, on 12/14/2017).

Deputy Hiran Gonçalves of the Social Security and Family Committee (CAS) endorsed:

"When creating the new system, it is necessary to establish minimum criteria and premises for its composition and functioning. Therefore, the Law must establish the minimum conditions, and the regulation must deal with its functioning. (Deputy Hiran Gonçalves, CT 0106 P01 n, on 10/29/2018).

The need for a law to affirm democracy and provide legal security to the rules safeguarding the rights and obligations of the parties involved was recognized. However, limits to the law were identified, which cannot foresee all the issues in the clinical research process.

For this reason, a law must establish general rules, and the Executive Power will issue administrative acts through its regulatory agencies to detail the rules in technical and specific matters.

The third type of translation described by Latour (1999) is close, and the opinion of deputies Afonso Mota – approved in full – followed by the endorsement of deputy Hiran Gonçalves, also approved by his peers, seems to have closed the first black box of this process.

However, a historical review of everything CEP/CONEP developed reopens the black box. It resumes the discussion on the relevance of a law for clinical research issued by the National Congress.

'Dirceu Greco+doctor+Brazilian Society of Bioethics' rescues the development process of the regulation that resulted in Resolution CNS 466/2021 and predicts that the law that will result from PL 7082/2017 will put an end to all this issue:

Since the adoption of the Declaration of Helsinki in 1964, which followed the Nuremberg Code of 1947, there has been an ongoing evolution in ethical guidelines, including UNESCO's Universal Declaration of Bioethics and Human Rights. The current legislative proposal (PL 7082/2017) aims to address these historical efforts but faces pressure due to its national implications stemming from international origins. Resolution 466 embodies independence and social control, but enacting a law formalizing these processes is complex and risky. Laws take significant time to adapt, as evidenced by delays since 2015 (Dirceu Bartolomeu Greco, UFMG, CT 01.06 P01 o, 12/06/2018).

Minister Barroso gives his opinion again to hold the critics' rejection towards PL 7082/2017:

The Clinical Research Alliance Brazil, comprising research entities and medical societies such as the Brazilian Medical Association and the Brazilian Society of Clinical Oncology, highlights a significant issue: the lack of a formal law governing the ethical review system in Brazil. Eminent jurist Luiz Roberto Barroso, now a Supreme Court Minister, noted that this absence undermines the system's legal validity by failing to allocate competencies and regulate behavior properly. The CEP/CONEP bodies, while carrying out their roles, must adhere to constitutional principles guiding public administration (Jaderson Sócrates Lima, CT0106 P01 p, 12/06/2018).

However, for the first time, Minister Barroso's opinion is challenged by new relations that change the direction of the actant´s behavior.

The opinion, even supported by the acting "minister," was produced by a lawyer, the same Luiz Barroso mentioned above. Later, after becoming a minister, he would have judged several processes in which the right to medicines was recognized, although not registered in the country. This shows the possibility of attributing rights even if no law exists.

Dr. Barroso's initial opinion as a lawyer highlighted legal fragility in using resolutions, which made sense then. However, numerous court decisions have since relied on these resolutions, supported by the Attorney of the Republic and other legal opinions. Later, as a Supreme Court Minister, Barroso acknowledged during patient judicialization discussions that unregistered medicines could be prescribed by court order, provided there was medical backing and specific conditions were met (Jorge Venâncio, CT0106 P01 h, 12/06/2018).

This is as saying that the opinion that endorses the position of those in favor of PL 7082/2017 is the product of the "lawyer+Barroso" network actor, while the "judge+Barroso" would be more inclined to make relative the legal certainty thesis, benefiting those who consider the current regime to be the more appropriate.

'The need for a law' thesis as a requirement for predictability in the regulatory process and demonstration of the democratic legitimacy of the attribution of rights and obligations to those involved in clinical research was rejected by the 'incompleteness of the law' to rule complex and dynamic situations, which would only be feasible if such attribution was exercised by experts, based on good practices developed by international standards and which inspired the construction of the current regulation of the CEP/CONEP system.

So, the performativity of international norms on current infra-legal acts would be enough to balance the Brazilian regulation that seeks to reconcile safety and agility in clinical research.

However, new actants are recognized in the relations of power that arise from the associations that constitute the CONEP. This Council would have been taken over by political interests that abuse regulatory power, although these same interests would have been justified precisely to confer popular participation in its constitution.

A new controversy arises regarding whether the regulatory legal framework (to be produced by PL 7082/2017) (a) reinforces the democratic principle of establishing an artifact made by those who represent the popular will or (b) suppresses social control by setting stricter rules on who will integrate the CONEP.

The controversy is over the type of democracy, participatory or representative, that one wants to establish in regulating clinical research.

These are, after all, the interests behind the issue regarding whether or not a law is necessary.

More than that, controversy becomes a political debate about who or what will make the compatibility between interests in developing pharmaceutical technology and safeguarding research participants.

Regulation abandons the spectrum of relations that regard it as the "good guy" or the "bad guy" and becomes an agencement supported by network actors who seek to resolve the recalcitrance of their ambivalence to reconcile interests through devices that control the development process of drugs against adverse effects of their use.

Controversies come to discuss not only (a) what dosage of regulation one wants to exert on clinical research but, above all, (b) who will make this decision.

The ethical instance of regulation, headed by the CEP/CONEP system, is reassembled.

7. FINAL CONSIDERATIONS

This study defended the cartography of controversies as a powerful extension of Actor-Network Theory (ANT) to explore complex realities empirically in a methodological way. Through a case study on regulating clinical trials in Brazil, involving legislative debates surrounding PLS 200/2015 and PL 7082/2017, we demonstrated how cartography aids in visualizing regulation as a socio-technical construct formed by heterogeneous networks. These networks operate within clinical research and are dynamically influenced by procedural and performative legislative processes.

By employing controversy mapping, this research identified actants, controversies, and network associations, highlighting that regulation shapes and is shaped by interactions mediated through diverse, symmetrically coexisting human and non—human elements. This dual nature of regulation and its constitutive processes underscores the performative effects within the network, emphasizing the mutual influence between regulatory frameworks and the actants they govern.

As outlined by Venturini (2010a), the methodological contribution of cartography proved effective in capturing the fluid, evolving processes inherent in ANT's procedural ontology. This technique provides actionable insights into how actor networks engage in translations that drive reality's continuous reconfiguration despite being subject to performative constraints. Although applying cartography presents inherent challenges—particularly maintaining fidelity to ANT's foundational principles—it has shown itself as a viable approach for advancing empirical studies of complex, relational phenomena.

Our findings suggest that the application of controversy mapping, even with its operational hurdles, is a valuable tool for scholars aiming to deepen their understanding of socio-technical dynamics. The example of Brazilian clinical trial regulation is a testament to its utility, encouraging further exploration into fluid and relational contexts where traditional methodologies might falter. By integrating this approach, researchers can advance the descriptive power of ANT, offering detailed, nuanced interpretations that resonate with contemporary academic and practical discourses.

This paper advocates for the more widespread use of cartography in research settings that require mapping complex actor networks. Such an expansion will refine methodological practices and ensure that ANT's theoretical principles are translated into robust, empirically grounded explanations.

8. REFERENCES

Alcadipani, R., & Hassard, J. (2010). Actor-Network Theory, organizations and critique: Towards a politics of organizing. *Organization*, *17*(4), 419–435. DOI: 10.1177/1350508410364441

Alcadipani, R., & Tureta, C. (2009). Actor-Network Theory and organizational analysis: Contributions and research possibilities in Brazil. *Organizações & Sociedade*, *16*, 647–664. DOI: 10.1590/S1984-92302009000400003

Bloomfield, B. P., & Vurdubakis, T. (1999). The outer limits: Monsters, actor networks and the writing of displacement. *Organization*, *4*(4), 625–647. DOI: 10.1177/135050849964004

Bloor, D. (1991). *Knowledge and social imagery*. University of Chicago Press.

Butler, J. (1997). *The psychic life of power: Theories in subjection*. Stanford University Press. DOI: 10.1515/9781503616295

Cabantous, L., & Sergi, V. (2018). Seeing the potentialities at the intersection: A reflection on performativity and processuality mindsets. *M@n@gement*, 21(4), 1229–1243.

Callon, M. (1984). Some elements of a sociology of translation: Domestication of the scallops and the fishermen of St Brieuc Bay. *The Sociological Review*, *32*(1, suppl), 196–233. DOI: 10.1111/j.1467-954X.1984.tb00113.x

Callon, M. (1987). Society in the making: The study of technology as a tool for sociological analysis. In Bijker, W. E., Hughes, T. P., & Pinch, T. J. (Eds.), *The social construction of technological systems: New directions in the sociology and history of technology* (pp. 83–103). MIT Press.

Callon, M. (1998). *The laws of the markets*. Sociological Review Monographs Series. Blackwell Publishers.

Callon, M. (1999). Actor-network theory—The market test. *The Sociological Review*, *47*(1, suppl), 181–195. DOI: 10.1111/j.1467-954X.1999.tb03488.x

Callon, M. (2010). Performativity, misfires, and politics. *Journal of Cultural Economics*, *3*(2), 163–169. DOI: 10.1080/17530350.2010.494119

Callon, M., & Latour, B. (1981). They were unscrewing the big Leviathan: How actors macro-structure reality and how sociologists help them. In Knorr-Cetina, K., & Cicourel, A. V. (Eds.), *Advances in social theory and methodology: Toward integrating micro-and macro-sociologies* (pp. 277–303). Routledge & Kegan Paul.

Callon, M., & Rip, A. (1992). Humans, non-humans: Morality of coexistence. In *La Terre Outragée. Les experts sont formels* (pp. 97–112). Autrement.

Camillis, P. K. D., & Antonello, C. S. (2016). From translation to enactment: Contributions of Actor-Network Theory to the processual approach to organizations. *Cadernos EBAPE.BR, 14*(1), 61–82. DOI: 10.1590/1679-395131412

Cerretto, C., & Domenico, S. M. R. D. (2016). Change and Actor-Network Theory: Humans and non-humans in controversies during the implementation of a shared services center. *Cadernos EBAPE.BR, 14*, 83–115. DOI: 10.1590/1679-395131385

Czarniawska, B., & Hernes, T. (2005). *Actor-network theory and organizing*. Liber.

Duarte, M. D. F., & Alcadipani, R. (2016). Contributions of organizing to organizational studies. *Organizações & Sociedade, 23*, 57–72. DOI: 10.1590/1984-9230763

Emirbayer, M. (1997). Manifesto for a relational sociology. *American Journal of Sociology, 103*(2), 281–317. DOI: 10.1086/231209

Gad, C., & Bruun Jensen, C. (2010). On the consequences of post-ANT. *Science, Technology & Human Values, 35*(1), 55–80. DOI: 10.1177/0162243908329567

Gomes, R. D. P., Pimentel, V. P., Landim, A. B., & Pieroni, J. P. (2012). Clinical trials in Brazil: International competitiveness and challenges. *BNDES Sectorial, 36*, 45–84.

Harman, G. (2009). *Prince of networks: Bruno Latour and metaphysics*. Re.press.

Hussenot, A. (2008). Between structuration and translation: An approach to ICT appropriation. *Journal of Organizational Change Management, 21*(3), 335–347. DOI: 10.1108/09534810810874813

Hussenot, A. (2014). Analyzing organization through disagreements: The concept of managerial controversy. *Journal of Organizational Change Management, 27*(3), 373–390. DOI: 10.1108/JOCM-01-2012-0006

Hussenot, A., & Missonier, S. (2010). A deeper understanding of the evolution of the role of the object in the organizational process: The concept of "mediation object.". *Journal of Organizational Change Management, 23*(3), 269–286. DOI: 10.1108/09534811011049608

Kanger, L. (2017). Mapping 'the ANT multiple': A comparative, critical and reflexive analysis. *Journal for the Theory of Social Behaviour, 47*(4), 435–462. DOI: 10.1111/jtsb.12141

Kanger, L. (2017). Mapping 'the ANT multiple': A comparative, critical and reflexive analysis. *Journal for the Theory of Social Behaviour*, *47*(4), 435–462. DOI: 10.1111/jtsb.12141

Laffont, J. J., & Maskin, E. (1982). Theory of incentives: An overview (part 1). In *Advances in Economic Theory* (pp. 31–93). Cambridge University Press.

Langley, A. (1999). Strategies for theorizing from process data. *Academy of Management Review*, *24*(4), 691–710. DOI: 10.2307/259349

Langley, A., & Tsoukas, H. (Eds.). (2016). *The SAGE handbook of process organization studies*. SAGE. DOI: 10.4135/9781473957954

Latour, B. (1984). The powers of association. *The Sociological Review*, *32*(1, suppl), 264–280. DOI: 10.1111/j.1467-954X.1984.tb00115.x

Latour, B. (1993). *The pasteurization of France*. Harvard University Press.

Latour, B. (1994). *We have never been modern*. Editora 34.

Latour, B. (1996). On actor-network theory: A few clarifications. *Soziale Welt*, •••, 369–381.

Latour, B. (2000). *Science in action: How to follow scientists and engineers through society*. Unesp.

Latour, B. (2007). *Reassembling the social: An introduction to actor-network theory*. OUP Oxford.

Latour, B. (2010). The making of law: An ethnography of the Conseil d'Etat. *Polity*.

Latour, B. (2019). *Inquiry into modes of existence: An anthropology of the moderns*. Editora Vozes.

Law, J. (1992). Notes on actor-network theory: Ordering, strategy, and heterogeneity. *Systems Practice*, *5*(4), 379–393. DOI: 10.1007/BF01059830

Law, J. (1999). After ANT: Complexity, naming, and topology. *The Sociological Review*, *47*(1_suppl, S1), 1–14. DOI: 10.1111/j.1467-954X.1999.tb03479.x

Law, J. (2004). *After method: Mess in social science research*. Routledge. DOI: 10.4324/9780203481141

Law, J., Bijker, W. E., Hughes, T. P., & Pinch, T. (2012). Technology and heterogeneous engineering: The case of Portuguese expansion. In *The social construction of technological systems: New directions in the sociology and history of technology* (pp. 105–127). MIT Press.

Law, J., & Hassard, J. (1999). *Actor-network theory and after*. Blackwell.

Mattos, P. T. L. (2006). Decisional autonomy, administrative discretion and legitimacy of the State's regulatory function in Brazilian legal debate. In de Aragão, A. S. (Ed.), *The normative power of regulatory agencies* (pp. 333–364). Forense.

Mitev, N. (2009). In and out of actor-network theory: A necessary but insufficient journey. *Information Technology & People*, *22*(1), 9–25. DOI: 10.1108/09593840910937463

Mol, A. (1999). Ontological politics: A word and some questions. *The Sociological Review*, *47*(1, suppl), 74–89. DOI: 10.1111/j.1467-954X.1999.tb03483.x

Mol, A. (2010). Actor-network theory: Sensitive terms and enduring tensions. *Kölner Zeitschrift für Soziologie und Sozialpsychologie*, *50*(1), 253–269.

Myles, G. D. (1995). *Public economics*. Cambridge University Press. DOI: 10.1017/CBO9781139170949

Nardacchione, G., & Acevedo, M. H. (2013). Pragmatic-pragmatist sociologies tested in Latin America.

Nobre, J. C. D. A., & Pedro, R. M. L. R. (2010). Reflections on methodological possibilities of Actor-Network Theory. *Cadernos UniFOA*, *5*(14), 47–56. DOI: 10.47385/cadunifoa.v5.n14.1018

Orlikowski, W. J., & Scott, S. V. (2008). Sociomateriality: Challenging the separation of technology, work, and organization. *The Academy of Management Annals*, *2*(1), 433–474. DOI: 10.5465/19416520802211644

Redes, A. C. S. (2008). Networks and controversies: Tools for mapping psychosocial dynamics.

Tonelli, D. F. (2016). Origins and epistemological affiliations of Actor-Network Theory: Implications for organizational analysis. *Cadernos EBAPE.BR*, *14*, 377–390. DOI: 10.1590/1679-395141596

Tsoukas, H. (2010). Practice, strategy making, and intentionality: A Heideggerian onto-epistemology for strategy as practice. In Golsorkhi, D., Rouleau, L., Seidl, D., & Vaara, E. (Eds.), *Cambridge Handbook of Strategy as Practice* (pp. 47–62). Cambridge University Press. DOI: 10.1017/CBO9780511777882.004

Tsoukas, H., & Chia, R. (2002). On organizational becoming: Rethinking organizational change. *Organization Science*, *13*(5), 567–582. DOI: 10.1287/orsc.13.5.567.7810

Tureta, C., Américo, B. L., & Clegg, S. (2021). Controversies as a method for ANTi-history: An inquiry into public administration practices. *Organization*, *28*(6), 1018–1035. DOI: 10.1177/13505084211015375

Venturini, T. (2010). Diving in magma: How to explore controversies with actor-network theory. *Public Understanding of Science (Bristol, England)*, *19*(3), 258–273. DOI: 10.1177/0963662509102694

Venturini, T. (2012). Building on faults: How to represent controversies with digital methods. *Public Understanding of Science (Bristol, England)*, *21*(7), 796–812. DOI: 10.1177/0963662510387558 PMID: 23832559

Venturini, T., Ricci, D., Mauri, M., Kimbell, L., & Meunier, A. (2015). Designing controversies and their publics. *Design Issues*, *31*(3), 74–87. DOI: 10.1162/DESI_a_00340

Villar, E. G., & Roglio, K. D. D. (2019). The alignment between Actor-Network Theory and relational sociology: An onto-epistemological discussion for organizational studies. [TPA]. *Theory and Practice in Administration*, *9*(1), 77–91.

Walsham, G. (1997). Actor-network theory and IS research: Current status and future prospects. *Information Systems and Qualitative Research*, 466-480.

Chapter 11
Qualitative Methodology Techniques and Approaches of Business Research in the Digital Era:
Perceptions and Way Forward

Mrigakshi Das
Exeed College, Westford University, India

ABSTRACT

Qualitative research has been focused on accumulating data for providing a realistic view of phenomena under study and form understanding from the perspectives of respondents. Uncertain situations such as COVID-19 pandemic has seen extraordinary social distancing measures thereby becoming a barrier in the data collection process. However, the progression of a IT showed favorable repercussions. Virtual methods have rapidly gained momentum. Limited research comprehensively explores the techniques for qualitative data collection and analysis through digital mode, opportunities and challenges faced. The PRISMA framework was used for presentation. The review was initiated with a scoping search to identify the existing articles. Articles were reviewed deeply to determine if the inclusion conditions were met. Conducting qualitative research through the usage of digital technologies could be perplexing for both parties and the debate on ethics and reliability have found mixed results calling paving the ground for future scope of the research.

DOI: 10.4018/979-8-3373-4432-4.ch011

1. INTRODUCTION

It's wonderful existing in an era when information is readily available at our fingertips. However, this can lead us to contemplate that as if there is no question that has ever been asked and no problem which has not been solved. Whatever may-be the topic sparking our interest, there always appears to be a massive amount of information readily available, and for the majority of us there is never enough time to read. This results in the unescapable feeling that whatever it is we have in mind must have been done before (Annik, 2017; Creswell and Creswell, 2017). The truth is maybe it has been but maybe not in the way we would intend to do it. But that doesn't mean that we are unable to confront it. One of the most popular business research topics is "Impact of leadership on enhancing employee performance". 'What is the impact of leadership style employed on enhancing organizational performance?' is a frequently asked question. The textbooks comprise of generic answers to this question, and even university libraries are full of project reports seeking answers to the same question. However, there are multiple managers in organizations who need to understand the matter in respect of their own organization. For instance, the qualitative approach depends on the kind of intelligence that machines (currently) lack, since phenomenon like positive associations with a brand, management trustworthiness, customer satisfaction, ability of a company to achieve competitive advantage, and cultural shifts are difficult, arguably impossible, to capture with numerical inputs (Basias and Pollalis, 2018). In other words, even the most well researched topic can be explored through numerous and specific situations. The understanding cannot be built based on assumptions or intuitions. This calls for business research.

Research is a procedure having its own language. The Research approaches (or methodologies) are procedures that span the ladders from broad assumptions to comprehensive methods of data collection, analysis, and interpretation. The plan of research involves numerous decisions and serve as the heart for the research process. Enlightening this decision should be grounded in the philosophical assumptions the researcher brings to the study; actions of inquiry (called research designs); and precise methods of data collection, analysis, and interpretation (Saunders and Lewis, 2017; Basias and Pollalis, 2018). The selection of a research approach includes the issue being addressed, the personal experiences of researchers, and the respondents chosen for conducting the study.

Qualitative research methods validate a different approach to scholarly inquiry (Huarang et al., 2018). There is mounting consensus that qualitative research contributes positively to the diversity of academic inquiry in business research, advancing richer and novel insights about the organizational phenomena (Javadian et al., 2020). Qualitative data is beyond numbers and includes motivations, opinions, and feelings. The approach relies on personal observations and perceptions of the respondents that needs to be recorded by taking consent of prior consent, has unique steps in analysis of data, and draws on diverse procedures (Huarang et al., 2018). It's best used when a researcher is seeking to understand the 'why' behind a market trend or wants to explore complex concepts with depth without limiting responses to predefined categories (Crick, 2021). Qualitative techniques in business research has been historically using stories by capturing real life experiences to describe and understand actual conduct or behavior in factual settings (Gephart and Saylors, 2020; Javadian et al., 2020). The heart of qualitative research focuses initially on accumulating data for providing a realistic view of phenomena under study and form understanding from the point of view of respondents or actors forming part of the research study in factual settings. Data is accumulated in qualitative research through observations of the factual phenomena and record the descriptions of observations, often through field notes (Javadian et al., 2020). Established methods for description include participant and non-participant observation, focus groups discussions, individual in-depth interviews, and collection of archival data (Rahman et al., 2021). The qualitative data analysis process spans from looking over the data to generating themes and a story line. The qualitative research methods require researchers to use multiple strategies to validate their findings, and the findings needs to be presented creatively (Crick, 2021; Creswell and Creswell, 2017). The outcomes of qualitative research are narratives of the actor's activities in factual settings, detailed descriptions of their behavior, discoveries of novel phenomena, and creating or testing of established theories.

Uncertain situations such as the inception of the COVID-19 pandemic has seen the enactment of extraordinary social distancing measures, confining social interactions thereby the possibility on conducting face-to-face qualitative research (Rahman et al., 2021; Rashid and Jamil, 2021). However, the progression of information technology and the broad use of smartphones have had favorable repercussions on qualitative research during the pandemic (Rashid and Jamil, 2021). COVID-19 has necessitated innovation in qualitative research as in-person qualitative data collection has been difficult by the restraints of social distancing and the prioritization of safety of the participants and researchers. As a consequence, virtual methods have rapidly gained momentum and the use of video communication technologies at work and at home is increasingly common (De Villiers et al., 2021). In this new digital era, we inhabit and prefer using virtual spaces and future social, economic, and environmental crises may further offshoot change (De Villiers et al., 2021). Due to advancement of technology, there is an increasing trend in qualitative researchers often perform their work through virtual platforms. However, there is a dearth in research that comprehensively explores the techniques and approaches, range of perceptions on the opportunities and challenges faced during conduct on qualitative research virtually with range of interviewees, and how the reliability, validity, and ethical considerations that arise when engaging in virtual qualitative research taken into consideration especially in the context of business research (Rehman et al., 2020; Roberts et al., 2021; Thunberg and Arnell, 2022).

With the growing technological advancement, maybe future researchers might get an option to select the most appropriate data collection tool for themselves without decreasing the quality of the research. However, literature tells that the relationship between these technologies and the various theoretical perspectives of qualitative research remain unexplored (Vivek, 2022). Historically, data analysis software (eg ATLAS, Nvivo) have been rapidly used for analysis of qualitative analysis of business data. However, this further leads to the question that whether data generated through newer technologies could be used for analysis by the existing softwares (Vivek, 2022).

2. RESEARCH QUESTIONS

The landscape of qualitative research has witnessed transformations, evolving from the days of notetaking to leveraging the sophisticated capabilities of technological advancements. This journey has been accompanied with manifold paradigm shifts in way researchers approach, analyze, and deduce qualitative data. With this backdrop, this research is driven by the following questions in the context of business research-

1) What critical developments have taken place in qualitative research techniques employed by business researchers to collect data virtually?
2) What are the numerous approaches utilized by qualitative researchers to encourage the respondents towards a desired answer or outcome in a virtual setting?
3) What are the different ways of ensuring reliability of findings based on the qualitative data collected through virtual techniques?
4) What data analysis tools are being used by qualitative business researchers for analyzing the data collected through virtual modes?

3. RESEARCH METHODOLOGY

The methodology for undertaking the study was initiated with a scoping search to identify the existing articles/publications. The objective of the scoping search was to identify the quantity and quality of the relevant primary studies published in the proposed area of study. The scoping search made it clear that the studies concerning the digital qualitative research techniques are quite broad and consisted of a general search for articles describing rapid data collection and analysis techniques of research catering to several academic fields, such as, medicine, social work, sociology, political science, etc. A total of 121 articles were collected and reviewed. The articles collected were from different databases (such as, Taylor and Francis, EBSCO, Proquest, Wiley, etc) and other general sources (Google Scholar, Academia, Research Gate, etc). The articles were screened by the researcher/author to determine if the articles met the inclusion conditions and to identify any additional terms used to describe virtual/digital data collection and analysis techniques (Bachir et al., 2020; Slablewska and Kubacki, 2023). After the screening of the articles, the duplicate copies of the articles were removed and articles which were considered non-relevant were removed. The articles needed to be peer-reviewed and able to be read in full, as the author emphasized on similarities with in-person qualitative research techniques despite the difference in the medium (Slablewska and Kubacki, 2023). The articles not published in peer reviewed journals, and editorial articles were excluded. Additionally, book review based articles were also excluded as the book reviews are not peer reviewed articles. The articles published in the last decade were considered for the review as digital techniques for qualitative research have gained momentum after the global pandemic. After the screening process, 20 non-relevant articles for this study and the duplicate copies of 9 articles were removed. Each article was reviewed in depth to explain the digital techniques used by researchers for data collection and analysis in research. Further, 13 more articles were removed due to reasons such as, incompleteness in reporting of the findings, published in languages other than English, etc. Therefore, this led to 78 articles that formed part of the systematic review process.

Identification
Screening
Eligibility
Included

Figure 1. PRISMA framework explaining the inclusion and exclusion criteria for the review

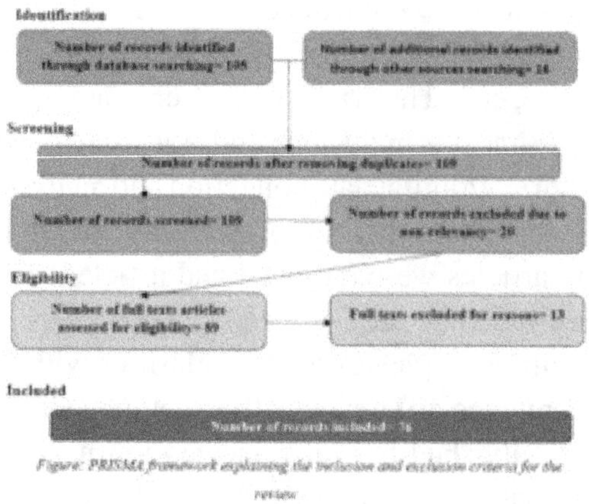

Though a broad scholarly perceptions were gathered, but the general search was too broad to describe the techniques employed by researchers to understand the approaches used especially by the business and management researchers and how the qualitative researchers ensures getting the desired outcome in a virtual setting. Therefore, the search was further refined to enhance understanding on the approaches utilized by researchers through review of business case studies and other publications. The search key words included terms like, usefulness and successful conduct of digital interviews, digital focus group discussions, video conferencing, etc. Further search was also carried on in understanding the how qualitative business researchers ensure reliability and validity of the findings from the data collected through digital modes.

Adding to the perceptions gathered on the data collection techniques/approaches/methods, further review was carried on to understand the transformation in the data analysis techniques accompanying the development in the data collection techniques/approached/methods due to the advent of digital age (Archibald et al., 2019).

4. PRIMARY CONSTITUENTS OF A QUALITATIVE INQUIRY IN BUSINESS RESEARCH

The qualitative inquiry in business research is driven by generating deep understanding and thick description of the phenomenon under study. The qualitative inquiry is driven by the thought processes of the participants. This necessitates a researcher to consciously decide whether the select participants are knowledgeable enough and is having the requisite skills required for undertaking of a research project. In the entire process of qualitative research inquiry, the researchers emphasize on learning and understanding the meaning that the participants embrace about the problem under study, and not the meaning that the researchers bring to the research. Historically, qualitative researchers have collected the data required for getting answers to the phenomenon under study in the field or at the site where participants experience the issue or problem under study. The data is gathered directly by talking directly to research participants and seeing them behave within their context has always remained a major characteristic of qualitative research. In the natural setting, the researchers interact face to face, and these interactions often extends over a prolonged period. Historically one of the grounds of criticism of qualitative research has been dependence on perceptions of participants. Therefore, qualitative researchers typically gather multiple forms of data, such as interviews, observations, archival documents, and audiovisual or social media information (Archibald et al., 2017). Consequently, the researchers review the entire data collected, make sense out of it, and organize it into patterns and themes that cut across all data sources. These are all open-ended forms of data in which the participants share their ideas freely.

The key idea behind qualitative research is to understand a problem or issue from the standpoint of the participants and to address the research to obtain that information. Therefore, as a researcher starts delving deeper into the topic of the phenomenon under study, some of the phases of the process may change or shift after the researcher begins to collect data. For instance, the questions for the interviewees may change, and the initial list of participants studied may change and the organizations from where data was originally planned to be collected may subject to change. Historically, the qualitative research inquiry techniques have been predominantly dependent on the researchers visit to the site or organizations for collecting the data. This leads us to the question is there a shift in the qualitative research techniques in the age which is heavily driven by digital technologies.

5. DEVELOPMENTS IN THE QUALITATIVE RESEARCH TECHNIQUES FOR BUSINESS RESEARCH

Qualitative inquiry has historically been seeking to discover and to describe narratively what specific people do in their day to day lives and what their actions mean to them. It generates meaning of relevant kinds of things in the world, such as, types of people, types of actions, types of beliefs and interests, centering on differences in the forms of things that makes an alteration for meaning. In qualitative inquiry, usually the researcher has always formed part of the environment of those the research is being conducted and responding to the ways in which they carry out their usual activities. Those who agree to be observed are usually called informants/participants (Denzin et al., 2023). Participant Observation technique has been widely used as a method to collect data throughout our everyday lives. However, the level of observation can vary with the requirements of the different research study (Archibald et al., 2019; Saunders et al., 2019).

If research questions and objectives are concerned with what people do and how they interact, an obvious way to collect data is to directly participate, watch and listen to them do it. The directly participant observation technique reflects the dual purpose of the qualitative research method: to take part in the event being studied by participating as a member, and to observe the social world of those in the work group, organization or community. The researcher through his/her experiencing of the event both as an insider and outsider, develops explicit awareness through wide observation being reflective and recording carefully all observations. The observation process as a research technique has its roots in the field of social anthropology studies, but it was the University of Chicago that changed its focus by using ethnographic methods to examine the lives of street gangs in Boston (Saunders et al., 2019). This approach involved researchers living among those whom they studied, to observe and talk to them to produce detailed accounts of their shared beliefs, interactions pattern, behaviors, language, rituals and the events that shaped their lives (Cunliffe 2010; Van Maanen 2011). The high level of researcher immersion in the research setting is a key strength of observation technique as the researcher become a member of the group within which observation is conducted, generating an in-depth understanding on the symbolic world of the informants and their perceptions about their social situation. The interactions happening between the researchers and the participants are unstructured.

Another technique of qualitative inquiry is the structured observation technique. This technique has a long history in the business and management research. Structured observation is uses a high level of predetermined structure and a formalized process, often to quantify what is observed systematically. Such observation involves the systematic viewing, recording, description, analysis and interpretation of people's behavior in a given setting. Structured observation is particularly effective at revealing 'the mundane, routine activities that collectively make up those practices of everyday life' such as the incidence of particular behaviors, interactions or events (Sauders et al., 2019). Structured observation may form only a part of a researcher's data collection process because its functions to find out which things happen, how often and possibly when and in what sequence rather than why they happen. For instance, the technique has been used over several decades to analyses how factory workers carry out their tasks and to measure the times that it took to complete these. Such 'time-and-motion' studies were used by employers to increase their control over the way work was conducted, and reduce the time required to undertake different tasks. Structured observation requires adopting a passive or detached observational role. The researcher taking an active part means the researcher communicating or interacting with people whose behaviour, interactions or activities are under study. The interactions might contaminate the observation by the researcher.

The third technique is the research interview method. It is a meaningful discussion between two or more people, during which the interviewer asks succinct and definite questions and listens attentively to the interviewee's responses. The interviews rely on establishing a rapport between the interviewer and the interviewee with the objective of generating valid and reliable data. The interviewees are selected as a part of the research technique because they have been witness to the phenomenon under study. Therefore, interviews are considered as an effective, unproblematic and reasonably structured means to gather valid data, provided the researchers gets access to the appropriate participants. The interviewer has a central role in this techniques as they ask questions to collect opinions or attitudes, responds to the participant's views, and interprets the resulting data (Saunders et al; 2023). The interviews conducted could be structured, semi-structured, or unstructured in nature. In structured interviews, each participant is asked a predetermined set of questions and in the same tone of voice. Responses to each question are recorded on a standardized schedule, usually with pre-coded answers. In semi-structured interviews, based on the research topic a list of pre-determined themes and initial questions are being used to guide the conduct of each interview. The flexibility is provided to the interviewees to freely express their views on the questions being inquired by the researcher. In-depth interviews are informal and used to explore in depth a general area. The researcher does not use predetermined questions to structure or guide the course of the interview, although the researcher needs to have some idea of the topic, event, experience, or aspect that is wished to explore with participants, giving them the opportunity to talk freely. These interviews are non-directive, the participant guides the interview process. An insight of one's own experiences, motivations, and feelings can be gained through the application of the interview technique most effectively (Roulston, 2024).

The next widely used technique is the focus group discussion. A focus group, sometimes called a 'focus group interview', is a group interview that focuses upon a particular issue, product, service or topic by encouraging discussion among participants and the sharing of perceptions in an open and tolerant environment (Krueger and Casey 2015; Saunders et al., 2023). Participant interaction is a key feature of focus group design. Focus groups discussions comprises of talks that are led by a researcher and take place among a limited number of participants. Such interactions can lead to a highly productive discussion as participants respond to the questions asked by the researcher. The researcher gets to explore collective perspectives and social dynamics provided participation from each member is encouraged.

Through the use of focus groups, researchers are able to study how opinions are formed, challenged, and negotiated within the context of a social setting (Croix et al., 2018). As a result of the group setting, individuals are frequently encouraged to offer ideas that they might not have considered on their own, which results in a wider variety of insights being provided. Focus groups are able to capture the larger social dynamics and group consensus, whilst interviews are able to provide precise individual insights. In order to properly use these methods, it is necessary to engage in cautious preparation, proficient moderation, and a comprehensive comprehension of the strengths and limitations associated with each of them.

6. DEVELOPMENTS IN THE DATA COLLECTION TECHNIQUES IN THE DIGITAL ERA

Evolution of Data Collection Strategies

As the process of collection of the data is the heart of research, in qualitative research the process occupies generally a longer time than other activities in the research timeline. However, phenomenon like COVID-19 pandemic significantly hindered qualitative research activities, putting them on hold. Such unexpected and abrupt phenomenon lead to researchers experiencing difficulties specifically during the collection of data process. The major difficulties have been delayed projects due to the control of movement order levied by the government (Becker et al., 2019; Rashid et al., 2021). However, the progression of information technology and the extensive use of smart devices have had favorable implications on qualitative research during the pandemic. In-person interviews have been observed as the 'gold standard' for qualitative research because interviews pave the path for collecting both verbal and non-verbal cues. However, in-person interviews may not always be possible because of topographical distance, ill health or, as situation like global pandemic that necessitates the need for social distancing.

During the 1990s when people started using computers, synchronous and asynchronous interviews through e-mail or chat forums started getting introduced and used as alternatives. Telephone and email interviews have been considered as an alternative historically, especially for research participants residing geographically far away from the researcher. An email interview is generally described as asynchronous because it is not necessary to ask questions and answer these sequentially without any time gaps. However, it may be possible to conduct an email interview in one period, where the interviewee responds immediately to each question and emails continue to be exchanged until the interviewer draws it to a close and thanks the interviewee for her/his participation. An electronic interview conducted by text-based instant messaging is synchronous because the technology uses real-time transmission. However, there still may be time gaps between the interviewer asking a question and the interviewee providing a response, so it may extend over several periods when both are online. For research topics where the researcher wishes to ask personal or sensitive questions, the anonymity offered by the interviewer and interviewee typing synchronously to each other is likely to produce reliable and useful data. The weakness in the method has been considered as a lack of non-verbal cues such as body language. This has resulted in telephone interviews being viewed as an alternative to, but not a replacement for, in-person interviews (Thunberg and Arnell, 2021). However, there is a lack in evidence on data getting lost or becoming poor due to the lack of non-verbal cues (Cecec et al., 2020; Renosa et al., 2021; Vivek, 2021). However, the risk of non-verbal cues being missed in addition to the missing voice of the interviewee to provide guidance continue to exist (Nair, 2021). However, there is a shift that has been observed especially post the global pandemic phase.

Due to Covid-19, the year 2020 saw enormous changes in the living style of people and mode of comprising of conduct of research. People were bound to learn use digital technology that only a year before were not considered equivalent to in-person options. The usage of digital technology paves the way for incorporation of both audio and visual communication increasing the possibility of reduction of the negative aspects associated with telephone and online text interviews, as both verbal and non-verbal cues can be studied (Fritz and Vandermouse, 2018; Hawkins, 2018; Robert et al., 2021; Santosh et al., 2021; Thunberg and Arnell, 2021; Vivek, 2022). Additionally, the researcher can record both the visual and audio interactions of the interview through simple software (Skype, Zoom Meetings, Microsoft Teams, Google Meet, etc) downloaded onto their computer (Renosa et al., 2021). Based on the researcher's experience, these platforms are user friendly, inexpensive, and equipped with vital security features and the ability to back up recordings unswervingly to the host's cloud (Rashid et al., 2021; Santosh et al., 2021; Sipes et al., 2022). Thus, in the globalized and digitalized world, researchers are having the boon of digital methods as it is making hard to reach people or groups possible to include as participants in research projects. The advantage that most of the studies have raised is due to the ability to reach out to participants in geographical distant or remote areas, this could lead to cutting down of travelling costs and time for the researcher or participants (Santhosh et al., 2021). Traditionally, the travel costs would also have been quite high for the researchers, as they would have had to travel to interview sites from even more remote areas (Santosh et al., 2021). All these costs were reduced by using digital interviewing and focus group discussions. The field journal constructed by the researcher to take note of the researcher's reflections using an online note-taking application (such as Evernote or OneNote). It is significant to set up a system to organize the observations from the commencement of the research.

Labels should include rudimentary information such as the time and place of the reflection. A consistent approach makes analysis stage simpler.

Additionally, another innovative online technique is becoming popular amongst the qualitative business researchers, namely, Google Doc (a commonly used easily available software for online work processing) for facilitating research interviews conducted online (Opara et al., 2023). The researchers highlighted that Google Doc allows researchers to conduct online interviews in written form, removing the need for transcription (Opara et al., 2023). Further, Google Doc does not require the researcher to download any software or create profiles. Google Docs can serve as an alternative to face-to-face interviews allowing the opportunity of reaching out to individuals or groups who are either geographically or culturally apart (Rashid et al., 2021; Opara et al., 2023). Further, Google Docs allows the researcher to observe what the other person is writing while the writing take place paving the ground for psedo-anonymity. The researchers noted that the research participants were able to read, re-read, reflect and rewrite their thoughts/insights on the google document making Google Doc as a preferred choice for sensitive or personal (for instance, abuse, vulnerable groups, violence, corruption, etc), while allowing for appropriate understanding to be constructed between researcher and interviewees (Sipes et al., 2022; Opara et al., 2023). The researchers noted that Google Doc as a digital technique could pave the way for increasing the sample of interviews which otherwise would lead to increased cost and time in case of offline interviews (Opara et al., 2023).

In business and management research, background data is often found in archival documents (e.g. organizational documents and reports, rules and regulations, policy documents, standards, etc.). Researchers have found that employees are more comfortable to undertake video interviews in the accustomed surroundings of their own workplace, probably because of its better privacy and suitability (Crick, 2021). The location provides an opportunity for interviewees to access and refer to files and documents (crucial contextual data) available at their workplaces, which would be challenging in a face-to-face interview in a café or conference room. Additionally, video interviews permit interviewees to "screen share" relevant documents. This convenience could overcome interviewees' hesitancy in emailing complete files and rather allow them to display specific pages. This enhances ability of the researcher to understand information in archived documenets stimulating creation or addition of new interview questions, thereby allowing the researcher to collect more indepth data. However, researchers need experience with video conferencing channels, research topic, and participants.

However, another body of literature have stated that during the interview process, people often use numerous signs and symbols. Due to lack of pre-determined arrangement about the usage of language and jargon in such online interviews, resulting in misunderstanding among participants and researchers posing risk to the study's rigour (Santosh et al., 2021; Sipes et al., 2022; Vivek, 2022). Additionally, the process could be more stressful for elderly or ailing or disabled participants who are mostly unfamiliar with the digital technologies. Further, time lag due to internet problems or any other technical issue can break the flow of conversation (Salmons, 2022: Opara et al., 2023). This could lead to the problem of engaging the participants in the interview process in the digital mode.

Participant engagement in Data Collection in the Digital Era

Thoughtful questioning and meaningful exchange of thoughts are central to the qualitative research (Roberts et al, 2021; Salmons, 2022). Given the fact that the researchers and participants are not acquainted with each other, building rapport could be particularly challenging. Therefore, qualitative researchers should connect with the participants to make them feel comfortable for sharing their problems and feelings throughout the qualitative collection stage. Whether the data is collected through online or offline mode, the interviewer needs to prepare in advance on the contributions by the participants to achieve the purpose of the research (Maher et al., 2018). The researchers have experienced that usage of an interview guide keeps the interviews on track, providing ample information in answering the questions (Rahman et al., 2021; Rashid et al., 2021). The questions asked if progresses from general to specific it paves the ground for participants to respond from their own perspective and becomes more of a conversation or dialogue flow (Thunberg and Arnell, 2021). This has been considered as one of the ideal ways for ensuring participant engagement. There could be instances when the participants do not understand certain questions, those questions need rephrasing.

In the digital mode, the researchers are of the view that the interviewer needs to possess good interviewing skills, including showing professionalism in the conduct, ability to build rapport, being an engaged listener and not interjecting when the participants are sharing their perspectives. Active listening heartens the interviewee for opening up with their experiences (Konecki, 2019; McGranth et al., 2019). Therefore, this necessitates the researcher for being attentive throughout the interview process, and nod with a smile and look at the screen, to demonstrate that the researcher is interested in the interviewee's perspectives. A proactive facilitator or researcher asking follow up questions may promote the quality of dialogues elicited. This keeps the participant motivated to keep sharing their insights (Konecki,2019; Javadian et al., 2021). Additionally, quiet spaces may give digital record of the expected quality of the research interview, but if the interviewees and the researcher are into spaces that are not natural, the content may not be as per the expectation (Saarijarvi and Bratt, 2021).

Reliability and Ethical Considerations during Online Data Collection Process

While digital techniques have made it easier to create bridge between different parts of the globe separated by geographical boundaries, ethics remain pivotal accompanied by unique data privacy and transparency challenges (Amankwaa, 2016; Bryda and Costa, 2023). Digital qualitative techniques introduce challenges in management of data accompanied by vital queries concerning the genuineness and reliability of data obtained through online mode (Ghafouri and Ofoghi, 2016; Rose and Johnson, 2020). Achieving a subtle balance between capitalizing on digital benefits and perpetuation of research integrity necessitates a reflective understanding and implementation of these digital techniques. This requires the researchers to continuously strive for reconciling cutting-edge digital technologies with fundamental research ethics, research principles and qualitative data security (Rose and Johnson, 2020; Riazi et al., 2023). Professionals needs to maintain integrity and stringent ethical standards to ensure reliability of the data collected through online mode.

Qualitative researchers have faced challenges on the ground of unwillingness of certain research participants to share data using online platform on the grounds of lack of trust and disclosure of confidential information (Kang and Hwang, 2021; Roberts et al., 2021; Vivek, 2022). Additionally, virtual methods present privacy concerns, as researchers could be interfering into participants' personal space, especially if participants are not aware of using virtual background or have access to headphones (Sy et al., 2020; Rose and Johnson, 2020; Riazi et al., 2023). This can significantly have an impact on the reliability of the findings from the study. Therefore, the commitment to ongoing learning with the new norms needs to accompany. Professionals needs to stay informed via workshops and webinars sessions as the digital landscape evolves to ensure smooth teamwork, even when researchers and participants are globally dispersed (Kang and Hwang, 2021). In addition to the knowledge on digital tools, digital literacy involves deciding upon which tool would be most suitable for a specific task and evaluation of the credibility of online platform. Professionals needs to be proficient in rereading AI-generated outputs, and identifying potential errors. Understanding of the essentials of cybersecurity is of supreme importance given the increasing cybercrime in the digital-centric world. Professionals should be versed in best practices related to data encryption, secure data storage, and safe data transmission (Bryda and Costa, 2023).

The researcher needs to provide the commitment for protecting the privacy of research participants for ensuring their safety and job security, because "online video/audio interviews by default, include identifiable material" (Kang and Hwang, 2021; Villiers et al, 2021). Ensuring robust digital data security has become crucial in the era of digitization such as strong cybersecurity measures and established techniques for management of data and training of researchers in ethical data handling (Bryda and Costa, 2023). This necessitates that modern professionals must be conversant with various platforms and skillful at leveraging the convoluted functionalities.

Further, the approach of researchers to questioning could minimize the scope for biasness thereby increasing the reliability of the information obtained. The researchers have found that the questions needed to be phrased evidently for the participants to understand them, and needs to be asked in a neutral voice tone (Rashid et al., 2021; Saarijarvia and Bratt, 2021; Salmons, 2022). Further, rapport between the researcher and interviewee can enhance value to the process as it comprises a feeling of comfort and confidence in the interviewee (Khan and MacEachen, 2022). Research if conducted through mode like Google Docs paved the way for adding link to archival data available in online repository which paved the way for triangulation of data at low cost and time (Kakar et al., 2023; Opara et al., 2023). Further, research has revealed that despite of much advancement in the qualitative research through digital techniques, but it may not be suitable for all kinds of sensitive topics (Heath et al., 2018; Thunberg and Arnell, 2021). For instance, for a researcher who is trying to gather perspectives from participants who are generally looked down by the society, the researcher will not always be able to read the visual and non-verbal cues that participants express. The major criticism is whether or not visual cues can be identified. Although more visual cues can probably be read in an in-person setting, but the camera usually only captures the head and upper body. So, certain studies have raised the ethical concerns with regards to visual cues, especially if the researcher has not created a mechanism of determining if the participant is distressed during communication. This could be considered as a violation of ethical norms.

However, certain researchers have found that research conducted on minorities or vulnerable groups such as transgender groups or indigenous communities, interviews through online setting has been a boon as the discomfort of participants in sharing information was lesser and increased the emotional connection between the researcher and participant which served paramount importance. This suggests that the approach of interviewer to ethically convince a participant for soliciting their internal interpretations on a topic is an art. This further needs to be supported by software programs available for conducting online qualitative data collection. It is imperative to choose software following the local and national security regulations. The preferences of interviewees need to be respected when recording, as it might be considered sensitive for participants who has apprehensions for being interviewed virtually. If encryption is needed to safeguard security and confidentiality during the video interview, Microsoft Teams are reliable options if Zoom is not considered safe enough (Johnson et al., 2021; Saarijarvi and Bratt, 2021; Bryda and Costa, 2023; Maia, 2024).

7. DATA ANALYSIS

Qualitative research is employed by a researcher for generating a profound understanding of a given situation providing an array of findings from generating findings to inform practice to providing in-depth descriptions of a practical problem under research (Mertens et al., 2017; Lester et al., 2020; Alam, 2021). The conduct of qualitative research is notable and is dependent upon researcher's ability to do rigorous analysis. Generally, the qualitative data analysis involves a 5 step process of analysis (Fritz and Vandermause, 2018; Akinyode and Khan, 2021; Bingham, 2023). It starts with organizing of data based on the themes generated from the literature review, followed by the process of sorting of the data based upon the research questions or objectives, reading and re-reading of the transcripts to understand the data and generate new themes or ideas emerging out of the narratives, followed by in depth interpretation of the data and finally involves explanation of the data which involves connecting with the relevant theories or develop new theoretical understanding emerging out from the data (Maher et al., 2018; Alam, 2021).

In the first step of organizing of the data collected, the researcher is involved in gathering together of the transcripts, archives or secondary reports, or audiovisual items that are included in the data of the research. The data must be developed and structured in a way that can be accessed rapidly for examination of data (Mertens et al., 2017). This phase lays the foundation for the following the data analysis and answering of the research questions or addressing of the research aim.

The second step is getting comfortable with data already collected by paying attention to the texts in the transcripts, examination or scrutiny of the visuals, and listen to audios multiple times. Once the researcher becomes acquainted with the data collected and conducts the preliminary assessment, the research develops the idea that if the researcher has collected sufficient data for answering the research questions. While the researcher goes through the data collected, it is crucial to pen down the key emerging questions or thoughts or topics that come to mind (Fritz and Vandermause, 2018; Alam, 2021; Creswell and Creswell, 2023). If the researcher feels that there are certain loose ends and more data is required to be collected to fill in the gaps, the researcher needs to collect more data or revisit the same interviewee(s) for seeking answers to the newly emerging gaps or questions (Elliot, 2018; Vears and William, 2022).

The third step comprise of the procedure of breaking down the entire data collected into smaller and manageable chunks that are more relevant to the situation. The process of breaking down of the data into different categories identified in the literature serves as a foundation for the preliminary analysis of data (Elliot, 2018; Scjott et al., 2019). As the researcher interacts with each individual piece of content, labels or codes needs to be assigned as per the themes or categories identified in the literature which can provide a summary of the data. When the data collected is organized and put into different classes, the method is known as axial coding (Elliot, 2018; Lineberg and Korsgard, 2019; Alam, 2021; Vears and William, 2022). This process includes assessing the data according to the theoretical associations between them. With the help of this process of classifying the data, the researcher creates a theoretical framework which considers attention to the ideas that are overarching and linkages that are predominant in the information (Bingham, 2023; Nickmanis, 2024).

The previous step allows the researcher to simplify the data gathered into a narrative that is consistent and clear and in alignment with the theory or literature. The researcher reviews the data that was coded in the third phase, looking for patterns that emerge within and across data sources in order to develop themes. This process is sometimes referred to as pattern coding, a method of collapsing the codes created during open/initial coding to separate the data into fewer analytic concepts (Scottjet et al., 2019; Locke et al., 2022). In the process of pattern coding, some of the inductive codes developed during open coding in phase 3 as well as any pattern codes applied or created throughout this phase may be subsumed, revised, or eliminated altogether (Elliot, 2018; Locke et al., 2022). In this process of identifying patterns and collapsing codes, the researcher can begin to identify themes. Themes are developed by synthesizing the remaining inductive codes into theme statements that describe the patterns seen in the data (Elliot, 2018; Lindren et al., 2020). With the utilization of categories and codes, the researcher has ability to choose the information that serve as the foundation for the ultimate interpretation and analysis.

Assessment of the patterns and themes includes the next phase in the process (Birmingham, 2023). To find the formulation of patterns and themes, enhanced classifications and codes must be used. The data includes different insights and ideas that are assessed. It is fundamental to consider the relevance, prevalence, and association between different categories and codes (Lindren et al., 2020). In addition to this, it is essential to discern patterns and relationships, as well as to recognize complicated patterns and linkages that exist between different categories and topics. It is essential that the researcher has an open mind for allowing and facilitating the development of new themes or areas naturally. In order to identify the underlying themes and patterns that might not be immediately apparent, the researcher needs to give closer attention to the particulars in this entire process that includes higher attention to detail. Understanding the intricacies of qualitative sets of data, such as interview and focus group discussions transcripts and archives and other secondary reports, could present substantial challenge for researchers (Maia, 2024; Nickmanis, 2024).

Interpreting and reporting the findings constitutes the next step in the process. The data presentation needs to be done through thorough investigation into the ways in which the various codes and categories contribute to the overall image. The researcher needs to make an assessment on how each theme relates to the objectives of the research and how they are connected to previously published works of literature, current society trends, or historical influences (Toutkoshian et al., 2022; Vears and Gilliam, 2022). The presentation of the analysis needs to be done through a script that is not only lucid and substantial but also successfully explains the outcomes. A researcher is expected to utilize descriptive language, bits and pieces by citing the participants, and graphic elements in order to provide an explanation of important concepts (Viasmoradi et al., 2016). One of the objectives is to convey the concepts in a way that is both engaging and easy to fully comprehend. Increasing the dependability and validity of the system is the focus of the presentation. To get results that are believable and trustworthy, it is essential to make certain that qualitative research is both valid and reliable (Chandra et al., 2019; Creswell and Creswell, 2023). It is essential to use triangulation to improve the validity of the findings. Through the collection of data from a variety of sources, the utilization of a wide range of research methodologies, and the collaboration with a number of researchers, it improves the conclusions produced. It is crucial to validate the results by receiving input from participants and fellow researchers. Additionally, member checking and peer review are two additional significant considerations that should be taken into account. In addition, another factor to take into account is the evaluation of the researcher's potential for bias (Cole, 2023).

Techniques of Qualitative Data Analysis

Different data analysis techniques for qualitative research have evolved over the years.

Narrative Analysis involves the tales that people tell which serves as the primary subject of narrative analysis. The researcher examines how participants convey their experiences, and self-identities through the medium of words instead of simply assessing the constituent of their words. The researcher tries to delve deeper into the meaning that lies beyond the surface of the narratives. When the researcher is interested in understanding the drivers, motives, and lived experiences of the users, this approach is useful since it provides a more comprehensive explanation (Salmon, 2022). For successfully undertaking the narrative analysis, the researcher needs to have empathy and open-mindedness when examining the stories (Bingham and Witskwosky, 2021; Nickmanis, 2024).

Thematic analysis is another technique of qualitative research which is considered to be flexible. Its primary objective is to identify, scrutinize, and state on repetitive themes or arrays within the data. Thematic analysis allows themes to emerge naturally from the data rather than forcing a predetermined theoretical framework on the study (Braun and Clarke, 2021; Bingham, 2024). It is beneficial for the researcher to use this method when they are interested in examining the length and breadth of findings from the qualitative data without restricting themselves by specific coding or theories from the literature reviewed (Viasmoradi et al., 2016; Braun and Clarke, 2021; Bingham and Witskwosky, 2021). This technique paves the way for more flexibility as it allows the researcher to discover unexpected insights and in building an increasingly comprehensive and circumstantial understanding of the phenomenon as the theme generation is allowed naturally (Lester et al., 2020; Salmona and Kaczynski, 2024). This technique of data analysis enables researchers to uncover themes concerning the complex social and managerial dynamics that shape behaviors, and the unintended consequences of a phenomenon under study (Viasmoradi et al., 2016). The ability to maintain an open mind and involve in iterative processes throughout the data collection and analysis stage is essential in conducting an effective thematic analysis (Bingham and Witskwosky, 2021; Naganathan et al., 2022; Nickmanis, 2024). This allows the researcher to uninterruptedly refine their theoretical framework as new insights come to light (Braun and Clarke, 2021).

Grounded Theory Approach is a rigorous and systematic approach to qualitative research. The objective of this approach is to produce new ideas and conceptual models that are developed from the data itself (Elliot, 2018; Pearse, 2021; Salmona and Kaczynski, 2024). Grounded theory analysis is does not begin with a pre-determined premise or any literature reviewed but reasonably involves the process of creating new theory or theories through the iterative process of data collection, observations and reflections emerging out of the ongoing analysis (Eliott, 2018). When the researcher ventures into an unaccustomed domain or area of work and is endeavoring to acquire a reflective, thoughtful and comprehensive grasp of the phenomenon under study, this technique is enormously advantageous (StPierre and Jackson, 2014; Salmona and Kaczynski, 2024). In the process of letting the theory to cultivate naturally from the data collected, the researcher has the opportunity to discover inimitable insights that challenge the beliefs that have been made previously and paves the ground for new investigations (Bingham and Witskwosky, 2021). The key to doing an operative grounded theory analysis is to dodging the impulse of utilizing existing theoretical frameworks and maintain an openness to unimagined arrays and relations within the data (Lester et al., 2020; Bringham, 2024; Salmona and Kaczynski, 2024).

Content analysis, initially a quantitative technique for identifying patterns in qualitative data, has evolved into a widely used qualitative method (Nickmanis et al., 2023; Nickmanis, 2024). Content Analysis provides the ground for carrying an in-depth assessment of qualitative data that might comprise of graphics, arguments, dialogues, and concepts. Through the process of observing and reading, the key points can be assessed, the key issues can be categorized, and repeated patterns can be identified that can otherwise be missed because of the gaps in the information

(Nickmanis et al., 2023). The researcher has the competence to reveal deep layers through the utilization of this approach adding to an additional vigorous assessment of the phenomenon and helps to enhance the process of comprehending the issue (Vears and William, 2022; Nickmanis, 2024). When the researcher commences the task of content analysis, the theoretical framework is used to provide assistance and guidance in the process. This technique serves as a guide to the researcher through the assistance of the procedure of evaluating meaningful insights from the data (Nickmanis et al., 2023). Through compliance with the rules, the researcher needs to have confidence that the examination or evaluation or analysis of the data is well-organized, comprehensive, and capable of minimize the obscured or masked amounts of the text or data.

Softwares for Qualitative Data Analysis

Qualitative Data Analysis Software packages delivers a range of techniques and tools for providing aid to researchers in a variety of research based activities such as collecting and managing data, transcribing of the recordings, analysis of data, writing and presentation of findings. The three most robust and well-known packages are NVivo, MAXQDA and ATLAS.ti (Estrada and Koolen, 2018; Evers, 2018; Bingham and Witskwosky, 2021; Paulus, 2023). The five activities that can be supported by software for qualitative data analysis comprises of integration of sources of data and analytical approaches; exploration of the content and structure of the data; organization of materials and ideas; reflecting and drawing interpretations on data; and retrieving, reviewing and rethinking ideas about the data.

QDAS packages enables digitalization of different data sources (e.g., audio and video recordings, field notes, documents, social media posts, images, interviews, responses to open-ended survey questions). The software packages also take into consideration other relevant project files related to the research that are originally in electronic format (e.g., ethical board approval forms, journal entries by researcher during data collection, etc). Data and documents that are not already in electronic form can be digitized and then stored and organized in ways that are relevant for enhancing the focus of the research project. A digital flow of work within the QDAS platforms support clearness and teamwork online in ways that the traditional system cannot (Aguinuis and Solarino, 2019; Ronzani et al., 2020; Neidbalski and Slezak, 2022). It is extremely easier to share, review and merge copies of analyzed software files than it is to share tons of highlighted data segments in an attempt to engage in collaborative research.

Data can be generated and maintained within the QDAS platform such as importing of tweets, generation of images and videos and or text data or document which can be easily accessible by the researcher from anywhere across the globe. Transcripts can be created elsewhere, and uploaded to the QDAS platform, and then harmonized with the audio/video originally recorded. After that the researcher can straight away click on the transcript and listen to that part of the recorded data file to make better sense of how something was actually said rather than solely relying typed transcript. Transcripts which are automatically generated via platforms such as Trint or Otter. ai can also be imported and matched with the recordings. QDAS packages support direct coding of audio and video files without even transcribing them.

These platforms provide a variety of writing tools that can be used to interpret the data while reading, or listening, or simply viewing the files. Coding tools can be used to put a tag on the segments of text, or graphics in logically meaningful ways. Once all related segments are marked or categorized with codes, the retrieval can happen together and can be further reviewed for consistency and accuracy. OAT-LAS.ti, NVivo and MAXQDA also has certain features enabling some automated analysis tools (such as sentiment analysis) (Johnson et al., 2021; Naganathan et al., 2022). These additional features provide the researchers to have a birds-eye view of the content of their data collected.

The software also provides the feature of drawing visual representations of the coded data and relationships which can be exported as image files and directly form part in the research reports. QDAS writing tools can be used to combine the coded data into narrative form. The findings are generated alongside the sources of data, interpretations and the sources of literature (Paulus, 2023).

8. DISCUSSION

The medium of communication, way of working and living have transformed with the advent of the global pandemic drastically. In this globalized and digitalized world, the usage of virtual spaces has increased enormously (Archibald et al., 2019). The global pandemic encouraged countries for imposing lockdowns to contain the spread of the virus forcing many to work from home and a substantial segment of the population had to familiarize themselves with video communication technologies. Qualitative researchers have traditionally been dependent on face to face interactions with the research participants, paying visits to the site of the data has been considered vital for ensuring the authenticity and reliability of the findings. However, due to such restrictions the qualitative researchers had to start using digital tools and resume working in the new environment for their data collection.

The current systematic review scrutinized and bought together existing research to identify potentials and boundaries of using digital interviews especially within the business and management disciplines. Generally, there is somewhat of an agreement that digital interviews can be a righteous approach for collecting data when the participants are living geographically away from the researcher and while research is happening on sensitive topics or on vulnerable groups, assuming that ethical issue are reflected upon and handled, and the internet infrastructure is robust enough. This approach influences the ability to get diversity of clues, which allows the collection of more varied and rich data. Ever since the global pandemic, organizations increasingly have been encouraging flexible working, formation of virtual teams and work from home arrangements, enabling the video conferencing/ online meetings the new the norm. Video interviewing such managers could possibly yield prolific results, especially considering the fact that the new norm gives their physical environment less importance to understand the context of their work-life This will also be useful where it is not possible for researchers to physically visit the participant's environment either due to reasons or restrictions. In addition, physically disabled researchers may prefer to conduct video interviews. Furthermore, experience with the channel, research topic, art of communication, and the selection of the research participants are influential in shaping the observations made from the video communication channel.

The research projects where researchers need to take ethical approval for conducting the research will mandatorily have to abide by the requirements which probably might include providing explanations to participants' rights to them, getting the consent of the participants to video record the interviews, and respecting the laws or regulations for protecting the privacy applicable in the jurisdictions of both the researcher and participants. While participants may be unwilling to provide consent to recording of their image or video either could be due to job security issues or trust factor on the digital mode of communication. This necessitates the commitment from researcher to provide protection of the image and privacy of participants for ensuring the safety There has been a consensus on the fact that rapport between researcher and interviewee is crucial to promote better understanding. However, on the contrary for several individuals find a video interview with a stranger challenging. Although video communication technologies are helping people in staying connected, but this review has revealed that researchers experience multiple challenges, due to issues distinctive to this mode of communication, having the possibility of making it more stressful than face-to-face communication. These includes, bigger effort or attentiveness required to understand the non-verbal cues, a little delay in either sending or receiving a message could create a perception of the interviewee or interviewer being less friendly, anxiety during the video calls due to lack of trust in the medium or due to a sentiment of being examined (i.e. being self-conscious) causing stress. Certain individuals may not feel comfortable due to introversion and have less confidence in front of the camera or a stranger. Conducting the interviews in a group (panel or a focus group discussion) may help in disabling this anxiety, due to the presence of the rest of the group members and could normalize the switching on of cameras. However, the researcher needs to keep a check or oversee how participants take turns to speak. Software, such as MS Teams, offers all the participants the option of virtually raising their hand, but this needs to be explained to everyone so that each participant is aware about the protocol to be followed at the outset. This necessitates the parties have digital literacy. During the conversation the interviewer could occasionally nod their head, give a smile and use different and appropriate expressions and provide tokens of acknowledgement like 'hmm', 'yep', 'that's interesting' to let the interviewee know that the interviewer is paying attention to what they are saying. This also could lead to an impression that the interviewer can clearly listen and understand the thoughts or perspectives of the interviewee and also could give an impression that the interviewer is enjoying the conversation. Therefore, using a diversity of cues, personalization, and language variety represent useful techniques in maintaining rapport with our interviewees.

The digital mode of communication or research is still developing its norms, and needs to be considered as something unique than in-person interviews. The richness of data, as well as ethical issues can be connected to the possibility for the interviewer and the interviewee to see and hear each other appropriately, and also establish their own norms and rules for the communication. Thus, it is significant to provide superior ethical consideration to each research project, as the digital medium raises ethical concerns that might not be an issue in in-person interviews. However, looking at the quality of the data, there seems to be not much of a difference between the in-person interviews and interviews conducted using digital mode. As long as the technology works as planned, conducting interviews with audio/visual software seems to be a good option. Additionally, audio/visual digital interviews are quite new, but are steadily increasing with the development of computer software. It is pertinent to mention that this article has achieved its primary objective by opening new windows that researchers could think of in their subsequent qualitative research. Concerning data quality, the studies show different results, with some arguing that in-person options are still to be preferred, and others that digital interviews are just as good and in some respects even better. Technical problems are mentioned, and although the software has improved over time and become more user-friendly, it is not uncommon to experience problems related to the technology, such as disconnections or lags, which is also confirmed.

9. CONCLUSION

Online interviews may be of benefit to academicians or researchers in terms of convenience and viability. However, conducting a qualitative research through the usage of digital technologies could be perplexing for both parties: the researcher and the participants. The challenges include poor internet connection, lack of funds to purchase expensive software, digital illiteracy or inability to work comfortably on an online platform, participants' inability and difficulty to use the applications, and limited visual cues that may affect the quality of data obtained. Therefore, researchers need to consider several aspects during the selection of appropriate interview modalities including ethical issues. Researchers are required to possess adequate interviewing skills, needs to be good listener, should be able to accommodate with the needs of participation and understand the same, and should be proactive in initiating a back-up plan for smooth running of the process. Researchers have observed that there's not much difference in the analysis of qualitative data collected through either offline or digital modes. However, the researchers' experiences provide the insight that into the prominence of digital literacy and people's willingness to use information technology when providing insights matters. There is still much debate about the degree to which modern techniques can produce qualitative analysis or purely contribute to its development by human researchers.

10. MANAGERIAL IMPLICATIONS

The study seeks to serve as a guide for business and management researchers, managers, practitioners who seeks to use the qualitative techniques for conducting in-depth inquires on business phenomenon needing deep investigation. The study would serve as a guide for the researchers in choosing between numerous techniques available, to make the adequate preparations needed for conducting the data collection virtually which would best serve the purpose of the study. The researchers would also develop understanding on the approaches for keeping the research participants motivated to stick with the ongoing enacted study and resist the urge of clicking on the window closed when some discomfort arises. Additionally, the proposed chapter would serve as a guide for the business researchers to ensure the reliability and validity of the qualitative data collected using virtual mode and the developments in the qualitative analysis techniques in the digital age for interpreting qualitative data collected through digital mode.

REFERENCES

Aguinis, H., & Solarino, A. M. (2019). Transparency and replicability in qualitative research: The case of interviews with elite informants. *Strategic Management Journal, 40*(8), 1291–1315. DOI: 10.1002/smj.3015

Akinyode, B. F., & Khan, T. H. (2018). Step by step approach for qualitative data analysis. *International Journal of built environment and sustainability, 5*(3).

Alam, M. K. (2021). A systematic qualitative case study: Questions, data collection, NVivo analysis and saturation. *Qualitative Research in Organizations and Management, 16*(1), 1–31. DOI: 10.1108/QROM-09-2019-1825

Amankwaa, L. (2016). Creating protocols for trustworthiness in qualitative research. *Journal of Cultural Diversity, 23*(3). PMID: 29694754

Annink, A. (2017). Using the research journal during qualitative data collection in a cross-cultural context. *Entrepreneurship Research Journal, 7*(1), 20150063. DOI: 10.1515/erj-2015-0063

Archibald, M. M., Ambagtsheer, R. C., Casey, M. G., & Lawless, M. (2019). Using zoom videoconferencing for qualitative data collection: Perceptions and experiences of researchers and participants. *International Journal of Qualitative Methods, 18*, 1609406919874596. DOI: 10.1177/1609406919874596

Bashir, M., Naqshbandi, M. M., & Farooq, R. (2020). Business model innovation: A systematic review and future research directions. *International Journal of Innovation Science, 12*(4), 457–476. DOI: 10.1108/IJIS-06-2020-0081

Basias, N., & Pollalis, Y. (2018). Quantitative and qualitative research in business & technology: Justifying a suitable research methodology. *Review of Integrative Business and Economics Research, 7*, 91–105.

Becker, K. M. (2019). Beyond researcher as instrument: Researcher with instrument: musicking in qualitative data collection. *Qualitative Research Journal, 19*(4), 426–437. DOI: 10.1108/QRJ-02-2019-0021

Bingham, A. J. (2023). From data management to actionable findings: A five-phase process of qualitative data analysis. *International Journal of Qualitative Methods, 22*, 16094069231183620. DOI: 10.1177/16094069231183620

Bingham, A. J., & Witkowsky, P. (2021). Deductive and inductive approaches to qualitative data analysis. *Analyzing and interpreting qualitative data: After the interview, 1*, 133-146.

Braun, V., & Clarke, V. (2021). Can I use TA? Should I use TA? Should I not use TA? Comparing reflexive thematic analysis and other pattern-based qualitative analytic approaches. *Counselling & Psychotherapy Research, 21*(1), 37–47. DOI: 10.1002/capr.12360

Braun, V., & Clarke, V. (2021). One size fits all? What counts as quality practice in (reflexive) thematic analysis? *Qualitative Research in Psychology, 18*(3), 328–352. DOI: 10.1080/14780887.2020.1769238

Bryda, G., & Costa, A. P. (2023). Qualitative research in digital era: Innovations, methodologies and collaborations. *Social Sciences (Basel, Switzerland), 12*(10), 570. DOI: 10.3390/socsci12100570

Cecez-Kecmanovic, D., Davison, R. M., Fernandez, W., Finnegan, P., Pan, S. L., & Sarker, S. (2020). Advancing qualitative IS research methodologies: Expanding horizons and seeking new paths. *Journal of the Association for Information Systems, 21*(1), 1. DOI: 10.17705/1jais.00599

Chandra, Y., Shang, L., Chandra, Y., & Shang, L. (2019). Computer-assisted qualitative research: An overview. *Qualitative Research Using R: A Systematic Approach*, 21-31.

Cole, R. (2023). Inter-rater reliability methods in qualitative case study research. *Sociological Methods & Research*, •••, 00491241231156971.

Creswell, J. W., & Creswell, J. D. (2017). *Research design: Qualitative, quantitative, and mixed methods approaches*. Sage publications.

Crick, J. M. (2021). Qualitative research in marketing: What can academics do better? *Journal of Strategic Marketing, 29*(5), 390–429. DOI: 10.1080/0965254X.2020.1743738

Cunliffe, A. L. (2010). Retelling tales of the field: In search of organizational ethnography 20 years on. *Organizational Research Methods, 13*(2), 224–239. DOI: 10.1177/1094428109340041

Cypress, B. S. (2019). Data analysis software in qualitative research: Preconceptions, expectations, and adoption. *Dimensions of Critical Care Nursing, 38*(4), 213–220. DOI: 10.1097/DCC.0000000000000363 PMID: 31145168

De Villiers, C., Farooq, M. B., & Molinari, M. (2022). Qualitative research interviews using online video technology–challenges and opportunities. *Meditari Accountancy Research, 30*(6), 1764–1782. DOI: 10.1108/MEDAR-03-2021-1252

Denzin, N. K., Lincoln, Y. S., Giardina, M. D., & Cannella, G. S. (Eds.). (2023). *The Sage handbook of qualitative research*. Sage publications.

Elliott, V. (2018). Thinking about the coding process in qualitative data analysis. *The Qualitative Report, 23*(11). Advance online publication. DOI: 10.46743/2160-3715/2018.3560

Estrada, L. M., & Koolen, M. (2018). Audiovisual media annotation using qualitative data analysis software: A comparative analysis. *The Qualitative Report, 23*(13), 40–60.

Evers, J. C. (2018). Current issues in qualitative data analysis software (QDAS): A user and developer perspective. *The Qualitative Report, 23*(13), 61–73. DOI: 10.46743/2160-3715/2018.3205

Fritz, R. L., & Vandermause, R. (2018). Data collection via in-depth email interviewing: Lessons from the field. *Qualitative Health Research, 28*(10), 1640–1649. DOI: 10.1177/1049732316689067 PMID: 29298576

Gephart, R. P., & Saylors, R. (2020). Qualitative designs and methodologies for business, management, and organizational research. In *Oxford research Encyclopedia of Business and management*. DOI: 10.1093/acrefore/9780190224851.013.230

Ghafouri, R., & Ofoghi, S. (2016). Trustworth and rigor in qualitative research. *International Journal of Advanced Biotechnology and Research, 7*(4), 1914–1922.

Hawkins, J. E. (2018). The practical utility and suitability of email interviews in qualitative research. *The Qualitative Report, 23*(2). Advance online publication. DOI: 10.46743/2160-3715/2018.3266

Heath, J., Williamson, H., Williams, L., & Harcourt, D. (2018). "It's just more personal": Using multiple methods of qualitative data collection to facilitate participation in research focusing on sensitive subjects. *Applied Nursing Research, 43*, 30–35. DOI: 10.1016/j.apnr.2018.06.015 PMID: 30220360

Huarng, K. H., Rey-Martí, A., & Miquel-Romero, M. J. (2018). Quantitative and qualitative comparative analysis in business. *Journal of Business Research, 89*, 171–174. DOI: 10.1016/j.jbusres.2018.02.032

Javadian, G., Dobratz, C., Gupta, A., Gupta, V. K., & Martin, J. A. (2020). Qualitative research in entrepreneurship studies: A state-of-science. *The Journal of Entrepreneurship, 29*(2), 223–258. DOI: 10.1177/0971355720930564

Javadian, G., Dobratz, C., Gupta, A., Gupta, V. K., & Martin, J. A. (2020). Qualitative research in entrepreneurship studies: A state-of-science. *The Journal of Entrepreneurship, 29*(2), 223–258. DOI: 10.1177/0971355720930564

Johnson, D. R., Scheitle, C. P., & Ecklund, E. H. (2021). Beyond the in-person interview? How interview quality varies across in-person, telephone, and Skype interviews. *Social Science Computer Review*, *39*(6), 1142–1158. DOI: 10.1177/0894439319893612

Kakar, Z. U. H., Rasheed, R., Rashid, A., & Akhter, S. (2023). Criteria for assessing and ensuring the trustworthiness in qualitative research.

Kang, E., & Hwang, H. J. (2021). Ethical conducts in qualitative research methodology: Participant observation and interview process. *Journal of Research and Publication Ethics*, *2*(2), 5–10.

Khan, T. H., & MacEachen, E. (2022). An alternative method of interviewing: Critical reflections on videoconference interviews for qualitative data collection. *International Journal of Qualitative Methods*, *21*, 16094069221090063. DOI: 10.1177/16094069221090063

Konecki, K. T. (2019). Creative thinking in qualitative research and analysis. *Qualitative Sociology Review*, *15*(3), 6–25. DOI: 10.18778/1733-8077.15.3.01

Lester, J. N., Cho, Y., & Lochmiller, C. R. (2020). Learning to do qualitative data analysis: A starting point. *Human Resource Development Review*, *19*(1), 94–106. DOI: 10.1177/1534484320903890

Lindgren, B. M., Lundman, B., & Graneheim, U. H. (2020). Abstraction and interpretation during the qualitative content analysis process. *International Journal of Nursing Studies*, *108*, 103632. DOI: 10.1016/j.ijnurstu.2020.103632 PMID: 32505813

Linneberg, M. S., & Korsgaard, S. (2019). Coding qualitative data: A synthesis guiding the novice. *Qualitative Research Journal*, *19*(3), 259–270. DOI: 10.1108/QRJ-12-2018-0012

Locke, K., Feldman, M., & Golden-Biddle, K. (2022). Coding practices and iterativity: Beyond templates for analyzing qualitative data. *Organizational Research Methods*, *25*(2), 262–284. DOI: 10.1177/1094428120948600

Maher, C., Hadfield, M., Hutchings, M., & De Eyto, A. (2018). Ensuring rigor in qualitative data analysis: A design research approach to coding combining NVivo with traditional material methods. *International Journal of Qualitative Methods*, *17*(1), 1609406918786362. DOI: 10.1177/1609406918786362

Maia, I. (2024). How to overcome the secretiveness of a group: Opportunities of online interviews. *Sociological Research Online*, *29*(1), 171–183. DOI: 10.1177/13607804231168253

McGrath, C., Palmgren, P. J., & Liljedahl, M. (2019). Twelve tips for conducting qualitative research interviews. *Medical Teacher*, *41*(9), 1002–1006. DOI: 10.1080/0142159X.2018.1497149 PMID: 30261797

Mertens, E., Heylighen, A., Declercq, A., Hannes, K., Truyen, F., Denier, Y., & Dierckx de Casterlé, B. (2017). QualiBuddy: An online tool to improve research skills in qualitative data analysis. *Qualitative Research Journal*, *17*(4), 306–318. DOI: 10.1108/QRJ-06-2016-0034

Naganathan, G., Srikanthan, S., Balachandran, A., Gladdy, A., & Shanmuganathan, V. (2022). Collaborative zoom coding—A novel approach to qualitative analysis. *International Journal of Qualitative Methods*, *21*, 16094069221075862. DOI: 10.1177/16094069221075862

Nair, L. B. (2021). From 'whodunit'to 'how': Detective stories and auditability in qualitative business ethics research. *Journal of Business Ethics*, *172*(2), 195–209. DOI: 10.1007/s10551-020-04479-4

Nicmanis, M. (2024). Reflexive Content Analysis: An approach to qualitative data analysis, reduction, and description. *International Journal of Qualitative Methods*, *23*, 16094069241236603. DOI: 10.1177/16094069241236603

Nicmanis, M., Chur-Hansen, A., & Linehan, K. (2023). The information needs and experiences of people living with cardiac implantable electronic devices: Qualitative content analysis of Reddit posts. *JMIR Cardio*, *7*, e46296. DOI: 10.2196/46296 PMID: 37766632

Niedbalski, J., & Ślęzak, I. (2022). Encounters with CAQDAS: Advice for beginner users of computer software for qualitative research. *The Qualitative Report*, *4*, 22–2022. DOI: 10.46743/2160-3715/2022.4770

Opara, V., Spangsdorf, S., & Ryan, M. K. (2023). Reflecting on the use of Google Docs for online interviews: Innovation in qualitative data collection. *Qualitative Research*, *23*(3), 561–578. DOI: 10.1177/14687941211045192

Paulus, T. M. (2023). Using qualitative data analysis software to support digital research workflows. *Human Resource Development Review*, *22*(1), 139–148. DOI: 10.1177/15344843221138381

Pearse, N. (2021). Guidelines for theory development using qualitative research approaches. *Electronic Journal of Business Research Methods*, *19*(2), 95–103. DOI: 10.34190/ejbrm.19.2.2512

Rahman, S. A., Tuckerman, L., Vorley, T., & Gherhes, C. (2021). Resilient research in the field: Insights and lessons from adapting qualitative research projects during the COVID-19 pandemic. *International Journal of Qualitative Methods, 20,* 16094069211016106. DOI: 10.1177/16094069211016106

Reñosa, M. D. C., Mwamba, C., Meghani, A., West, N. S., Hariyani, S., Ddaaki, W., Sharma, A., Beres, L. K., & McMahon, S. (2021). Selfie consents, remote rapport, and Zoom debriefings: Collecting qualitative data amid a pandemic in four resource-constrained settings. *BMJ Global Health, 6*(1), e004193. DOI: 10.1136/bmjgh-2020-004193 PMID: 33419929

Riazi, A. M., Rezvani, R., & Ghanbar, H. (2023). Trustworthiness in L2 writing research: A review and analysis of qualitative articles in the Journal of Second Language Writing. *Research Methods in Applied Linguistics, 2*(3), 100065. DOI: 10.1016/j.rmal.2023.100065

Roberts, J. K., Pavlakis, A. E., & Richards, M. P. (2021). It's more complicated than it seems: Virtual qualitative research in the COVID-19 era. *International Journal of Qualitative Methods, 20,* 16094069211002959. DOI: 10.1177/16094069211002959

Ronzani, C. M., da Costa, P. R., da Silva, L. F., Pigola, A., & de Paiva, E. M. (2020). Qualitative methods of analysis: An example of Atlas. TI™ Software usage. *Revista Gestão & Tecnologia, 20*(4), 284–311. DOI: 10.20397/2177-6652/2020.v20i4.1994

Rose, J., & Johnson, C. W. (2020). Contextualizing reliability and validity in qualitative research: Toward more rigorous and trustworthy qualitative social science in leisure research. *Journal of Leisure Research, 51*(4), 432–451. DOI: 10.1080/00222216.2020.1722042

Saarijärvi, M., & Bratt, E. L. (2021). When face-to-face interviews are not possible: tips and tricks for video, telephone, online chat, and email interviews in qualitative research.

Salmona, M., & Kaczynski, D. (2024). Qualitative data analysis strategies. In *How to conduct qualitative research in finance* (pp. 80–96). Edward Elgar Publishing. DOI: 10.4337/9781803927008.00012

Salmons, J. (2021). *Doing qualitative research online*. Sage Publications.

Santhosh, L., Rojas, J. C., & Lyons, P. G. (2021). Zooming into Focus Groups: Strategies for Qualitative Research in the Era of Social. *ATS Scholar, 2*(2), 176–184. DOI: 10.34197/ats-scholar.2020-0127PS PMID: 34409412

Saunders, M., & Lewis, P. (2017). *Doing research in business and management.* Pearson.

Saunders, M., Lewis, P., & Adrian, T. (2019). *Qualitative Research in Business*. Pearson.

Shree, D., Singh, R. K., Paul, J., Hao, A., & Xu, S. (2021). Digital platforms for business-to-business markets: A systematic review and future research agenda. *Journal of Business Research, 137*, 354–365. DOI: 10.1016/j.jbusres.2021.08.031

Sipes, J. B., Roberts, L. D., & Mullan, B. (2022). Voice-only Skype for use in researching sensitive topics: A research note. *Qualitative Research in Psychology, 19*(1), 204–220. DOI: 10.1080/14780887.2019.1577518

Skjott Linneberg, M., & Korsgaard, S. (2019). Coding qualitative data: A synthesis guiding the novice. *Qualitative Research Journal, 19*(3), 259–270. DOI: 10.1108/QRJ-12-2018-0012

St. Pierre, E. A., & Jackson, A. Y. (2014). Qualitative data analysis after coding. *Qualitative Inquiry, 20*(6), 715–719. DOI: 10.1177/1077800414532435

Sy, M., O'Leary, N., Nagraj, S., El-Awaisi, A., O'Carroll, V., & Xyrichis, A. (2020). Doing interprofessional research in the COVID-19 era. *Discussion Paper (Deutsches Institut für Entwicklungspolitik)*. PMID: 32718262

Szablewska, N., & Kubacki, K. (2023). Empirical business research on modern slavery in supply chains: A systematic review. *Journal of Business Research, 164*, 113988. DOI: 10.1016/j.jbusres.2023.113988

Thunberg, S., & Arnell, L. (2022). Pioneering the use of technologies in qualitative research–A research review of the use of digital interviews. *International Journal of Social Research Methodology, 25*(6), 757–768. DOI: 10.1080/13645579.2021.1935565

Toutkoushian, R., Riffe, K., Sanford, P., & Ness, E. (2022). How do faculty select between defined benefit and defined contribution retirement plans? A qualitative investigation. *Journal of Education Finance, 47*(3), 250–274. DOI: 10.1353/jef.2022.a852709

Vaismoradi, M., Jones, J., Turunen, H., & Snelgrove, S. (2016). Theme development in qualitative content analysis and thematic analysis.

Van Maanen, J. (2011). *Tales of the field: On writing ethnography*. University of Chicago Press. DOI: 10.7208/chicago/9780226849638.001.0001

Vears, D. F., & Gillam, L. (2022). Inductive content analysis: A guide for beginning qualitative researchers. Focus on Health Professional Education: A Multi-Disciplinary Journal, 23(1),111–127.

Vindrola-Padros, C., & Johnson, G. A. (2020). Rapid techniques in qualitative research: A critical review of the literature. *Qualitative Health Research*, *30*(10), 1596–1604. DOI: 10.1177/1049732320921835 PMID: 32667277

Vivek, R. (2022). Systematic review of modern techniques in qualitative research. *Management and Entrepreneurship: Trends of Development*, *4*(22), 34–49. DOI: 10.26661/2522-1566/2022-4/22-03

ADDITIONAL READING

Abdul Rashid, N. A., Lee, K., & Jamil, N. A. (2021). Conducting qualitative research in the new norms: Are we ready? *Nursing & Health Sciences*, *23*(4), 967–973. DOI: 10.1111/nhs.12872 PMID: 34370379

Chapter 12
Content Analysis, Triangulation of Results Through a Protocol Based on the Combination of the Pre-Analysis Material Exploration and the Coding Cycles

Adriano Pinto Teixeira
https://orcid.org/0000-0001-6378-5018
Mackenzie Presbyterian University, Brazil

Claudio Parisi
Mackenzie Presbyterian University, Brazil

ABSTRACT

The research is explanatory, using the "Engaged Scholarship" framework (Van de Ven, 2007). Data collection was divided into evidence from primary sources, obtained through semi-structured interviews and focus group sessions with experts from Brazil's National Financial System (NFS), and secondary sources, based on documentary analysis. The data was processed using the Atlas.ti computer tool (Friese, 2014), applying a protocol based on the proposals of Saldaña (2016) and Bardin (2016), triangulating the results (Yin, 2014). The analysis of the results, aided by automated Atlas.ti reports, was carried out based on the content analysis of each data source in relation to the assumptions of the study's theoretical approach. The

DOI: 10.4018/979-8-3373-4432-4.ch012

methodological platform applied in the thesis proved to be innovative due to the use of the Engaged Scholarship approach, supported by the suggestion of a protocol for analyzing and triangulating evidence, with the aid of a computer platform.

INTRODUCTION

On October 9, 2019, Mackenzie Presbyterian University launched the professional doctorate program in controllership and finance, the first in this modality and theme in Brazil, with the aim of training professionals to seek solutions to complex problems in organizations, supported by scientific methods, bringing the practical closer to academia (Mackenzie, 2019). And in this program, doctoral research was carried out by Teixeira (2023) whose methodological choice was the qualitative approach. The study made theoretical, methodological and practical contributions, with emphasis on updating the framework "Institutional Contradictions, Praxis, and Institutional Change: A Dialectical Perspective" by Seo and Creed (2002), considering the influence of conflicting logics (Thornton, Ocasio, & Lounsbury, 2012) in organizational fields (DiMaggio & Powell, 1983); the empirical finding of the effectiveness of applying the Engaged Scholarship; as well as the creation of a protocol for content analysis combining the proposals of Bardin (2016) and Saldaña (2016), supported by the "Atlas.ti" application (Friese, 2014). A conclusive technical report and recommendations were drawn up for the Basel Committee (BCBS), the Central Bank of Brazil (CBB) and the banks, represented by the Brazilian Federation of Banks (FEBRABAN), in order to encourage alignment between the interests of the supervisory body and the financial industry.

BACKGROUND

On the one hand, they work to meet the interests of their stakeholders, who demand efficiency and a return on their investments, and on the other, they need to comply with the determinations of regulatory and supervisory bodies, setting aside substantial amounts of capital to withstand possible losses related to credit, market and operational risks. In this challenging scenario, banks can choose to use models adjusted to the profile of the institutions (internal models - IM) because they are developed internally or follow the conservative rules standardized by the CBB for calculating capital requirements. Regulators and supervisors are also in a delicate position, where conservative measures impact the supply of credit and the country's economic growth, and the softening of measures can affect the resilience of the National Financial System (NFS) and cause catastrophic economic effects, such as the consequences of the 2008 global crisis (Ramirez, 2017).

The contradictory environment is formed when decision-makers have divergent interests, opening up the space for the formation of institutional logics in organizational fields, directly affecting the institutionalization of new practices. The process of institutionalizing internal models, or choosing to use the parameters standardized by the CBB, is a decision-making environment made up of at least two logics: the market logic, represented by the banks, and the regulatory logic of the supervisory body itself. The market logic supports and justifies the creation of more competitive and efficient financial services (Lounsbury, 2002), in which the use of IM would lead to better capital management. The regulatory logic, on the other hand, aims to protect investors and the NFS, and reflects the imposition of regulatory restrictions, such as the requirement for minimum requirements to authorize the use of internal risk assessment approaches, aimed at limiting leverage in order to contain financial exposure, as well as monitoring and comparing banks' ability to withstand possible liquidity crises in market stress scenarios (Lounsbury, 2002).

In this chapter, we will demonstrate the methodology applied in the study carried out by Teixeira (2023), which aimed to analyze how conflicting logics and institutional contradictions influence the application for the use of internal models to determine the capital requirements of banks in Brazil, and the research question "How do conflicting logics and institutional contradictions influence the application for the use of internal models to determine the capital requirements of banks in Brazil?"

The research is explanatory, using the "Engaged Scholarship" framework by Van de Ven (2007), which proposes a method for studying complex social problems that exceed our ability to carry them out individually. For the analysis, the author proposed the creation of a protocol based on the proposals of Saldaña (2016) and Bardin (2016), connecting the stages of content analysis and the coding cycles. Finally, the researcher used the computer support of Atlas.ti (Friese, 2014), which catalogs documents, quotes, video transcripts, remote meetings (Microsoft Teams), notes, memos, calls, in order to assist in the content analysis.

ENGAGED SCHOLARSHIP MODEL

The "engaged scholarship" model is divided into four steps: problem formulation, theory construction, research design and problem solution (Van de Ven, 2007). In Figure 1, the author summarized in the "diamand model" how the elements of the methodology can increase knowledge and understanding of the phenomenon being studied by engaging other participants who may be relevant in each of the research activities. In the context of the study, FEBRABAN plays a relevant role in the interaction and engagement of the banks' risk specialists with the researcher, especially during the intervention in the field through the various data collection techniques (semi-structured interviews, focus groups and document analysis).

Figure 1. Engaged Scholarship - diamond model

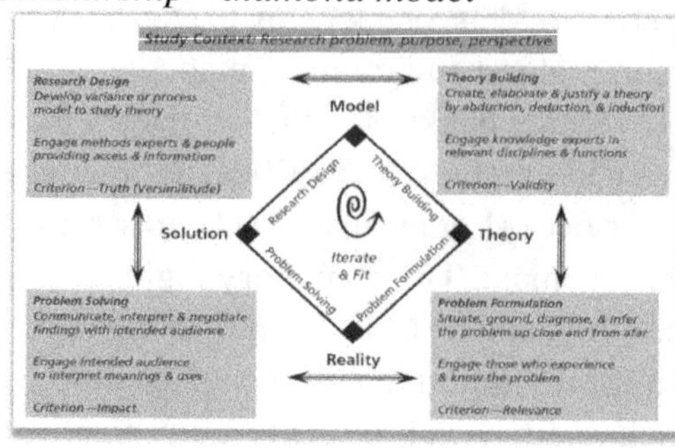

Note. *Source: Van de Ven (2007, p. 10)*

Problem formulation

According to Van de Ven (2007), the formulation of the problem is often the most important task in the *engaged scholarship* model, and is considered crucial in grounding the phenomenon in reality, as well as directly affecting the conception of the theoretical model that aided the investigation. Formulating the problem involved carrying out four correlated activities (Van de Ven, 2007): (1) recognizing and contextualizing the problem; (2) identifying evidence to substantiate and frame the problem; (3) diagnosing the information gathered to determine the characteristics or symptoms of the problem; and (4) defining the actions or guiding question of the research, with the aim of proposing a solution to the problem. The study problem was formulated in accordance with the following activities:

1) Understanding the context of the problem, considering the scenario with the small number of financial institutions that participated in applications to use internal models to calculate capital requirements in Brazil; evaluation of the normative publications made by the banking regulatory and supervisory bodies in Brazil; and the survey of international best practices recommended by the BCBS;
2) To carry out a systematic literature review in order to identify studies that show positive factors in the use of internal models, elucidating the discussion of efficiency in the management of financial institutions' capital, in addition to the

scarcity of research that has addressed the decision-making process of applying for the use of internal models;

3) Draw up diagnoses resulting from the literature review of both the practical aspects surrounding the calculation and management of capital requirements in banks and the variables that can affect the rational decision-making of agents, in scenarios impacted by institutional logics and contradictions; and
4) To propose a solution to the study's problem based on the formulation and practical application of the theoretical model, which takes into account the logics, organizational fields, institutional contradictions and human praxis of the banks' decision-making process when choosing the approach for determining capital requirements, based on updating the theoretical framework of the seminal study by Seo and Creed (2002).

Building the theoretical model

The central aim of theory building was to develop plausible and interesting conceptual models that help answer the research question, promoting the central interaction between theory and practice (Van de Ven, 2007), as shown in Figure 1.

The study was based on institutional theory, whose field of application is broad and has been classified differently by various authors (Russo & Guerreiro, 2017). The authors Burns and Scapens (2000) suggested a threefold classification: *New Institutional Economics (NIE*), *Old* Institutional Economics (*OIE*) and *New Institutional Sociology* (*NIS*), the latter being the approach adopted in this study. The authors Russo and Guerreiro (2017) point out that *NIE* focuses on macro institutions and studies the relationships between organizations and their environment, *OIE* is related to the micro process of institutionalization (habits that become routines) and *NIS* focuses on analyzing the effects of extra-organizational institutions (social, economic and political).

The *NIS* has been used in recent years in various studies evaluating the process of organizational change, with an emphasis on the extent to which tensions arising from the external environment could cause movements in response to these pressures ("from the outside in"), which most of the time were motivated more strongly by the search for legitimacy in the eyes of their *stakeholders*, such as: regulators; competitors; investors; even disregarding or not prioritizing aspects related to operational efficiency.

DiMaggio and Powell (1983) explored how agents rationalize changes in organizations, given that they were becoming increasingly similar. They also argue that organizations in a field that is already institutionally structured respond to environments that others are already responding to, so agents' decision-making rationalization is directly impacted by the external environment. Finally, the researchers presented the three mechanisms of institutional isomorphic change: coercive isomorphism, which derives from political and regulatory influences and the problem of legitimacy; mimetic isomorphism, which results from standardized responses to uncertainty; and normative isomorphism, which is associated with the category of professionalization.

The proposal to revise and update the framework "Institutional Contradictions, Praxis, and Institutional Change: A Dialectical Perspective", by Seo and Creed (2002), made it possible to integrate the process of institutional change from the micro and macro levels, pointing out the influence of institutional logics (Thornton, Ocasio & Lounsbury, 2012; Thornton & Ocasio, 1999, 2008) in the formation of organizational fields (DiMaggio & Powell, 1983), as an antecedent and fomenter of the creation of contradictions, which can trigger the human praxis necessary to question the dominant and/or conflicting scenario and drive successful change.

The theoretical model was applied to the phenomenon that is the subject of this study. The application highlights the agents involved, the logics, the organizational fields and the institutional contradictions in the process of applying to use internal models to calculate capital requirements, listed in six phases of the proposed new framework.

The Figure 2 shows the application of the theoretical model to the research context, highlighting all the study variables such as: the agents involved, the institutional logics and contradictions of the application process and the use of internal models to calculate capital requirements.

Figure 2. Theoretical Model Applied - Decision to apply for the use of internal models to determine the capital requirements of banks in Brazil

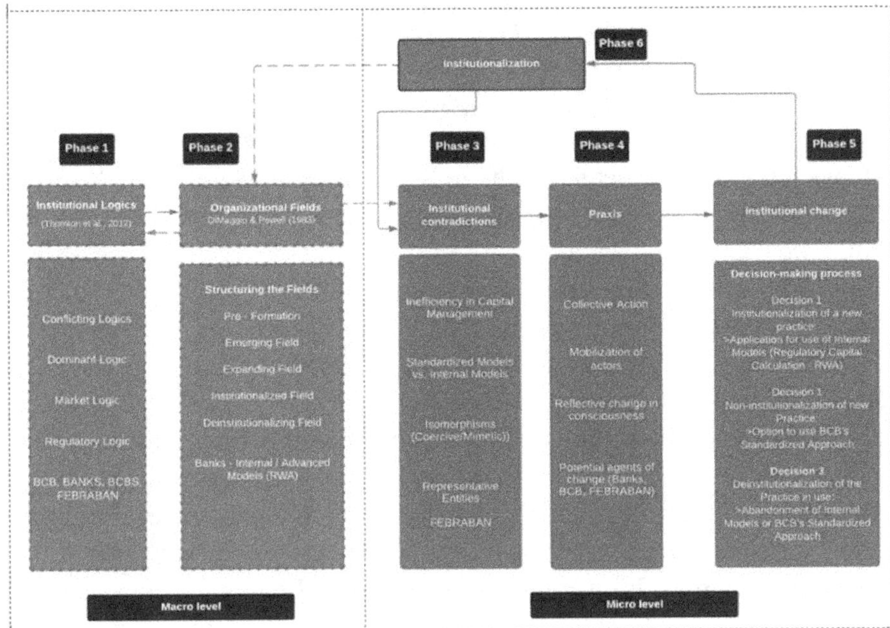

Note. *Prepared by the author.*

The process was divided into six phases, the first two at the macro organizational level and the others at the micro level, with the respective research assumptions.

Phase 1 - Institutional Logics

The authors Thornton and Ocasio (1999) proposed a broader definition of institutional logics, taking into account the definitions developed by Friedlande Alford (1991) and Jackall (2015). The new expanded definition conceptualizes institutional logics as the socially constructed pattern, based on the history of material practices, assumptions, values, beliefs and rules by which individuals produce and reproduce their material subsistence, organize time and space, and give meaning to their social reality. The study considered conflicting logics, dominant logics, market logics and regulatory logics, as follows:

I. Conflicting Logics: The creation of conflicting logics arises in scenarios formed by various organizational actors influenced by multiple social systems, such as the institutional orders of the family, religion, the state, the market, the professions and the community, where each order has its interests, its choices, its practices, its symbols, its principles that condition both the behavior of individuals and organizations, causing competitions and conflicts between each institutional logic of the decision-making agents (Thornton et al., 2012).

II. Dominant Logics: Dominant logics put decision-makers under pressure, which can have various consequences for organizations, whether through isomorphism, the ceremonial institutionalization of practices, or the creation of barriers to innovation (Cinar & Benneworth, 2021; Johansson, Malmström, Wincent, & Parida, 2021; Monios & Ng, 2021; Ocasio & Radoynovska, 2016; Rossoni, Poli, Sinay, & Araújo, 2020).

III. Regulatory Logics: Regulatory logics appear with reflections of power in the defense of the decision-makers who support their purposes. Scott (2014, p. 61) highlights the typical characteristics of the regulatory pillar of institutions and also conceptualizes very clearly how this logic is sustained: "[...] force, sanctions and expedited responses are central ingredients of the regulatory pillar, but they are often tempered by the existence of rules that justify the use of force". The author adds: "[...] when coercive power is both overcome and limited by rules, we enter the realm of authority". The author goes on to say: "[...] The regulatory emphasis is on compliance with rules: Legitimate organizations are those established and operating in accordance with legal or quasi-legal clearance requirements" (Scott, 2014, p. 74). In this scenario, there are regulatory and supervisory bodies, such as the Central Bank of Brazil (CBB), the supervisor of the National Financial System (NFS) in Brazil.

IV. Market Logic: Organizations that seek to institutionalize practices that favour operational efficiency are adept at market logic. According to Thornton and Ocasio (1999), market logic is associated with committing capital to a given investment in the search for the highest return.

This phase represents the banks' various interactions with the CBB (Supervisor) and their stakeholders (clients, investors, supervisors, regulators and self-regulators) in the outside world (macro level), which puts pressure on the FIs' risk managers to rationalize their decisions regarding the use of internal models or those standardized by the regulator for calculating capital requirements. Both the external and internal pressures for efficiency vis-à-vis the supervisor's demand to maintain capital cushions to deal with the risks to which they are exposed, place decision-makers in a real dilemma, considering that the process of applying to use internal models may not represent an efficiency gain, given the number of requirements and the intervention of the CBB in the process.

On the other hand, the CBB is concerned with preserving the soundness of the NFS, preventing FIs from irregularly using proprietary approaches to reduce capital amounts, with the practice of regulatory arbitrage. Banks have the option of choosing between internal and standardized models to calculate their risk-weighted assets and constitute their capital amounts, but the presence of conflicting logics and/or influence by a dominant logic of the supervisory body can interfere with the rational decision-making of agents and also in the formation and/or alteration of organizational fields (internal/advanced models and standardized models). This presents the first four assumptions of the study at the macro organizational level:

V. Assumption 1: The regulator's dominant institutional logics influence the institutionalization of the organizational field by inducing financial institutions to adopt standardized models for calculating regulatory capital requirements.
VI. Assumption 2: Conflicting institutional logics influence the formation of emerging organizational fields in financial institutions that use internal models to determine regulatory capital requirements.
VII. Assumption 3: The regulatory logic induces the institutionalization of the organizational fields of financial institutions that use the regulators' standardized models for calculating regulatory capital requirements, stimulating the isomorphism of the methodology.
VIII. Assumption 4: The market logic of financial institutions influences the pre-formation of fields from institutions that use internal models to determine regulatory capital requirements, encouraging new entrants to apply to the regulator for authorization of internal models.

Phase 2 - Organizational Fields

Organizational fields are made up of institutions that together constitute a recognized area of institutional life, comprising and often sharing the same key suppliers, offering similar products and services, as well as being subject to the supervision of the same regulatory bodies and compliance with common regulatory frameworks (DiMaggio & Powell, 1983).

An organizational field, at its initial stage, is made up of isolated and specialized organizations, while, over time, the structuring of the field makes it possible to share values, beliefs and practices, recognizing the importance of each other, either by strengthening relationships or increasing interactions (Holanda, 2003; Milano, Vieira, & Carvalho, 2003). According to Holanda (2003), the formation of organizational fields is divided into four stages: **1) Pre-Formation**, where organizations are isolated and independent; **2) Emerging Field**, with the presence of inter-organizational links and concentration; **3) Expanding Field**, due to the increase in relationships and converging values; and **4) Institutionalized Field**, which shows the high level of linkage in inter-organizational relationships and the sharing of values.

Standardized or internal approaches to determining regulatory capital requirements are shared within the organizational fields of financial institutions, which in turn are supported by institutional logics. The action of these institutional logics in the fields can cause contradictions raised by the agents in search of innovations or changes that can induce the formation of new fields with the entry of new practices or the opening of fronts that will question the historically consolidated ones. Assumptions 5 and 6 present the scenarios that correlate fields and institutional logics:

IX. Assumption 5: Emerging fields formed from conflicting market logics influence the decision to apply for the use of internal models.
X. Assumption 6: Fields institutionalized by regulatory logic induce the adoption of isomorphic practices based on standardized models for calculating regulatory capital requirements.

Phase 3 - Institutional contradictions

Drawing on Benson's (1977) dialectical perspective, Seo and Creed's (2002) proposal introduced mechanisms that explain how institutional arrangements generate various inconsistencies and tensions within and between social systems, known as contradictions. They also explored how these contradictions affect agents of change, with a focus on theorizing a specific type of human action, called 'praxis'.

I. The legitimacy that compromises functional efficiency

The first source of contradiction questions whether gaining legitimacy justifies a lack of efficiency, accompanied by the need to become isomorphic. Along these lines, one of the central premises of institutional theory is anchored, which is that organizational success depends on factors other than technical efficiency; considering that organizations gain legitimacy and the necessary resources by becoming isomorphic with their institutional environments (DiMaggio & Powell, 1983; Meyer & Rowan, 1977). Meyer and Rowan (1977) argue that although institutional rules are capable of conflicting with the logic of efficiency, technical activities require diverse and personalized solutions. Seo and Creed (2002) conclude that even in situations where institutionalized organizations make decisions that improve both legitimacy and technical efficiency in the short term, these decisions become easily questionable if new "optimal" solutions are not continually pursued and adopted.

II. Adaptation that undermines the capacity for adaptability

According to Seo and Creed (2002), institutional isomorphism that increases legitimacy is an adaptive movement for survival, but this occurs not only because the perceived benefits associated with familiarity easily outweigh the expected gains associated with flexibility, but also because a considerable amount of economic resources invested are "shielded internally". They state that although institutionalization is an adaptive process, it is likely that entities are both psychologically and economically impeded and isolated, not responding to the need for change, creating space for contradictions to develop.

III. Intra-institutional conformity that creates inter-institutional incompatibilities

For Seo and Creed (2002), compliance with certain institutional arrangements within a given level or sector can cause conflicts or inconsistencies with the institutional arrangements of different internal levels or sectors, especially if the prescriptions for action are strongly incoherent, all supported by rationalized myths, often leading organizations to incorporate incompatible practices and procedures in the search for legitimacy and stability. The authors emphasize that individuals and organizations are increasingly exposed to multiple and contradictory institutional arrangements and prescriptions.

IV. Isomorphism that conflicts with divergent interests

Seo and Creed (2002) state that the formation and reproduction of institutional arrangements is unlikely to satisfy the divergent interests of all participants, especially those of the least powerful. The authors argue that the formation and reproduction of social arrangements are basically political processes involving several participants who have divergent interests and unequal power, to the extent that this misalignment constitutes a strong source of contradiction.

If the logics of the agents provide strong stimuli in the organizational fields to drive the sources of contradiction with high intensity (inefficiency, non-adaptability, inter-institutional incompatibilities and misaligned interests between agents), can create movements among the banks' risk managers or through the main entity representing Financial Institutions (FIs) in the NFS, FEBRABAN, aimed at creating actions, the so-called "Praxis", to unleash fronts of discussion on the paths or proposals for revising regulations, creating self-regulation or preparing questions for the supervisory body. Thus we have the assumptions that connect the organizational fields and the contradictions, correlating the macro and micro levels in the environment of institutional change:

XI. Assumption 7: Institutionalized fields with the regulator's standardized models negatively influence the creation of institutional contradictions that would lead to the decision to apply for the use of internal models.

XII. Assumption 8: Emerging fields influence the formation of strong contradictions that positively induce the formation of praxis for institutional change in the use of internal models.

Phase 4 - "Praxis":

Human 'praxis' are creative and flexible reconstructions of social patterns, resulting from an analysis that considers both the limits and the potential of existing social forms. In specific decision-making situations, people can become active agents in the reconstruction of their own social relations (Benson, 1977).

After the movements to discuss and question practices that have been legitimized but are inefficient or institutionalized in a ceremonial way, they will trigger individual or collective actions, if the banks choose to take action against FEBRABAN, when it involves more than one organization associated with the representative body, in order to finally begin the process of institutional change.

Phase 5 - Institutional Change:

With the whole scenario set for challenging the use of obsolete or inefficient processes, even if they are legitimized internally and externally, organizations can decide to remain inert, opting to maintain the CBB's standardized approach, or start the process of institutional change by applying to use internal models to calculate regulatory capital requirements (RWA), with a view to evolving its governance structures, systemic infrastructure, the specialization of employees in the areas of risk management, internal auditing, BackOffice and model validation, which can result in the expected efficiency of regulatory capital management (Teixeira, 2017). However, the practices in use (internal or standardized models), when affected by contradictions arising from dominant and/or conflicting logics, can come under pressure to be deinstitutionalized.

Phase 6 - Institutionalization:

The last stage, after going through the previous cycles, comes with the successful institutionalization of the use of internal models, after approval by the CBB supervisor. Risk managers, supported internally and externally by the implementation of sophisticated risk assessment and regulatory capital mechanisms, have the challenge of demonstrating over time the assertiveness of the decision that brought about the successful change, especially when it comes to the efficient management of capital amounts, taking into account the maintenance of the amounts needed to withstand possible risks arising from economic and banking crises. The last phase brings us to the last three assumptions of the study, which connect the micro and macro levels:

XIII. Assumption 9: The successful institutionalization of the use of internal models to determine regulatory capital requirements positively stimulates the formation of organizational fields emerging from the new practice.
XIV. Assumption 10: The institutionalization of regulators' standardized models for determining regulatory capital requirements influences the stability of the institutionalized organizational field.
XV. Assumption 11: The institutionalization of regulators' standardized models for calculating regulatory capital requirements encourages the de-institutionalization of the use of internal models in emerging fields.

The research instruments were used according to their origin, whether primary or secondary. A tying matrix was drawn up (Mazzon, 1983; Telles, 2001), which takes into account: the research assumptions, the variables of the theoretical model, the references to previous studies, obtained from systematic literature reviews (Kitchenham (2007), with computer support from the systemic tool StArt (State of the Art through Systematic Review (Fabbri *et al., 2016*; LAPES - UFSCAR, 2021), the correlation with the groupings of data sources, as shown in the Content Analysis and Triangulation of Results protocol proposed in the study (Figure 3).

Methods used for data collection

The methods used to collect data for the research were subdivided into secondary and primary data sources. Data from primary sources was obtained through semi-structured interviews and focus group sessions, while secondary data was extracted from documentary analysis of regulations, technical articles, video transcripts of lectures at technical congresses in the banking industry, as well as documents provided by bank representative association practitioners in Brazil.

Primary data collection was carried out with the support of FEBRABAN, which helped with the research by appointing experts from the FIs, sending out the call for meetings, and the field intervention was divided into three stages:

i. **Stage 1** (Semi-structured interviews): An empirical study was carried out with experts from the FIs, appointed by FEBRABAN, to identify and confirm the variables. Sixteen interviews were conducted between October 25 and December 15, 2022. The interviews lasted an average of 45 minutes, 15 of which were conducted remotely using the Teams application (Microsoft Office Package) and one in person. All the participants authorized the recording and transcription of the interviews, and the transcription was carried out by a specialist company hired by the researcher. The interviewees had extensive experience of the phenomenon under study:

I. 12 interviewees were from the S1 and S2 Segments, where these groups of FIs represent around 80% of the NFS's total capital requirement (CBB - 02/2022), with an average experience of 19 years in the areas of risk and capital management;
II. interviewees with an average experience of more than 28 years at the CBB, working to supervise the process and to standardize and regulate the application for the use of internal models;
III. The sum of the 16 interviewees' years of experience totals 345 years in the NFS.

ii. **Stage 2** (*Focus group*): Data was collected by conducting an *online focus group*. For Van de Ven (2007), the information obtained from brainstorming or focus group meetings often provides an overall gestalt appreciation of a problem that rarely emerges from individual interviews or survey techniques.

The protocol proposed by Gatti (2015) was used to carry out the *focus group*. According to the author, the method of working in focus groups shows rich possibilities for generating knowledge on various themes and problems, based on experiences, reported/observed experiences, expressions of behavior, attitudes, values and narratives of different oral and written natures. The protocol proposed by Gatti (2015, pp. 240-242) recommends that each session should last between an hour and a half and no more than three hours with groups of 6 to 12 people, with one or two meetings being sufficient to obtain a good analysis.

The focus group session was held on November 30, 2022, involving experts from FEBRABAN's Market and Liquidity Risk, Credit and Capital Risk Management and Operational Risk and Business Continuity Management advisory groups, which are subordinate to the entity's Risk Management Committee. Each advisory group has around 60 FIs members, with representatives from the NFS segments (Conselho Monetário Nacional, 2017).

iii. **Stage 3** (Documentary Analysis): Secondary data was collected based on surveys and research on the websites of local and international regulators, as well as information provided by FEBRABRAN and research carried out in journals that publish practical articles on a global scale, as listed below:

I. CBB (Period: 2019 to 2022) - DOCUMENTS: Emails replied by CBB to the researcher about the application for the use of internal models in Brazil, documents published on the CBB's website (Standards/Public Consultation Notices, presentations at congresses and meetings with the Financial Industry, as detailed in the Protocol;

II. FEBRABAN (Period: 2020 to 2022) - DOCUMENTS: (Letters/memos/presentations, Videos of the International Risk Congress (2020, 2021 and 2022), made available by FEBRABAN related to the application process for the use of internal models of the member banks, considering the discussions held in the industry with the supervisory body - CBB;

III. ECB / European Council (Period 2021 to 2022) - DOCUMENTS: E-mails answered by the European Council, reports published by the ECB and proposals for regulatory amendments;

IV. BIS (Period: 2017 to 2019) - DOCUMENTS: Documents published on the BIS website, which address the regulatory guidelines for the world's central banks on the best risk management practices for FIs;

V. MOODY'S (2022) - In November 2022, in its report to its clients and investors, entitled *"EU states propose watered down Basel* III *banking regulations, a credit negative for banks globally"*, which addressed the proposal of the European Council of Ministers which, among other matters, proposes to dilute and postpone for up to 10 years the compliance of the *output floor* limits;

VI. SS&C (2022) DOCUMENT: Material developed with lessons and challenges on the implementation of the new Fundamental Review of the Trading Book (FTRB), published on its website;

VII. PODCAST (2022) - NFS SPECIALIST: S1 segment FIs promoted a *podcast* with the aim of explaining to investors the importance and impacts of capital management. Access and the audio transcription process, which was transcribed by a specialized company; and

VIII. Practical Articles - Risk.net (Period: 2012 to 2022) - 11 articles published in the period 2012 to 2022 were collected, whose topics discussed address the international context on the practices adopted and to be implemented based on the Basel recommendations, considering the magazine's practical approach, which directly corroborates to compare the internal and external scenario to obtain evidence for the phenomenon studied in this research.

Data analysis methods

The analysis of the qualitative data collected in the field study stages was based on document coding (Saldaña, 2016) and content analysis (Bardin, 2016), which was processed using the Atlas.ti computer tool, which catalogs documents, quotes, video transcripts, remote meetings (Microsoft Teams), notes, memos, calls, with the aim of aiding content analysis (Friese, 2014).

According to Saldaña (2016), coding is one way of analyzing qualitative data, not the only way, and it is emphatic that the researcher himself must choose the most appropriate technique for his research project. The author advocates two cycles of coding and one of transition, where at each stage the researcher carries out a constant analysis in a non-linear way, be it code with code, data with data, data with code, code with categories, categories with categories, categories returning with the data, etc.

According to Bardin (2016), content analysis is a set of communication analysis techniques aimed at obtaining objective content description procedures that allow the inference of knowledge and the conditions of production of the message variables, comprising three chronological stages: pre-analysis; exploration of the material; and treatment of the results, inference and interpretation. The author states that treating the material means codifying it, transforming raw text data in order to achieve a representation of the content or its expression.

The Atlas.ti computer tool allows the researcher to code the evidence obtained during the field stage, correlating it with the variables of the theoretical model, where the researcher has greater control of their data, establishes patterns, focusing on their objective, establishing networks that provide a visualization and interpretation of the research corpus, which helped in the sequencing of the analyses (Bley & Carvalho, 2019).

Strategies to guarantee quality and scientific rigor

The results of the content analysis were triangulated, taking into account the various pieces of evidence collected from primary and secondary sources (Bardin, 2016; Saldaña, 2016; Yin, 2014). Saldaña (2016) and Bardin (2016) therefore approach the analysis process in the form of coding, the former through coding cycles and the latter through content analysis. Thus, considering that the models can be complementary, which allows for greater rigor and robustness in the evaluation of data and preparation of inputs for the triangulation of results, this study proposes the implementation of a protocol based on the proposals of Saldaña, (2016) and Bardin (2016), as shown in Figure 3 - Content Analysis and Triangulation of Results, which presents the junction of the stages of content analysis and the coding cycles:

Figure 3. Content Analysis and Triangulation of Results

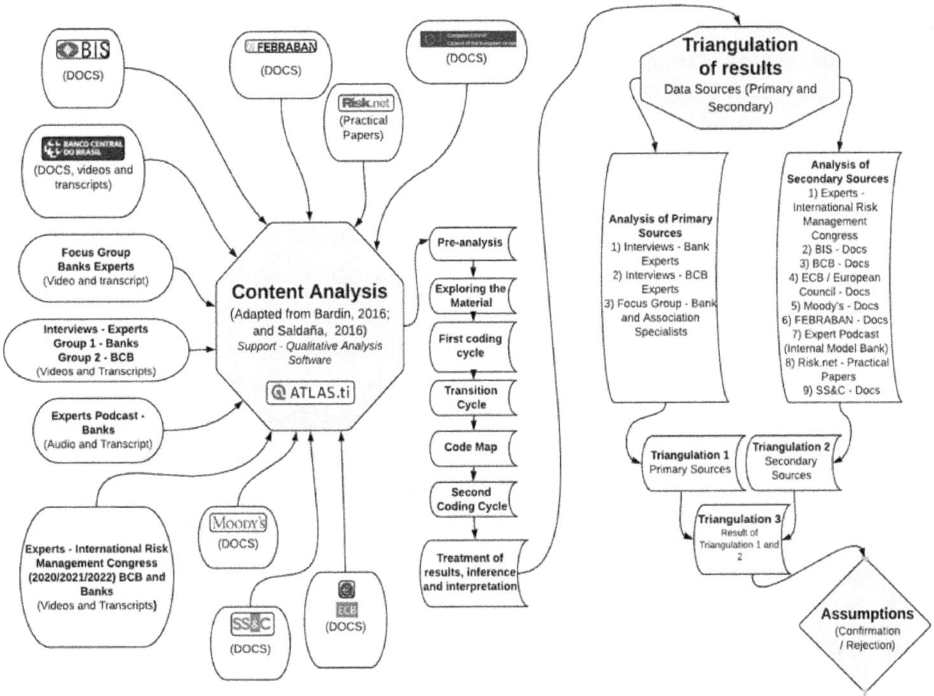

Note. *Prepared by the author.*

Step-by-step data analysis with Atlas.ti computer support and application of the protocol proposed in the study

Step 1 - Creating the Atlas.ti Project

The researcher must create a new project and add the various documents collected from the primary and secondary data sources obtained during the field research, creating codes and document groups (Figure 4).

Figure 4. Project creation in Atlas.ti

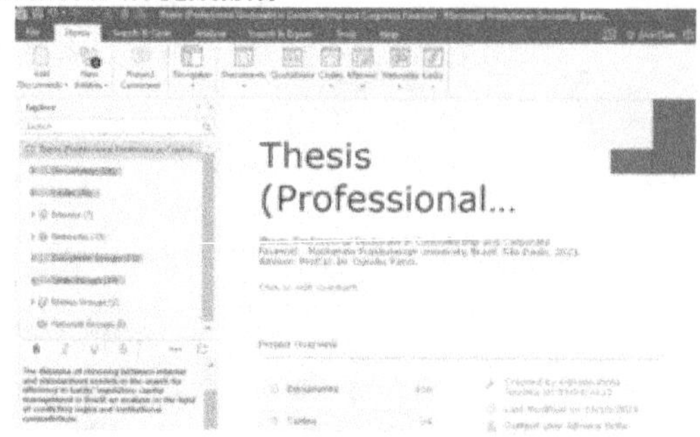

Note. *Prepared by the author.*

Step 2 - Creating the Codes (Variables of the theoretical model)

The codes were created according to the variables in the theoretical model applied (see Figure 5 - Creating the codes (Theoretical Model variables), as shown in Figure 2 - Theoretical Model Applied - Decision to apply for the use of internal models to determine the capital requirements of banks in Brazil).

Figure 5. Creating the codes (Theoretical Model variables)

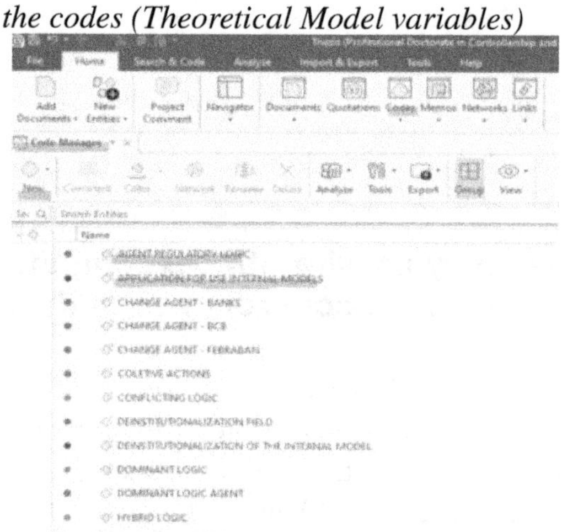

Note. *Prepared by the author.*

Step 3 - Creating Document Groups (Primary and Secondary Sources)

From the data sources obtained, groups of documents were created to enable analysis grouped according to the evidence, as shown in Figure 6.

Figure 6. Creating Document Groups (Data Sources)

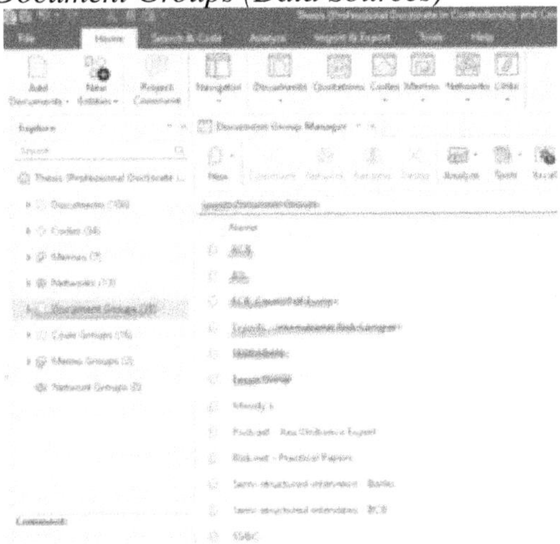

Note. *Prepared by the author.*

Step 4 - Creating the Code Groups (Study assumptions)

From the codes (variables of the theoretical model), create the groups of codes that represent each of the assumptions of the study.

Figure 7. Creation of code groups (study assumptions)

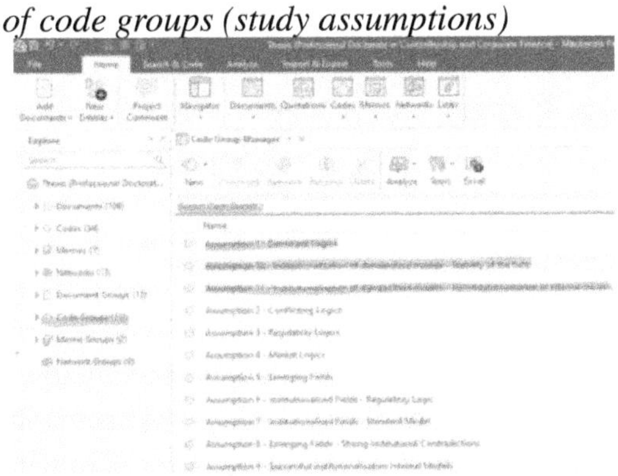

Note. *Prepared by the author.*

Step 5 - Start of the study protocol (pre-analysis and exploration of the material)

Once the codes and groups of codes and documents have been created, the protocol begins (see Figure 3 - Content Analysis and Triangulation of Results), starting with the pre-analysis and exploration of the material from the documents in each of the document groups, as shown in Figure 8.

Figure 8. Pre-analysis and exploration of the material

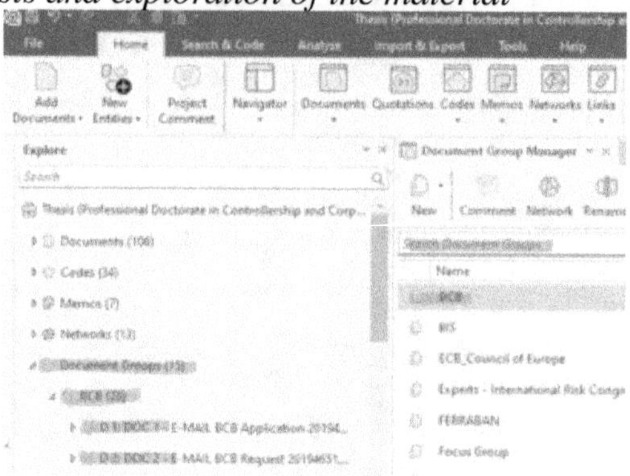

Note. *Prepared by the author.*

Step 6 - First Coding Cycle

The first coding cycle pro-rates all the documents in each of the groups, taking into account the codes created (variables from the theoretical model), as shown in Figure 9.

Figure 9. Coding - Cycle 1

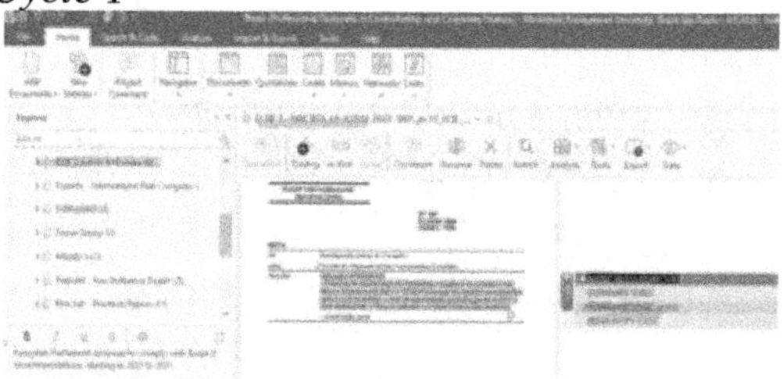

Note. *Prepared by the author.*

Step 7 - Transition cycle, code map and second coding cycle

Based on the code management functionality, the researcher can evaluate the relationship between variables and document sources, assumptions, build relationship networks between codes, groups of codes, citations and documents, as shown in Figure 10.

Figure 10. Transition cycle, code map and second coding cycle

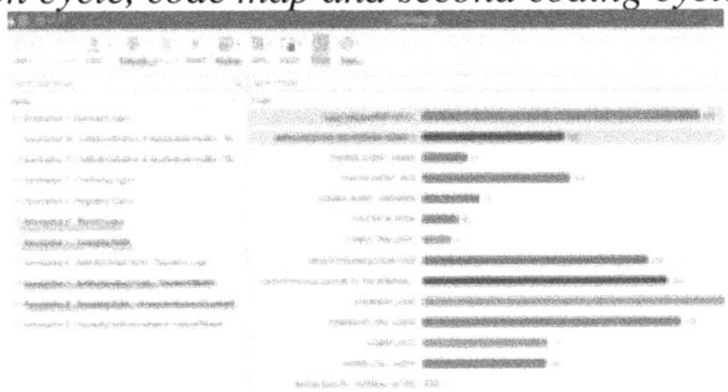

Note. *Prepared by the author.*

Step 8 - Treatment of results, inference and interpretation

The penultimate stage of the protocol before the triangulations was carried out on the basis of the analysis functions and Research tools, considering the primary and secondary data sources individually and the relationship with the study's assumptions, as shown in Figure 11.

Figure 11. Treatment of results, inference and interpretation

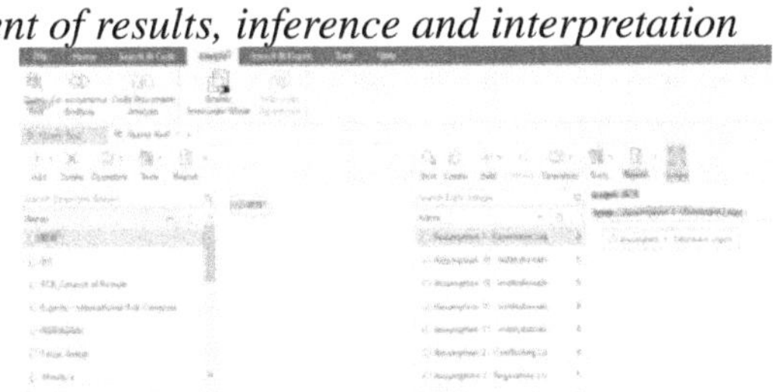

Note. *Prepared by the author.*

Step 9 - Triangulation of results

The results were triangulated, considering each document in the groups, classified and primary and secondary sources. The Atlas.ti tool offers the creation of automatic reports with the result of the data relationship (scope in Atlas.ti), coded with all the variables of the theoretical model and each assumption (Atlas.ti Code Group) of the study. Figure 12 demonstrates the application of the functionality that results in the report for the researcher's analysis to accept or reject the assumption, considering the analysis of each data source.

Figure 12. Triangulation of results – Atlas.ti Report

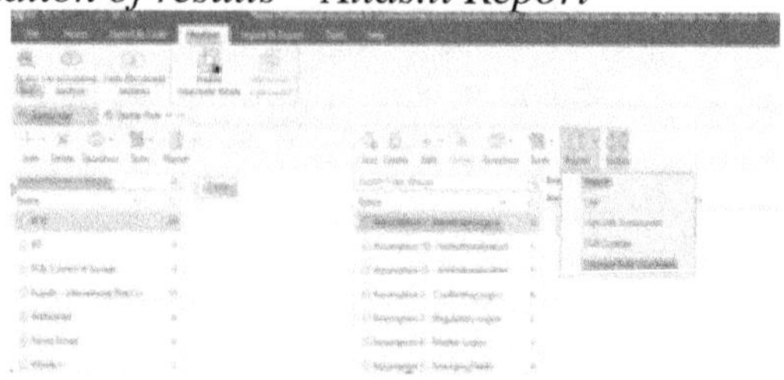

Note. *Prepared by the author.*

In Figure 13, we show the report that addresses the result of the relationship between Assumption 1 and the secondary data source (CBB), showing 40 quotes that were coded into the variables of the theoretical model. In the example, document number 1, quote 1:4, was coded with four codes (variables), helping the researcher to analyze the content directly.

Figure 13. Report extracted from Atlas.ti - Relationships between the research assumption and the data source - CBB

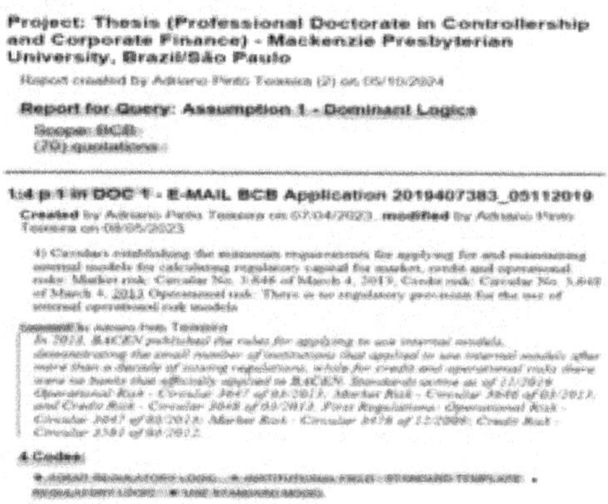

Project: Thesis (Professional Doctorate in Controllership and Corporate Finance) - Mackenzie Presbyterian University, Brazil/São Paulo

Report for Query: Assumption 1 - Dominant Logics

1:4 p 1 in DOC 1 - E-MAIL BCB Application 2019407383_05112019

Note. *Prepared by the author.*

Step 10 - Research assumptions - Confirmation / Rejection

After analyzing and cross-referencing the results obtained in the semi-structured interviews and the focus group section (primary sources), we found that “Assumptions 1, 3, 6, 7, 10 and 11” were confirmed, while “Assumptions 2, 4, 5, 8 and 9” were rejected.

After analyzing and cross-referencing the results obtained from each of the secondary sources, “Assumptions 1, 3, 6, 7, 10 and 11” are confirmed, while rejected “Assumptions 2, 4, 5, 8 and 9”, the last action in the protocol (See Figure 3: Content Analysis and Triangulation of Results).

Problem solving / results

For Van de Ven (2007), in solving the research problem, the researcher must communicate, interpret and apply the empirical results on which alternative models best answer the research question. At the end of the study, the research offered recommendations in order to analyze how conflicting logics and institutional contradictions influence the candidacy for using internal models to determine the capital requirements of banks in Brazil, enabling proposals for possible alignments between banks and the supervisor in the search for a solid and efficient financial system, with positive repercussions for all stakeholders (bank customers, investors, shareholders).

Recommendations - Basel Committee, CBB and BANKS

Based on the analysis of the evidence obtained in the field research, recommendations were made with the aim of promoting greater integration between the bodies responsible for banking supervision, the international entities that disseminate best risk management practices, the financial institutions, as well as the associations representing the financial industry. It should be emphasized that these actions are not exhaustive, considering the high complexity involved, the regulatory issues, the risks intrinsic to the operations of each institution, in addition to the legal responsibilities that each agency must follow, whose analysis and decision regarding the evaluation of the recommendations will be the sole responsibility of the institutions and/or regulatory, supervisory and/or representative bodies involved.

The coexistence of conflicting institutional logics has been the subject of studies, the results of which have shown the possibility of alignment, even considering the presence of dominant logics (Mahmood & Uddin, 2020; Maibom & Smith, 2016), whose pressures have a direct impact on the process of changing practices and, consequently, a possible opportunity to evolve and improve processes, as was found in this study, with the strong influence of CBB's regulatory logic on the decision-making process of banks, which in turn rely on the conflicting logic of the market in deciding whether to apply for the use of internal models to calculate the capital requirements of banks in Brazil.

In view of the above, the following are points for reflection by the agents involved (CBB; BCBS; banks; FEBRABAN), noting that the security of the NFS will have priority weight in the analysis of the points listed below:

- Carrying out impact studies with the aim of assessing the quality of the internal models used by the industry, regardless of whether they are banks already authorized to determine capital requirements (*RWA*) or which use these approaches for management purposes or within the scope of the ICAAP - Internal Capital Adequacy Assessment Process;

- Creation of permanent forums for prior discussion between the industry and the banking supervisor of the technical aspects and impacts when "tropicalizing" BCBS recommendations;
- Promoting industry participation in the Basel Committee's methodological discussions, supported by the regulatory and supervisory bodies of the National Financial System, so that local particularities and their impacts are no longer put on the table in advance of the conclusion of international agreements;
- Preparation of joint actions to speed up the dissemination of local rules when new BCBS guidelines are implemented, reducing the impact of uncertainties regarding changes to processes, systems, staff training for regulatory compliance, as well as the necessary investments and possible impacts on capital requirements;
- CBB's assessment of the possibility, if possible, of adapting and/or equalizing the regulatory requirements established for IFRS 9 - *Financial Instruments* (Expected Credit Losses) and the application to the IRB (Internal rating-based) internal models for calculating capital requirements (unexpected credit risk losses), with a view to providing efficiency gains in the implementation of both processes and reducing the cost of compliance for both the supervisor and the institutions, in line with a study carried out by the ECB - European Central Bank in 2017, entitled "Thematic analysis on IFRS 9 carried out within the scope of the Single Supervisory Mechanism" (ECB, 2017), which revealed greater ease of implementation for banks that were already authorized to use internal models (IRB);
- Revision of the procedures for evaluating applications for the use of internal models, considering the little or no flexibility and the dissonance between the discourse of CBB's leaders and the technical staff of analysts who carry out the inspections, where the public servant's concern about possible penalties in the event of failures or errors can be seen;
- Definition of CBB's actions in conjunction with the industry to promote closer ties with risk management areas and models with a view to enabling the supervisor to have a better understanding of internal approaches;
- Discussion with the industry regarding possible incentives to reduce capital, through the use of internal models, since this has been identified as the main *driver* for banks to decide whether to apply, considering the restrictive measures adopted locally, even more conservative than the Basel Committee itself, which sometimes does not justify the high investment required to enter

an application process, according to considerations observed in interactions with the experts interviewed;

- Creation of a special transitional process for banks authorized to use internal models for market risks to adapt to the new Basel 3 requirements (FRTB), given the complexity and relevance of the changes, as well as the high cost of implementation, in order to minimize the impacts and avoid a migration of these institutions to the standardized model, as observed in large European banks;
- CBB hires third parties to meet the need for resources or even promote pre-applications, including evaluations carried out by the 2nd and 3rd line areas of organizations (Internal Controls, Compliance, Independent Validation of Models and Internal Audit), reducing the effort of the supervisory body itself when evaluating applications;
- Institution of a pre-application process in which institutions could adapt to the requirements, receiving partial feedback and evaluating the cost-benefit ratio for maintaining the process, moving on to the actual application after the supervisor's prior evaluation;
- Use of the ICAAP as a pre-candidate in assessing the quality of the internal models applied in the calculation of economic capital, allowing CBB to get an impression of the assertiveness of the models in relation to the values calculated by standardized methods; and
- Evaluation of the risks related to the concentration of the use of the standardized methods of banking supervisors both locally and around the world, considering the possibility of underestimating the capital requirements to support the unexpected losses of the institutions.

RESEARCH METHODOLOGY LEARNING

The study provided relevant methodological contributions by verifying the effectiveness of the Engaged Scholarship application, which brings academia closer to market practitioners, demonstrating the depth of the analysis of the phenomenon studied, as well as the practical nature of the results achieved. Still in the methodological dimension, a protocol for content analysis was developed, supported by the "Atlas.ti" computer tool (Friese, 2014; Bley & Carvalho, 2019), combining the stages proposed by Bardin (2016) and the coding cycles of Saldaña (2016).

It is worth noting that the computational tool enabled the complete application of the proposed protocol, making it possible for other researchers to replicate it "step by step", demonstrating the robustness of qualitative methodology, especially in studies involving complex themes with the need to analyze a large number of pieces of evidence and research strategies.

ACKNOWLEDGMENT

With Special thanks to PhD Claudio Parisi for his continued support and advices.

REFERENCES

Bardin, L. (2016). Content analysis. 3rd Reprint. *São Paulo: Edições, 70.*

Benson, J. K. (1977). Organizations: A dialectical view. *Administrative Science Quarterly*, *22*(1), 1–21. DOI: 10.2307/2391741

BIS. (2017a). *Guide to the BIS archives*. 45. www.bis.org

BIS. (2017b). *Basel III: Finalizing post-crisis reforms*.https://www.bis.org/BCBS/publ/d424.pdf

Bley, D. H. P., & Carvalho, A. B. G. (2019). Ciclos de codificação e o software Atlas ti: uma parceria criativa para análise de dados qualitativos em pesquisas sobre o uso das tecnologias digitais no campo da Educação. *In Teia | Revista De Educação Matemática E Tecnológica Iberoamericana, 10*(1). DOI: 10.36397/emteia.v10i1.240016

Burns, J., & Scapens, R. W. (2000). Conceptualizing management accounting change: An institutional framework. *Management Accounting Research, 11*(1), 3–25. DOI: 10.1006/mare.1999.0119

Cinar, R., & Benneworth, P. (2021). Why do universities have little systemic impact with social innovation? An institutional logics perspective. *Growth and Change*, *52*(2), 751–769. DOI: 10.1111/grow.12367

Conselho Monetário Nacional. (2017). *Resolução CMN n° 4.553 de 30/1/2017 – Estabelece a segmentação do conjunto das instituições financeiras e demais instituições autorizadas a funcionar pelo Banco Central do Brasil para fins de aplicação proporcional da regulação prudencial.* Resolução Do Concelho Monetário Nacional. https://www.CBB.gov.br/estabilidadefinanceira/exibenormativo?tipo=Resolução&numero=4553

Conselho Monetário Nacional. (2021). *Resolução CMN n° 4.958 de 21/10/2021 – Dispõe sobre os requerimentos mínimos de Patrimônio de Referência (PR), de Nível I e de Capital Principal e sobre o Adicional de Capital Principal (ACP).* Resolução Do Concelho Monetário Nacional. https://www.CBB.gov.br/estabilidadefinanceira/exibenormativo?tipo=Resolução CMN&numero=4958

de Holanda, L. A. (2003). *Formação e Institucionalização do Campo Organizacional do Turismo no Recife-PE* [Master's Dissertation. Postgraduate Program in Administration, Federal University of Pernambuco]. https://repositorio.ufpe.br/handle/123456789/1157

DiMaggio, P. J., & Powell, W. W. (1983). The iron cage revisited: Institutional isomorphism and collective rationality in organizational fields. *American Sociological Review*, *48*(2), 147–160. DOI: 10.2307/2095101

European Central Bank. (2017). *SSM thematic review on IFRS 9 Assessment of institutions' preparedness for the implementation of IFRS 9*. European Central Bank., DOI: 10.2866/732554

Fabbri, S., Silva, C., Hernandes, E., Octaviano, F., Di Thommazo, A., & Belgamo, A. (2016). Improvements in the StArt tool to better support the systematic review process. In *Proceedings of the 20th International Conference on Evaluation and Assessment in Software Engineering (EASE '16)*. Association for Computing Machinery, New York, NY, USA, Article 21, 1-5. https://doi.org/DOI: 10.1145/2915970.2916013

FEBRABAN. (2020). *2020 - LIVE: 10º International Risk Management Congress (1º day) - YouTube*. November 17, 2020. https://www.youtube.com/watch?v=GLJDypipDuU FEBRABAN (2021). *2021 - LIVE: 11º International Risk Management Congress – Day 14 (Portuguese)* October 14, 2021 - *YouTube*. https://www.youtube.com/watch?v=i59w9sO4Qz8

FEBRABAN. (2022a). *2022 - LIVE | 12º International Risk Management Congress (G- Risc) - Day 9 (Portuguese)* November 9, 2022 - *YouTube*.https://www.youtube.com/watch?app=desktop&v=eYFhpRGDouI

FEBRABAN. (2022b). *2022 - LIVE | 12º International Risk Management Congress (G-Risc) - Day 10 - (Portuguese) November 10, 2022 - YouTube*.https://www.youtube.com/watch?v=AnOde_eyS5g

FEBRABAN. (2022c). *Febraban - FEBRABAN*.https://portal.febraban.org.br/paginas/9/pt- br/institucional

Friedland, R., & Alford, R. R. (1991). Bringing Society Back. In Powell, I. W. W., & DiMaggio, P. J. (Eds.), *Symbols, Practices and Institutional Contradictions* (pp. 232–263). The New Institutionalism in Organizational Analysis.

Friese, S. (2014). *Qualitative Data Analysis with ATLAS.ti* (2. Ed).

Gatti, B. A. (2015). Grupo Focal: fundamentos, perspectivas e procedimentos. In M. Tavares & R. J. Ruchardson. (Org.), *Metodologias Qualitativas: teoria e prática* – Editora CRV (231- 257). DOI: 10.24824/978854440558.1

Jackall, R. (2015). Moral Mazes. *Wiley Encyclopedia of Management*, 1-2. https://doi.org/DOI: 10.1002/9781118785317.weom020150

Johansson, J., Malmström, M., Wincent, J., & Parida, V. (2021). How individual cognitions overshadow regulations and group norms: A study of government venture capital decisions. *Small Business Economics*, *56*(2), 857–876. DOI: 10.1007/s11187-019-00273-3

Kitchenham, B. (2007). Guidelines for performing Systematic Literature Reviews in software engineering. Reviews in software engineering. EBSE Technical Report. 2007 (1). https://www.researchgate.net/publication/258968007_Kitchenham_B_Guidelines_for_performing_Systematic_Literature_Reviews_in_software_engineering_EBSE_Technical_ Report_EBSE-2007-01

LAPES - UFSCAR. (2021). *StArt - State of the Art through Systematic Review.* http://lapes.dc.ufscar.br/tools/start_tool

Lounsbury, M. (2002). Institutional Transformation and Status Mobility: The Professionalization of the Field of Finance. *Academy of Management Journal*, *45*(1), 255–266. DOI: 10.2307/3069295

Mackenzie, U. P. (2019). Mackenzie launches unprecedented professional doctorate in controllership and finance Mackenzie. https://www.mackenzie.br/noticias/artigo/n/a/i/mackenzie-lanca-doutorado-profissional-inedito-em-controladoria-e-financas

Mahmood, Z., & Uddin, S. (2020). Institutional logics and practice variations in sustainability reporting: Evidence from an emerging field. *Accounting, Auditing & Accountability Journal*, *34*(5), 1163–1189. DOI: 10.1108/AAAJ-07-2019-4086

Maibom, C., & Smith, P. (2016). Symbiosis across institutional logics in a social enterprise. *Social Enterprise Journal*, *12*(3), 260–280. DOI: 10.1108/SEJ-02-2016-0002

Mazzon, J. A. (1983). Avaliação do programa de alimentação do trabalhador. Subsídios à formulação de um plano de marketing social. *Revista ADM*, *18*(4), 33–58. DOI: 10.1016/rausp.v18i4.167313

Meyer, J. W., & Rowan, B. (1977). Institutionalized organizations: Formal structure as myth and ceremony. *American Journal of Sociology*, *83*(2), 340–363. DOI: 10.1086/226550

Milano, M., Vieira, F., & Carvalho, C. A. (2003). *Campos Organizacionais: de wallpaper à construção histórica do contexto de organizações culturais em Porto Alegre e em Recife.*

Monios, J., & Ng, A. K. Y. (2021). Competing institutional logics and institutional erosion in environmental governance of maritime transport. *Journal of Transport Geography*, *94*, 103114. DOI: 10.1016/j.jtrangeo.2021.103114

Moody's Corporation. (2022). *About us*. https://www.moodys.com/web/en/us/about/who-we- are/about-us.html

Moody's Corporation. (2022). *About us*.https://www.moodys.com/web/en/us/about/who-we- are/about-us.html

Ocasio, W., & Radoynovska, N. (2016). Strategy and commitments to institutional logics: Organizational heterogeneity in business models and governance. *Strategic Organization*, *14*(4), 287–309. DOI: 10.1177/1476127015625040

Ramirez, J. (2017). *Handbook of Basel III Capital* (First edit). John Wiley & Sons Ltd.

Rossoni, L., Poli, I. T., de Sinay, M. C. F., & de Araújo, G. A. (2020). Materiality of sustainable practices and the institutional logics of adoption: A comparative study of chemical road transportation companies. *Journal of Cleaner Production*, *246*, 119058. DOI: 10.1016/j.jclepro.2019.119058

Russo, P. T., & Guerreiro, R. (2017). Perceptions about the sociomateriality of management accounting practices. *R*AE-Revista de Administração de Empresas | FGV EAESP, *57*(6), 567-584. https://doi.org/DOI: 10.1590/s0034-759020170605

Saldaña, J. (2016). *The Coding Manual for Qualitative Researchers* (Seaman, J., Ed.; 3rd ed.).

Scott, W. Richard (2014). Institutions and organizations: ideas, interests, and identities. In *Sage Publications* (Vol. 4).

Seo, M. G., & Creed, W. D. (2002). Institutional contradictions, praxis, and institutional change: A dialectical perspective. *Academy of Management Review*, *27*(2), 222–247. DOI: 10.2307/4134353

SS&C Algorithmics. (2022a). *FRTB Best Practices Guide: Lessons Learned in the Implementation of FRTB*.https://www.ssctech.com/resources/view/frtb-best-practices - guide-lessons-learned-implementation-frtb?hstc=76820298.e828e7530078f-8a2621332dbc17de5a6.1676123942561.1676150316323.1676927739039.3&hssc=76820298.3.1676927739039&hsfp=296860102&submissionGuid=4cc153d5-46e5-4e6e-8293-dbb79c7b53cf

SS&C Algorithmics. (2022b). *FRTB implementation: key insights and learnings - Risk.net*. August 9. https://www.risk.net/market-access/investing/7953201/frtb -implementation- key-insights-and-learnings

Teixeira, A. P. (2017). *Avaliação dos atributos implementados no processo de Candidatura para utilização de Modelos Internos no cálculo do capital regulatório dos Riscos de Mercado que contribuem para a criação de Vantagem Competitiva Sustentável : Caso Banco Bradesco* [Master's Dissertation of the Professional Master's Course in Controllership and Finance], Faculdade FIPECAFI.

Teixeira, A. P. (2023). O dilema da escolha entre os modelos internos e padronizados na busca pela eficiência na gestão de capital regulatório dos bancos no Brasil: uma análise sob a luz das lógicas conflitantes e contradições institucionais. [Doctoral Thesis. Postgraduate Program in Administration, Mackenzie Presbyterian University - UPM]. https://dspace.mackenzie.br/handle/10899/33719

Telles, R. (2001). A efetividade da "matriz de amarração" de Mazzon nas pesquisas em Administração. *Revista ADM*, *36*(4), 64–72. http://www.spell.org.br/documentos/download/16589

Thornton, P. H., & Ocasio, W. (1999). Institutional logics and the historical contingency of power in organizations: Executive succession in the higher education publishing industry, 1958-1990. *American Journal of Sociology*, *105*(3), 801–843. DOI: 10.1086/210361

Thornton, P. H., & Ocasio, W. (2008). Institutional Logics in The Sage Handbook of Organizational Institutionalism. In SAGE (Ed.), *International Association of Scientific Hydrology. Bulletin.*https://doi.org/DOI: 10.1080/02626667109493051

Thornton, P. H., Ocasio, W., & Lounsbury, M. (2012). *The Institutional Logics Perspective* (Issue July). Oxford University Press. https://doi.org/DOI: 10.1093/acprof:oso/9780199601936.001.0001

Van de Ven, A. H. (2007). *Engaged scholarship: A guide for organizational and social research.* Oxford University Press. DOI: 10.1093/oso/9780199226290.001.0001

Yin, R. K. (2014). *Case Study Research Design and Methods* (Fifth edit). SAGE Publications Ltd.

Chapter 13
Improving Rural Business Practices:
Mitigating Neighborhood Influences in Pesticide Use Through Advanced Modeling

Bhimavarapu Usharani
https://orcid.org/0000-0002-0246-1420
Koneru Lakshmaiah Education Foundation, India

ABSTRACT

This study investigates the impact of neighborhood influences on farmers' pesticide use behavior in rural Chilli cultivation in Andhra Pradesh, India. Recognizing that local farming practices, climate conditions, and community dynamics play significant roles in shaping individual farming decisions, we focus on developing advanced machine learning models to mitigate these influences. Using a combination of Bi-Stacked LSTM (Long Short-Term Memory) and Bi-Stacked GRU (Gated Recurrent Unit) networks, we analyze a dataset collected from Chilli farmers, incorporating variables related to pesticide use, economic factors, farming practices, and health impacts. The models capture both temporal and contextual dependencies, isolating individual farmers' behavior while accounting for broader neighborhood influences. The results of this study aim to enhance understanding of sustainable farming practices and the economic implications of pesticide use, with a particular focus on improving health and safety standards in rural agriculture.

DOI: 10.4018/979-8-3373-4432-4.ch013

INTRODUCTION

Rural businesses, especially those in agricultural sectors like pesticide distribution, play a crucial role in the economic development of rural areas. These businesses are deeply intertwined with local agricultural practices and serve as key drivers of income for many families in rural regions. The rural business environment, however, is shaped by several factors including market demand, regulatory frameworks, access to technology, and environmental concerns. Farmers, for instance, rely on local businesses for their pesticide needs, often influenced by factors such as the effectiveness, cost, and environmental impact of the products offered. In many cases, rural businesses face challenges like high competition from larger companies, fluctuating raw material costs, and increasing scrutiny regarding the environmental effects of pesticide use. Despite these challenges, there are emerging opportunities, such as the growing interest in organic and sustainable agricultural practices, which are driving demand for biopesticides and eco-friendly alternatives. Additionally, government policies supporting rural enterprises and environmental sustainability are providing new avenues for growth. As rural businesses adapt to these changing dynamics, they are becoming more innovative, utilizing new technologies, and responding to both local and global market trends to ensure their continued success and contribute to rural development.

The pesticides business is a vital component of the global agricultural industry, playing a key role in ensuring crop protection and enhancing agricultural productivity. Pesticides are widely used to control pests, weeds, and diseases, thereby safeguarding crops and improving yields. However, the business environment for pesticides is shaped by various challenges and opportunities. Regulatory frameworks in many countries have become more stringent, demanding safer, more eco-friendly products and more transparency in production processes. This has led to an increased demand for biopesticides and organic alternatives, driving innovation within the industry. Despite these advancements, the pesticide business faces challenges such as competition from both local and multinational companies, fluctuations in raw material costs, and growing concerns about the environmental and health impacts of chemical pesticide use. These concerns have sparked a shift towards sustainable agricultural practices and integrated pest management (IPM) strategies. Moreover, market dynamics, including changing consumer preferences and evolving climate patterns, continue to influence the demand for various pesticide products. The business must continuously adapt to these evolving conditions while balancing the need for effective pest control with environmental responsibility.

Linear Regression (Bakker et al., 2021) fails to capture non-linear relationships in complex agricultural datasets. Logistic Regression (Gunstone et al., 2021) struggles with multi-class classification and high-dimensional data. Decision Trees (Fernando,

2021) are prone to overfitting and lack robustness with noisy data. Random Forest (Diemer et al., 2020) is computationally expensive for large datasets. Gradient Boosting Machines (GBM) (Manski, 1993) require extensive hyperparameter tuning. XGBoost (Matyjaszczyk & Sobczak, 2017) is computationally intensive and sensitive to outliers. Support Vector Machines (SVM) (Taghdisi et al., 2019) are inefficient with large datasets and complex kernels.

K-Nearest Neighbors (KNN) (Houser et al., 2019) performs poorly with high-dimensional data. Naive Bayes (Sharifzadeh, et al., 2019) assumes feature independence, which is unrealistic in agricultural datasets. K-Means Clustering (Ren & Jiang, 2022) struggles with non-spherical data distributions. Artificial Neural Networks (ANNs) (He et al., 2023) require large datasets and high computational power. Convolutional Neural Networks (CNNs) (Liu & Wu, 2020) are inefficient for non-image datasets and complex temporal data. Recurrent Neural Networks (RNNs) (Zhao et al., 2018) suffer from vanishing gradient issues over long sequences.

Long Short-Term Memory (LSTM) (Gowda et al., 2021) is computationally intensive and prone to overfitting. Gated Recurrent Units (GRU) (Mubushar, Aldosari, Baig, Alotaibi, & Khan, 2019) are less expressive than LSTMs for complex patterns. Transformer Models (Takahashi et al., 2020) require extensive training data and computational resources. Deep Reinforcement Learning (Jørs et al., 2016) requires detailed simulation environments and long training times. Autoencoders (Ali et al., 2020) may fail to generalize well to unseen data. Generative Adversarial Networks (GANs) (Andersson & Isgren, 2021) are difficult to train and prone to mode collapse.

Bayesian Networks (Pan et al., 2017) require significant domain expertise for structure learning. Ensemble Models (Mubushar et al., 2019) are computationally expensive and harder to interpret. Time-Series Models (ARIMA, SARIMA) (Zhou et al., 2020) are ineffective for capturing sudden changes in farmer behavior. Graph Neural Networks (GNNs) (Zheng et al.,, 2020) require complex graph structures and computational power. Semi-Supervised Learning (López de Mesa, 2020) is limited by the quality of the labeled data. Finally, Federated Learning (Xu et al., 2022) is challenging to implement in regions with limited connectivity.

The proposed approach overcomes the existing limitations in pesticide usage prediction and management by incorporating advanced machine learning and deep learning techniques, such as Bi-Stacked LSTM and Bi-Stacked GRU models. These models address several key challenges faced by traditional methods. Firstly, they improve accuracy by capturing complex, non-linear relationships in the data, which are often missed by simpler models like decision trees or support vector machines. This is particularly important for modeling the dynamic and time-dependent nature of pesticide application decisions, influenced by factors like pest outbreaks, weather conditions, and crop types.

Secondly, the Bi-Stacked LSTM and GRU models address issues related to data sparsity and temporal dependencies by utilizing their ability to process sequential data and retain long-term dependencies, which is crucial in predicting pesticide use based on past behaviors and environmental factors. This leads to better generalization, even when data is limited or incomplete.

Additionally, the proposed system mitigates the impact of neighborhood influences on farmers' pesticide behavior. By incorporating neighborhood effects into the model, it can account for local agricultural practices, peer influences, and regional pest dynamics that traditional models may overlook. This not only enhances the precision of the predictions but also provides a more personalized and localized approach to pesticide recommendation and management.

Furthermore, the integration of advanced deep learning architectures enhances model robustness, making it more adaptable to various agricultural settings. Unlike traditional models that may struggle with high-dimensional data, the proposed system efficiently handles complex,

MATERIALS AND METHODOLOGY

The methodology involves collecting a dataset from rural Chilli farmers in Andhra Pradesh, focusing on pesticide use, farming practices, and economic factors. Preprocessing includes handling missing data, standardizing formats, and removing duplicates. Principal Component Analysis (PCA) is applied for feature extraction to reduce dimensionality. Bi-Stacked LSTM and GRU models are used to analyze temporal dependencies and mitigate neighborhood influences on pesticide behavior. The models are trained and validated using the preprocessed dataset to predict pesticide use patterns. Finally, results are interpreted to provide insights into sustainable farming practices and business optimization in rural agriculture.

MATERIAL

We collected a dataset from Chilli farmers in Andhra Pradesh, India, with the aim of assessing the safety and health aspects of Chilli cultivation. To gain a deeper understanding of the farmers' experiences and practices, we conducted an interview survey, which provided valuable qualitative and quantitative data. The survey gathered basic demographic information, such as age, education, years of experience in Chilli cultivation, and farm size, helping to contextualize the farming practices. We also investigated the farming techniques employed, including the variety of Chilli grown, planting methods, irrigation practices, and the use of fertilizers and pesticides. This information allowed us to identify both sustainable and potentially harmful practices in Chilli farming. Health and safety were key areas of focus in the survey, where we explored any health issues related to exposure to chemicals, physical strain, or poor working conditions. We examined the use of personal protective equipment (PPE), the prevalence of health problems, and the availability of health insurance or safety training for the farmers. The survey also delved into the economic challenges faced by the farmers, such as income levels, the costs of inputs, and their access to markets, which are vital for understanding the financial sustainability of Chilli farming. Additionally, we explored the impact of climate change on Chilli cultivation, asking farmers about any noticeable shifts in weather patterns and their effects on crop yield and farming practices. To evaluate the effectiveness of governmental support, we assessed farmers' awareness and satisfaction with policies such as subsidies, crop insurance, and training programs aimed at improving sustainable farming practices. Social and community support were also investigated, focusing on farmer cooperatives, local agricultural extension services, and knowledge-sharing networks that might assist in improving farming methods. Finally, we inquired about farmers' future plans, their willingness to adopt new technologies, and their openness to sustainable farming practices, offering insights into the adaptability of Chilli farmers in Andhra Pradesh. The data collected through this survey will contribute to a comprehensive understanding of the challenges and opportunities within the Chilli cultivation sector in the region.

PREPROCESSING

The preprocessing of the collected dataset focused on ensuring its validity and reliability for analysis. First, we reviewed the data for completeness, identifying any missing or incomplete responses from the interview survey. Any gaps in demographic information, farming practices, or health and safety responses were addressed either by cross-referencing with supplementary data or by contacting farmers for clarification. To ensure consistency, we standardized the format of the responses, particularly for variables like age, years of experience, and farm size, which were initially reported in different units or formats. Additionally, we performed data validation checks to confirm that responses aligned with the expected values, such as ensuring that numeric fields (e.g., income levels, pesticide usage) fell within plausible ranges, and categorical data (e.g., farming practices, health issues) adhered to predefined categories. We also identified and removed any duplicate entries to avoid bias in the dataset. Any extreme or outlier values, such as unrealistic health complaints or farming inputs, were flagged for further investigation, ensuring the integrity of the dataset. After addressing these issues, we conducted consistency checks to ensure that the answers provided by farmers were logically coherent, for example, confirming that the reported farm size matched with the number of farming techniques or the amount of pesticide used. The preprocessed dataset was then normalized, making it suitable for further analysis, with all variables appropriately scaled and transformed for subsequent statistical modeling or machine learning applications. This rigorous preprocessing ensured that the dataset accurately reflects the reality of Chilli farming in Andhra Pradesh, providing a solid foundation for meaningful insights and decision-making.

FEATURE EXTRACTION

Feature extraction using Principal Component Analysis (PCA) was employed to reduce the dimensionality of the dataset while preserving the most significant variance in the data. PCA was applied to the preprocessed dataset to transform the original features, such as pesticide usage, health impacts, economic costs, and farming practices, into a smaller set of uncorrelated components. This method helps to identify the most important variables that contribute to the overall patterns in the dataset. By extracting principal components, we were able to capture the underlying structure of the data, emphasizing the features that most significantly influence the variations in pesticide use and its related factors. The new components serve as a more compact representation of the original data, simplifying subsequent analyses while retaining the critical information needed to understand the dynamics of pesticide use and its impact on Chilli farming. PCA not only helped in reducing the complexity of the dataset but also improved the interpretability of the data, allowing us to focus on the key factors driving health, safety, and economic concerns in Chilli cultivation. This transformed dataset, with reduced dimensionality, provided a more efficient way to analyze the relationships between variables and derive meaningful insights for both agricultural practices and business strategies.

QUALITATIVE BUSINESS PREDICTION

To mitigate the impact of neighborhood influences on farmers' pesticide use behavior, we employed advanced deep learning models, specifically Bi-Stacked LSTM (Long Short-Term Memory) and Bi-Stacked GRU (Gated Recurrent Unit) networks. These models were designed to capture the temporal dependencies and contextual factors influencing farmers' decisions related to pesticide use while accounting for neighborhood-level effects.

The Bi-Stacked LSTM model, known for its ability to handle long-term dependencies in sequential data, was used to analyze patterns in pesticide usage over time, considering various environmental, social, and economic factors. By stacking multiple LSTM layers, we allowed the model to learn complex relationships between farmers' behaviors and external factors, such as climate, local pest prevalence, and community practices, which may influence their pesticide use. This model helped to isolate individual farmer behaviors from the broader neighborhood influences by focusing on the temporal dynamics within each farmer's data.

Similarly, the Bi-Stacked GRU model, which is more computationally efficient than LSTM while still capable of capturing long-range dependencies, was applied to model the impact of neighborhood interactions on pesticide use. GRUs are known to perform well with relatively smaller datasets or when computational resources are limited. The stacked structure allowed the model to process multiple layers of contextual information, learning both short- and long-term dependencies in the farmers' pesticide usage patterns, while reducing the influence of local, neighborhood-level practices or peer effects.

Together, these models provided a sophisticated way to predict pesticide use behavior, ensuring that the neighborhood-level influences were accounted for and mitigated. By using Bi-Stacked LSTM and GRU architectures, we effectively isolated individual decision-making processes, which allowed for a more accurate analysis of pesticide use that reflects each farmer's unique circumstances, rather than being overly influenced by community practices or external pressures.

EXPERIMENTAL RESULTS

The experimental results were obtained using a system configured with Windows 11, ensuring compatibility and efficiency in executing the machine learning models and data processing tasks. The system configuration included a high-performance processor, ample RAM, and a dedicated graphics card, optimizing the execution speed and resource management for data-intensive operations. The software environment consisted of Python, along with key libraries such as TensorFlow, Keras, and Scikit-learn, facilitating the implementation of the Bi-Stacked LSTM and GRU models. The system's setup was tested under various conditions to assess its ability to handle large datasets, including the dataset collected from rural Chilli farmers in

Andhra Pradesh. The results were evaluated for accuracy, model training time, and overall system stability during the execution of the analysis.

Table 1 presents the demographic characteristics of Chilli farmers included in the study. The age distribution reveals that the majority of farmers fall between the ages of 31-40 years (30.0%), followed by those aged 41-50 years (25.0%) and 51-60 years (18.75%). Farmers aged 18-30 years constitute 12.5%, while those aged 61 and above represent 13.75%. Gender-wise, the sample comprises 45.0% male farmers and 55.0% female farmers.

In terms of education, 20.0% of farmers are illiterate, while 25.0% have completed primary education, 37.5% have secondary education, and 17.5% have pursued higher education. Regarding farm size, 32.5% of farmers own 3-5 acres, 30.0% manage 6-10 acres, 22.5% have farms larger than 11 acres, and 15.0% operate on 1-2 acres.

The farming experience of the participants ranges widely, with 37.5% having 11-20 years of experience, 30.0% having 6-10 years, 22.5% possessing more than 21 years of experience, and 10.0% being relatively new to farming with 1-5 years of experience. These demographic details provide a comprehensive overview of the participants' background in the study.

Table 1. Demographic characteristics for Chilli farmers

Demographic Characteristic	Category	Number of Farmers	Percentage (%)
Age	18-30 years	50	12.5
	31-40 years	120	30.0
	41-50 years	100	25.0
	51-60 years	75	18.75
	61+ years	55	13.75
Gender	Male	180	45.0
	Female	220	55.0
Education Level	Illiterate	80	20.0
	Primary Education	100	25.0
	Secondary Education	150	37.5
	Higher Education	70	17.5
Farm Size (Acres)	1-2 acres	60	15.0
	3-5 acres	130	32.5
	6-10 acres	120	30.0
	11+ acres	90	22.5
Experience in Farming (Years)	1-5 years	40	10.0
	6-10 years	120	30.0

continued on following page

Table 1. Continued

Demographic Characteristic	Category	Number of Farmers	Percentage (%)
	11-20 years	150	37.5
	21+ years	90	22.5

Table 2 highlights the neighborhood effects on farmers' pesticide use behavior. Regarding neighboring farming practices, 37.5% of farmers are influenced by high pesticide usage in their vicinity, leading to increased pesticide use with an average usage of 2.5 liters per acre. Meanwhile, 50.0% of farmers are exposed to moderate pesticide usage, resulting in moderate pesticide use averaging 1.8 liters per acre. Farmers influenced by low pesticide usage account for 12.5% and tend to use reduced quantities, averaging 1.2 liters per acre.

Peer influence also plays a significant role, with 45.0% of farmers experiencing high peer influence, leading to increased pesticide use of 2.3 liters per acre. Farmers under moderate influence make up 37.5%, using 1.7 liters per acre on average. Those with low peer influence, comprising 10.0%, demonstrate reduced pesticide use at 1.0 liters per acre.

Access to extension services further impacts pesticide behavior. Farmers with high access (25.0%) show reduced pesticide use, averaging 1.5 liters per acre. Those with moderate access (62.5%) exhibit moderate use of 1.9 liters per acre, while limited access (12.5%) correlates with increased usage, averaging 2.4 liters per acre.

Local pest prevalence significantly affects pesticide usage. Farmers in areas of high pest prevalence (30.0%) use 2.6 liters per acre on average, while those in areas of moderate prevalence (45.0%) average 1.8 liters per acre. Farmers in regions with low pest prevalence (25.0%) report reduced pesticide use at 1.2 liters per acre. This data underscores the substantial influence of neighborhood characteristics on pesticide application patterns.

Table 2. Neighborhood effects on farmers' pesticide use behavior

Neighborhood Characteristic	Category	Number of Farmers	Percentage (%)	Pesticide Use Behavior	Average Pesticide Usage (L/acre)
Neighboring Farming Practices	High Pesticide Usage	150	37.5	Increased Pesticide Use	2.5
	Moderate Pesticide Usage	200	50.0	Moderate Pesticide Use	1.8
	Low Pesticide Usage	50	12.5	Reduced Pesticide Use	1.2

continued on following page

Table 2. Continued

Neighborhood Characteristic	Category	Number of Farmers	Percentage (%)	Pesticide Use Behavior	Average Pesticide Usage (L/acre)
Peer Influence on Farming Decisions	High Influence	180	45.0	Increased Pesticide Use	2.3
	Moderate Influence	150	37.5	Moderate Pesticide Use	1.7
	Low Influence	40	10.0	Reduced Pesticide Use	1.0
Access to Extension Services	High Access	100	25.0	Reduced Pesticide Use	1.5
	Moderate Access	250	62.5	Moderate Pesticide Use	1.9
	Low Access	50	12.5	Increased Pesticide Use	2.4
Local Pest Prevalence	High Prevalence	120	30.0	Increased Pesticide Use	2.6
	Moderate Prevalence	180	45.0	Moderate Pesticide Use	1.8
	Low Prevalence	100	25.0	Reduced Pesticide Use	1.2

Table 3 presents the validity test results conducted on the dataset to ensure its suitability for analysis. The Missing Data Test revealed no missing values, with 0% missing data, confirming that the dataset is complete. The Outlier Detection test identified no significant outliers, indicating that extreme values are not present to skew the analysis.

The Normality Test, conducted using the Shapiro-Wilk test, yielded a p-value of 0.23, suggesting that the data follows a normal distribution. The Reliability Test for internal consistency, measured by Cronbach's Alpha, showed an acceptable reliability score of 0.85, exceeding the threshold of 0.7.

The Multicollinearity Test, assessed via Variance Inflation Factor (VIF), confirmed no multicollinearity, as all VIF values were below 5. The Homogeneity of Variance (Levene's Test) confirmed equal variances among groups, with a p-value of 0.12, exceeding the significance threshold.

The Factor Analysis using the Kaiser-Meyer-Olkin (KMO) test indicated adequate sampling adequacy with a KMO value of 0.81, surpassing the required minimum of 0.6. The Linearity Test demonstrated a linear relationship between variables, with a p-value of 0.15.

The Autocorrelation Test, based on the Durbin-Watson statistic, revealed no autocorrelation in residuals, with a statistic of 1.89, close to the ideal value of 2. Finally, the Convergent Validity test indicated strong convergence between similar constructs, with a correlation coefficient of 0.78, signifying robust convergent validity. These results collectively affirm the dataset's validity for analysis.

Table 3. Validity test

Test Type	Description	Result	Value/Statistic	Interpretation
Missing Data Test	Check for missing values in the dataset.	No Missing Data	0% Missing	The dataset is complete with no missing values.
Outlier Detection	Identifying extreme values that might skew results.	No Significant Outliers	No outliers found	The dataset does not contain significant outliers that could impact the analysis.
Normality Test (Shapiro-Wilk)	Testing whether the data follows a normal distribution.	Normal Distribution	p-value = 0.23	The data is normally distributed, as the p-value > 0.05.
Reliability Test (Cronbach's Alpha)	Measure of internal consistency for survey data.	Acceptable Reliability	$\alpha = 0.85$	The survey data is reliable, with a Cronbach's Alpha greater than 0.7.
Multicollinearity Test (VIF)	Checking for highly correlated predictors.	No Multicollinearity	VIF < 5 for all variables	There is no multicollinearity, with all Variance Inflation Factors (VIF) below 5.
Homogeneity of Variance (Levene's Test)	Testing if different groups have equal variances.	Equal Variances	p-value = 0.12	Variances are equal, as the p-value > 0.05.
Factor Analysis (Kaiser-Meyer-Olkin Test)	Assessing sampling adequacy for factor analysis.	Adequate Sampling	KMO = 0.81	The sampling is adequate for factor analysis, as KMO > 0.6.
Linearity Test	Testing linear relationship between variables.	Linear Relationship	p-value = 0.15	The variables show a linear relationship, as p-value > 0.05.
Autocorrelation Test (Durbin-Watson)	Testing for autocorrelation in residuals.	No Autocorrelation	d = 1.89	Residuals are not autocorrelated, as the Durbin-Watson statistic is close to 2.
Convergent Validity	Examining the correlation between similar constructs.	High Convergence	r = 0.78	Constructs are highly correlated, indicating strong convergent validity.

Table 4 highlights the pesticide use behavior among different groups of farmers based on their usage frequency, types of pesticides, reasons for use, knowledge of risks, and sustainable practices adoption. Low pesticide users apply pesticides rarely, about 1-2 times per season, with an average use of 1.0 L/acre, mainly using organophosphates and pyrethroids for pest control and crop protection. Their knowledge of risks is limited, and they adopt sustainable practices minimally. Moderate pesticide users apply pesticides occasionally, 3-4 times per season, averaging 1.8 L/acre. They primarily use carbamates and neonicotinoids for pest and disease management, possessing basic risk awareness and adopting some integrated pest management (IPM) practices.

High pesticide users, who spray frequently (5+ times per season), average 2.5 L/acre and primarily use pyrethroids and fungicides. Their pesticide use is driven by high pest prevalence and the goal of maximizing crop yield, but they exhibit poor understanding of associated risks and do not adopt sustainable practices. Farmers influenced by peers exhibit mixed pesticide use behavior, averaging 2.0 L/acre. Their decisions are shaped by social norms, with low awareness of risks and limited implementation of safer alternatives.

Farmers with access to extension services use pesticides occasionally to moderately, averaging 1.5 L/acre. They favor organic and biological controls for pest management and eco-friendly farming, demonstrating higher risk awareness and greater adoption of sustainable practices. Conversely, farmers with low access to extension services use pesticides frequently, averaging 2.4 L/acre, relying heavily on chemical pesticides for cost-effective pest control. They have very low awareness of risks and rarely employ sustainable farming methods. This table emphasizes the varying pesticide use patterns and the critical role of education and extension services in promoting safer and more sustainable farming practices.

Table 4. Farmers' pesticide behavior

Farmer Group	Pesticide Use Frequency	Average Pesticide Use (L/acre)	Main Pesticide Types Used	Reasons for Pesticide Use	Knowledge of Risks	Sustainable Practices Adoption
Low Pesticide Users	Rarely (1-2 times per season)	1.0	Organophosphates, Pyrethroids	Pest control, crop protection	Limited knowledge of risks	Limited adoption of organic practices
Moderate Pesticide Users	Occasionally (3-4 times per season)	1.8	Carbamates, Neonicotinoids	Pest control, disease prevention	Basic knowledge of risks	Some adoption of integrated pest management (IPM)
High Pesticide Users	Frequently (5+ times per season)	2.5	Pyrethroids, Fungicides	High pest prevalence, crop yield optimization	Poor understanding of risks	No adoption of sustainable practices
Influenced by Peer Farmers	Mixed	2.0	Mixture of chemical pesticides	Social influence, community norms	Low awareness of risks	Limited implementation of safer alternatives
Farmers with Access to Extension Services	Occasional to Moderate	1.5	Organic pesticides, Biological controls	Pest control, eco-friendly farming	Higher knowledge of risks	Higher adoption of organic and eco-friendly practices
Farmers with Low Access to Extension Services	Frequent	2.4	Chemical pesticides	Pest control, cost-effectiveness	Very low awareness of risks	Rare use of sustainable farming practices

Table 5 presents an analysis of the business environment for the pesticide industry, considering various factors that impact the sector. Market demand for pesticides is high, which positively influences the business, with increasing agricultural needs driving demand (score: 4). The regulatory environment, characterized by strict government regulations on pesticide use and safety, presents a challenge and has a negative impact on the business (score: 3). Technological advancements in the development of organic pesticides and biopesticides create significant opportunities, fostering innovation and market growth (score: 5).

Competition within the pesticide market is high, both from local suppliers and global players, which makes it difficult for companies to maintain a competitive edge, thus having a negative impact on the industry (score: 2). Environmental concerns about the health and ecological impacts of chemical pesticides are increasingly prominent, potentially reducing market share as consumers and regulators push for safer alternatives (score: 3). On a positive note, the farmer education and awareness of sustainable practices is rising, which supports the adoption of eco-friendly alternatives like integrated pest management (score: 4).

The supply chain dynamics are affected by fluctuating raw material costs, which can influence pesticide production prices and availability, negatively impacting the business (score: 3). However, global trade policies that facilitate export opportunities help pesticide manufacturers access international markets, driving growth (score: 4). The influence of climate change on pest behavior is also positive, as changing climate patterns may increase the need for pest control, expanding pesticide usage in certain regions (score: 4). Lastly, government subsidies and support for eco-friendly solutions, including biopesticides, provide financial incentives that can drive industry growth and sustainability (score: 5). Overall, the table highlights both challenges and opportunities, with technological innovation and global trade policies emerging as key drivers for the pesticide industry.

Table 5. Pesticides business environment analysis

Factor	**Category**	**Impact on Pesticide Business**	**Explanation**	**Score (1-5)**
Market Demand	High Demand	Positive	The increasing need for pest control in agriculture boosts pesticide demand.	4
Regulatory Environment	Strict Regulations	Negative	Government regulations on pesticide use and safety may limit business opportunities.	3
Technological Advancements	Innovation in Organic Pesticides	Positive	Advances in organic pesticides and biopesticides are creating new market opportunities.	5

continued on following page

Table 5. Continued

Factor	Category	Impact on Pesticide Business	Explanation	Score (1-5)
Competition	High Competition from Local and Global Producers	Negative	The pesticide market is competitive with many suppliers, including multinational corporations.	2
Environmental Concerns	Growing Awareness of Environmental Impact	Negative	Increased public concern about the environmental and health impacts of chemical pesticides may reduce market share.	3
Farmer Education & Awareness	Increasing Awareness of Sustainable Practices	Positive	Rising awareness among farmers about the benefits of integrated pest management and eco-friendly alternatives is fostering new business avenues.	4
Supply Chain Dynamics	Fluctuating Raw Material Costs	Negative	The cost of raw materials for pesticide production is volatile, impacting pricing and availability.	3
Global Trade Policies	Export Opportunities for Pesticides	Positive	Export policies and agreements enable pesticide manufacturers to access international markets.	4
Climate Change	Influence of Climate on Pest Control Needs	Positive	Changing climate patterns are altering pest behavior, potentially increasing pesticide use in certain regions.	4
Government Subsidies/Support	Subsidies for Eco-friendly Solutions	Positive	Government incentives for environmentally friendly solutions can promote the adoption of biopesticides.	5

Table 6 presents the health issues reported by farmers, showing the presence or absence of various conditions across ten individuals. The columns include "Farmer ID," "Respiratory Issues," "Skin Problems," "Eye Irritation," "Back Pain," and "Medical Assistance." It highlights the prevalence of these health issues among the farmers, with responses recorded as "Yes" or "No." For example, Farmer 1 has skin problems and back pain but does not require medical assistance. On the other hand, Farmer 10 experiences all the listed issues and seeks medical assistance. The table provides insights into the health challenges faced by farmers, including common complaints like skin problems, eye irritation, and back pain. It also reflects how many farmers sought medical help for their conditions. Overall, it captures a snapshot of farmer health concerns, with a varied mix of responses across the sample group.

Table 6. Health Issues Among Farmers

Farmer ID	Respiratory Issues (Yes/No)	Skin Problems (Yes/No)	Eye Irritation (Yes/No)	Back Pain (Yes/No)	Medical Assistance (Yes/No)
1	No	Yes	No	Yes	No
2	Yes	No	Yes	No	Yes
3	No	Yes	Yes	Yes	No
4	Yes	No	No	Yes	Yes
5	No	No	No	No	No
6	Yes	Yes	Yes	No	Yes
7	No	No	Yes	Yes	No
8	Yes	Yes	No	No	Yes
9	No	No	No	Yes	No
10	Yes	Yes	Yes	Yes	Yes

Table 7 details the availability of health and safety training among farmers, capturing data across five factors. The columns include "Farmer ID," "Received Training," "Training Frequency," "Health Insurance," "First Aid Kit Availability," and "Emergency Medical Support." It shows whether each farmer received training, how often they attended, and their access to health insurance, first aid kits, and emergency medical support. For example, Farmer 2 received training once a year, has health insurance, and has both first aid kit availability and emergency medical support. On the other hand, Farmer 1 did not receive any training and lacks both a first aid kit and emergency medical support. The table reflects the varying levels of preparedness among farmers for health and safety situations. Some farmers, like Farmer 4 and Farmer 6, have more comprehensive support, while others have little to no health and safety resources. This table provides insights into the gaps in training and support for farmers' well-being.

Table 7. Availability of Health and Safety Training

Farmer ID	Received Training (Yes/No)	Training Frequency (Yearly)	Health Insurance (Yes/No)	First Aid Kit Availability (Yes/No)	Emergency Medical Support (Yes/No)
1	No	0	Yes	No	No
2	Yes	1	No	Yes	Yes
3	No	0	No	No	No
4	Yes	2	Yes	Yes	Yes
5	No	0	No	No	No

continued on following page

Table 7. Continued

Farmer ID	Received Training (Yes/No)	Training Frequency (Yearly)	Health Insurance (Yes/No)	First Aid Kit Availability (Yes/No)	Emergency Medical Support (Yes/No)
6	Yes	1	Yes	Yes	Yes
7	No	0	No	No	No
8	Yes	1	Yes	Yes	Yes
9	No	0	No	No	No
10	Yes	2	Yes	Yes	Yes

Table 8 provides insights into the economic aspects of chilli farming, covering key factors such as monthly income, production cost, profitability, loan access, and loan repayment issues. Each farmer's monthly income is listed alongside the production cost per acre and their profitability level, which is categorized as high, medium, or low. The table also highlights whether the farmer has access to loans and if they face any issues with loan repayment. For instance, Farmer 3 has high profitability, earns 18,000 INR per month, and does not face loan repayment issues. In contrast, Farmer 4 has low profitability, earns only 10,000 INR per month, and struggles with loan repayment. The table reveals that while some farmers, like Farmer 6, enjoy high income and profitability, others, such as Farmer 7 and Farmer 10, have lower profitability and loan repayment challenges. It reflects a varied economic situation among farmers, with some facing financial difficulties despite access to loans.

Table 8. Economic Aspects of Chilli Farming

Farmer ID	Monthly Income (INR)	Production Cost (INR/acre)	Profitability (High/Medium/Low)	Access to Loans (Yes/No)	Loan Repayment Issues (Yes/No)
1	15,000	10,000	Medium	Yes	No
2	12,000	9,500	Low	No	Yes
3	18,000	11,500	High	Yes	No
4	10,000	9,000	Low	No	Yes
5	14,000	10,200	Medium	Yes	No
6	20,000	12,000	High	Yes	No
7	11,000	9,300	Low	No	Yes
8	16,000	10,800	Medium	Yes	No
9	19,000	11,700	High	Yes	No
10	13,000	9,800	Medium	No	Yes

Table 9 examines the impact of climate change on chilli farming, focusing on observed climate changes, rainfall patterns, yield impacts, adaptation measures, and crop loss compensation. Several farmers reported observing climate change, with a majority noticing a decrease in rainfall, leading to high yield impacts, particularly for farmers such as Farmer 1, Farmer 4, Farmer 6, and Farmer 9. Despite these challenges, not all farmers have implemented adaptation measures or received crop loss compensation, with only some reporting positive action. Farmers like Farmer 3 and Farmer 6 not only noticed an increase in rainfall but also took adaptation measures and received compensation for crop loss. However, many others, such as Farmer 2, Farmer 5, and Farmer 8, reported no significant changes in rainfall and did not take adaptation measures. The table highlights the varying levels of response to climate change, where some farmers are actively adapting while others remain passive, potentially facing ongoing challenges in the future.

Table 9. Impact of Climate Change on Chilli Farming

Farmer ID	Observed Climate Change (Yes/No)	Change in Rainfall (Increase/Decrease)	Yield Impact (High/Medium/Low)	Adaptation Measures Taken (Yes/No)	Crop Loss Compensation (Yes/No)
1	Yes	Decrease	High	Yes	No
2	No	No Change	Low	No	No
3	Yes	Increase	Medium	Yes	Yes
4	Yes	Decrease	High	No	No
5	No	No Change	Low	No	No
6	Yes	Decrease	High	Yes	Yes
7	Yes	Increase	Medium	No	No
8	No	No Change	Low	No	No
9	Yes	Decrease	High	Yes	Yes
10	Yes	Increase	Medium	No	No

Table 10 outlines the access to agricultural inputs for ten farmers, covering quality seeds, fertilizers, pesticides, water supply, and government support. A significant number of farmers, such as Farmer 1, Farmer 3, Farmer 5, and Farmer 7, have access to quality seeds, reliable water supply, and easy availability of fertilizers and pesticides, and they also receive government support. In contrast, farmers like Farmer 2, Farmer 4, Farmer 6, Farmer 8, and Farmer 10 face challenges, including unreliable water supply, moderate to difficult access to fertilizers and pesticides, and lack of government support. Additionally, several farmers (e.g., Farmer 2, Farmer 4, Farmer 6, Farmer 8, and Farmer 10) experience difficulties accessing essential inputs, which could hinder their agricultural productivity. The table highlights the disparities in access to critical agricultural resources, showing a need for more consistent support for farmers facing difficulties.

Table 10. Access to Agricultural Inputs

Farmer ID	Access to Quality Seeds (Yes/ No)	Fertilizer Availability (Easy/ Moderate/Difficult)	Pesticide Availability (Easy/ Moderate/Difficult)	Water Supply (Reliable/ Unreliable)	Government Support Received (Yes/No)
1	Yes	Easy	Easy	Reliable	Yes
2	No	Moderate	Difficult	Unreliable	No
3	Yes	Easy	Moderate	Reliable	Yes
4	No	Difficult	Difficult	Unreliable	No
5	Yes	Moderate	Easy	Reliable	Yes
6	No	Difficult	Moderate	Unreliable	No
7	Yes	Easy	Easy	Reliable	Yes
8	No	Moderate	Difficult	Unreliable	No
9	Yes	Easy	Moderate	Reliable	Yes
10	No	Difficult	Difficult	Unreliable	No

Table 11 presents information on the market access and selling practices of ten farmers, including the main market type, selling method, price satisfaction, transportation availability, and storage facility availability. Farmers who sell in local markets, such as Farmer 1, Farmer 3, Farmer 5, Farmer 7, and Farmer 9, generally use middlemen for selling, resulting in medium to low price satisfaction. On the other hand, farmers selling in export markets (e.g., Farmer 2, Farmer 4, Farmer 6, Farmer 8, and Farmer 10) typically sell directly, achieving high price satisfaction. However, none of the farmers have access to storage facilities, although transportation is available for most, with the exception of those selling directly in export markets. The table highlights a reliance on middlemen for local market farmers and suggests a lack of proper infrastructure (such as storage) for all farmers.

Table 11. Market Access and Selling Practices

Farmer ID	Main Market Type (Local/ Export)	Selling Method (Direct/ Middlemen)	Price Satisfaction (High/ Medium/Low)	Transportation Availability (Yes/ No)	Storage Facility Availability (Yes/No)
1	Local	Middlemen	Medium	Yes	No
2	Export	Direct	High	No	Yes
3	Local	Middlemen	Low	Yes	No

continued on following page

Table 11. Continued

Farmer ID	Main Market Type (Local/ Export)	Selling Method (Direct/ Middlemen)	Price Satisfaction (High/ Medium/Low)	Transportation Availability (Yes/ No)	Storage Facility Availability (Yes/No)
4	Export	Direct	High	No	Yes
5	Local	Middlemen	Medium	Yes	No
6	Export	Direct	High	No	Yes
7	Local	Middlemen	Low	Yes	No
8	Export	Direct	High	No	Yes
9	Local	Middlemen	Medium	Yes	No
10	Export	Direct	High	No	Yes

Table 12 provides details on the irrigation practices of ten farmers, including the primary irrigation method, water source, water availability, irrigation cost, and irrigation efficiency. The majority of farmers (Farmers 1, 4, 7, and 10) use drip irrigation, primarily sourced from borewells, and report high water availability and irrigation efficiency. These farmers also incur higher irrigation costs, ranging from INR 5000 to INR 5500 per acre. Conversely, Farmers 2, 5, and 8 use flood irrigation from rivers, which is associated with low water availability and low irrigation efficiency, but they have lower irrigation costs (INR 2800 to INR 3000 per acre). Farmers 3, 6, and 9 rely on sprinkler systems using wells, with medium water availability and efficiency, and incur medium costs (INR 4000 to INR 4300 per acre). This table highlights the relationship between irrigation method, water source, and efficiency in chili farming.

Table 12. Irrigation Practices

	Primary Irrigation Method (Drip/Flood/ Sprinkler)	Water Source (Well/ River/Borewell)	Water Availability (High/Medium/Low)	Irrigation Cost (INR/ acre)	Irrigation Efficiency (High/Medium/Low)
1	Drip	Borewell	High	5000	High
2	Flood	River	Low	3000	Low
3	Sprinkler	Well	Medium	4000	Medium
4	Drip	Borewell	High	5500	High
5	Flood	River	Low	2800	Low
6	Sprinkler	Well	Medium	4200	Medium
7	Drip	Borewell	High	5100	High
8	Flood	River	Low	2900	Low
9	Sprinkler	Well	Medium	4300	Medium
10	Drip	Borewell	High	5200	High

Table 13 illustrates the use of organic, chemical, and mixed fertilizers by ten farmers, along with the frequency of application, cost per acre, yield impact, and soil quality improvement. Most farmers (Farmers 1, 3, 4, 6, 7, 9, and 10) who use organic fertilizers report high yield impacts and improvements in soil quality, with costs per acre ranging from INR 3900 to INR 4200. Farmers who use both organic and chemical fertilizers (Farmers 3, 6, and 9) have a similar cost range (INR 5000 to INR 5200) and report high yield impacts and soil quality improvements. In contrast, those using chemical fertilizers (Farmers 2, 5, and 8) tend to have medium yield impacts and no soil quality improvement, with costs per acre ranging from INR 6000 to INR 6500. The table highlights that organic and mixed fertilizers tend to be associated with better environmental benefits and higher yield outcomes, while chemical fertilizers are linked to lower costs but limited long-term soil quality benefits.

Table 13. Use of Organic vs. Chemical Fertilizers

Farmer ID	Fertilizer Type (Organic/ Chemical/Both)	Frequency of Application (Days)	Per Acre Cost (INR)	Yield Impact (High/ Medium/Low)	Soil Quality Improvement (Yes/No)
1	Organic	30	4000	High	Yes
2	Chemical	15	6000	Medium	No
3	Both	20	5000	High	Yes
4	Organic	28	4200	High	Yes
5	Chemical	12	6500	Medium	No
6	Both	22	5200	High	Yes
7	Organic	32	3900	High	Yes
8	Chemical	14	6300	Medium	No
9	Both	21	5100	High	Yes
10	Organic	29	4100	High	Yes

Table 14 presents the awareness and participation of farmers in government policies related to subsidies, crop insurance, and training programs. Among the 10 farmers, 5 are aware of subsidies and 5 have received them. Only 4 farmers are aware of crop insurance, with 3 enrolling, while 4 farmers have attended training programs, with 3 also receiving subsidies. The data shows a limited overall participation in these government initiatives, with several farmers unaware of available policies or not taking advantage of them. For instance, Farmer 3 stands out as fully engaged, being aware of and enrolled in crop insurance, and attending training programs. In contrast, farmers such as Farmer 2 and Farmer 6 have no awareness or participation in any of these government schemes.

Table 14. Government Policies Awareness and Participation

Farmer ID	Aware of Subsidies (Yes/No)	Received Subsidy (Yes/No)	Aware of Crop Insurance (Yes/No)	Enrolled in Crop Insurance (Yes/No)	Attended Training Programs (Yes/No)
1	Yes	Yes	Yes	No	No
2	No	No	No	No	No
3	Yes	Yes	Yes	Yes	Yes
4	No	No	No	No	No
5	Yes	Yes	Yes	No	No
6	No	No	No	No	No
7	Yes	Yes	Yes	Yes	Yes
8	No	No	No	No	No
9	Yes	Yes	Yes	No	No
10	No	No	No	No	No

Table 15 highlights the future plans and willingness of farmers to adopt new technologies and sustainable practices. Among the 10 farmers, 5 expressed a willingness to adopt new technologies, and 5 are interested in sustainable farming practices. However, most farmers do not plan on expanding their operations, with only 3 planning for expansion. Financial support is needed by 5 farmers, and all of them are aware of agri-tech solutions. For example, Farmer 3 is keen on adopting new technology, interested in sustainability, planning expansion, and requires financial support. On the other hand, Farmers 2, 4, 6, 8, and 10 are not interested in any of these aspects, showing a lack of awareness and participation in modern agricultural trends.

Table 15. Future Plans and Willingness to Adopt New Technology

Farmer ID	Willing to Adopt New Technology (Yes/No)	Interest in Sustainable Practices (Yes/No)	Planning Expansion (Yes/No)	Need Financial Support (Yes/No)	Knowledge of Agri-Tech Solutions (Yes/No)
1	Yes	Yes	No	Yes	Yes
2	No	No	No	No	No
3	Yes	Yes	Yes	Yes	Yes
4	No	No	No	No	No
5	Yes	Yes	No	Yes	Yes
6	No	No	No	No	No
7	Yes	Yes	Yes	Yes	Yes
8	No	No	No	No	No

continued on following page

Table 15. Continued

Farmer ID	Willing to Adopt New Technology (Yes/No)	Interest in Sustainable Practices (Yes/No)	Planning Expansion (Yes/No)	Need Financial Support (Yes/No)	Knowledge of Agri-Tech Solutions (Yes/No)
9	Yes	Yes	No	Yes	Yes
10	No	No	No	No	No

Table 16 provides an overview of the varieties and planting methods used by the farmers for growing chillies. The table includes details such as chilli variety, seed type, planting method, seed treatment, and growth duration. Among the 10 farmers, 6 use hybrid seeds, while 4 use local varieties. Transplanting is the most common planting method (7 farmers), while 3 farmers use direct sowing. Seed treatments vary, with some farmers using chemical treatments, others opting for organic treatments, and some not using any treatment. Growth duration ranges from 140 to 160 days, with Guntur Sannam and Teja varieties being planted by multiple farmers. For instance, Farmer 1 grows Guntur Sannam with a hybrid seed, using chemical treatment and transplanting, with a growth duration of 150 days, while Farmer 2 grows Byadgi with a local seed, applying organic treatments and direct sowing, taking 140 days to grow.

Table 16. Chilli Varieties and Planting Methods

Farmer ID	Chilli Variety	Seed Type	Planting Method	Seed Treatment	Growth Duration (Days)
1	Guntur Sannam	Hybrid	Transplanting	Chemical	150
2	Byadgi	Local	Direct Sowing	Organic	140
3	Teja	Hybrid	Transplanting	Chemical	160
4	Wonder Hot	Hybrid	Direct Sowing	None	155
5	334	Local	Transplanting	Organic	145
6	Guntur Sannam	Local	Direct Sowing	Chemical	150
7	Byadgi	Hybrid	Transplanting	Organic	148
8	Teja	Local	Direct Sowing	None	152
9	Wonder Hot	Hybrid	Transplanting	Chemical	158
10	334	Local	Direct Sowing	Organic	145

Table 17 summarizes the irrigation practices of the 10 farmers, including the irrigation type, water source, frequency of irrigation, use of drip irrigation, and rain dependency percentage. The table shows a variety of irrigation methods: drip irrigation is used by 4 farmers, flood irrigation by 3 farmers, and sprinkler irrigation by 3 farmers. Water sources include borewells, canals, and rivers. Farmers using drip irrigation tend to irrigate more frequently (every 4 to 6 days), while flood and sprinkler irrigation frequencies range from 7 to 12 days. Rain dependency varies, with farmers relying on rain anywhere between 20% to 60% of the time. For example, Farmer 1 uses drip irrigation from a borewell every 5 days, with only 30% rain dependency, while Farmer 5 uses flood irrigation from a borewell every 12 days, relying on rain 55% of the time.

Table 17. Irrigation Practices

Farmer ID	Irrigation Type	Water Source	Frequency (Days)	Drip Irrigation (Yes/No)	Rain Dependency (%)
1	Drip	Borewell	5	Yes	30
2	Flood	Canal	10	No	50
3	Sprinkler	Borewell	7	No	25
4	Drip	River	6	Yes	40
5	Flood	Borewell	12	No	55
6	Sprinkler	Canal	8	No	35
7	Drip	Borewell	4	Yes	20
8	Flood	River	9	No	60
9	Sprinkler	Borewell	7	No	25
10	Drip	Canal	5	Yes	30

Table 18 provides an overview of fertilizer usage among the 10 farmers, highlighting the type of fertilizer used, its organic status, application frequency, amount applied per acre, and associated costs. Farmers are divided between using organic and chemical fertilizers, with 5 using organic options such as compost, manure, or vermicompost, while the other 5 use chemical fertilizers like urea, DAP, or potash. Organic fertilizers tend to be applied less frequently, with intervals ranging from 42 to 60 days, while chemical fertilizers are applied more often, ranging from 22 to 30 days. The amount applied varies between 38 kg and 90 kg per acre, with costs ranging from INR 1800 to INR 3600 per acre. For example, Farmer 1 uses urea (a non-organic fertilizer) every 30 days, applying 50 kg per acre at a cost of INR 3000, whereas Farmer 4 uses manure (an organic fertilizer) every 60 days, applying 90 kg per acre at a cost of INR 1800.

Table 18. Fertilizer Usage

Farmer ID	Fertilizer Type	Organic (Yes/No)	Application Frequency (Days)	Amount (kg/acre)	Cost (INR/acre)
1	Urea	No	30	50	3000
2	Compost	Yes	45	80	2000
3	DAP	No	25	40	3500
4	Manure	Yes	60	90	1800
5	Potash	No	30	45	3200
6	Vermicompost	Yes	50	85	2200
7	Urea	No	28	55	3300
8	Compost	Yes	42	75	2500
9	DAP	No	22	38	3600
10	Manure	Yes	55	88	1900

Table 19 provides an overview of pesticide usage and safety measures among 10 farmers. The table categorizes the type of pesticide used, the application frequency, the use of personal protective equipment (PPE), reported health issues, and the cost per acre. The farmers use various pesticides, including insecticides, fungicides, and herbicides, with application frequencies ranging from 10 to 25 days. While some farmers use PPE during pesticide application (e.g., Farmer 1, 3, 5, 7, and 9), others do not (e.g., Farmer 2, 4, 6, 8, and 10). Health issues are reported by a few farmers, with some experiencing adverse effects after pesticide use. The cost per acre for pesticide application varies from INR 1800 to INR 3100, with the highest cost seen in Farmer 6 (using herbicide) and the lowest cost in Farmer 2 (using fungicide). For instance, Farmer 1 applies insecticide every 15 days, using PPE, and incurs a cost of INR 2500 per acre.

Table 19. Pesticide Usage and Safety Measures

Farmer ID	Pesticide Type	Application Frequency (Days)	PPE Usage (Yes/No)	Health Issues Reported	Cost (INR/acre)
1	Insecticide	15	Yes	No	2500
2	Fungicide	20	No	Yes	1800
3	Herbicide	25	Yes	No	3000
4	Insecticide	12	No	Yes	2800
5	Fungicide	18	Yes	No	2200
6	Herbicide	22	No	Yes	3100
7	Insecticide	10	Yes	No	2700

continued on following page

Table 19. Continued

Farmer ID	Pesticide Type	Application Frequency (Days)	PPE Usage (Yes/No)	Health Issues Reported	Cost (INR/acre)
8	Fungicide	14	No	Yes	2300
9	Herbicide	19	Yes	No	2900
10	Insecticide	13	No	Ye	2600

CONCLUSION

The application of Bi-Stacked LSTM and Bi-Stacked GRU models proved effective in mitigating the impact of neighborhood influences on farmers' pesticide use behavior in the rural Chilli farming sector. By isolating individual decision-making processes from community-level practices, these models provided more accurate predictions of pesticide usage, accounting for temporal dynamics and local factors. This approach not only advances the understanding of pesticide use patterns but also underscores the potential of machine learning in addressing rural business challenges, such as improving health and safety standards and optimizing economic decisions in agriculture. The findings suggest that incorporating such advanced analytical tools into rural business practices can foster more informed and sustainable decision-making. Ultimately, the research contributes to the development of tailored interventions aimed at reducing pesticide exposure, promoting safer agricultural practices, and supporting the financial sustainability of rural farmers.

REFERENCES

Ali, M. P., Kabir, M. M. M., Haque, S. S., Qin, X., Nasrin, S., Landis, D., Holmquist, B., & Ahmed, N. (2020). Farmer's behavior in pesticide use: Insights study from smallholder and intensive agricultural farms in Bangladesh. *The Science of the Total Environment, 747*, 141160. DOI: 10.1016/j.scitotenv.2020.141160 PMID: 32781314

Andersson, E., & Isgren, E. (2021). Gambling in the garden: Pesticide use and risk exposure in Ugandan smallholder farming. *Journal of Rural Studies, 82*, 76–86. DOI: 10.1016/j.jrurstud.2021.01.013

Ataei, P., Gholamrezai, S., Movahedi, R., & Aliabadi, V. (2021). An analysis of farmers' intention to use green pesticides: The application of the extended theory of planned behavior and health belief model. *Journal of Rural Studies, 81*, 374–384. DOI: 10.1016/j.jrurstud.2020.11.003

Bakker, L., Sok, J., Van Der Werf, W., & Bianchi, F. (2021). Kicking the habit: What makes and breaks farmers' intentions to reduce pesticide use? *Ecological Economics, 180*, 106868. DOI: 10.1016/j.ecolecon.2020.106868

Berni, I., Menouni, A., El, I. G., Duca, R.-C., Kestemont, M.-P., Godderis, L., & Jaafari, S. E. (2021). Understanding farmers' safety behavior regarding pesticide use in Morocco. *Sustainable Production and Consumption, 25*, 471–483. DOI: 10.1016/j.spc.2020.11.019

Carriger, J. F., Rand, G. M., Gardinali, P. R., Perry, W. B., Tompkins, M. S., & Fernandez, A. M. (2006). Pesticides of potential ecological concern in sediment from South Florida canals: An ecological risk prioritization for aquatic arthropods. *Soil & Sediment Contamination, 15*(1), 21–45. DOI: 10.1080/15320380500363095

Conley, T. G., & Udry, C. R. (2010). Learning about a new technology: Pineapple in Ghana. *The American Economic Review, 100*(1), 35–69. DOI: 10.1257/aer.100.1.35

Diemer, N., Staudacher, P., Atuhaire, A., Fuhrimann, S., & Inauen, J. (2020). Smallholder farmers' information behavior differs for organic versus conventional pest management strategies: A qualitative study in Uganda. *Journal of Cleaner Production, 257*, 120465. DOI: 10.1016/j.jclepro.2020.120465

Duzy, L. M., Campana, D. J., & Brain, R. (2023). Agroeconomic costs for meeting the Environmental Protection Agency's mitigation menu approach to pesticide regulation. *Agricultural & Environmental Letters, 8*(2), e20119. DOI: 10.1002/ael2.20119

Fernando, A. N. (2021). Seeking the treated: The impact of mobile extension on farmer information exchange in India. *Journal of Development Economics*, *153*, 102713. DOI: 10.1016/j.jdeveco.2021.102713

Gowda, B., Sendhil, R., Adak, T., Raghu, S., Patil, N., Mahendiran, A., Rath, P. C., Kumar, G., & Damalas, C. A. (2021). Determinants of rice farmers' intention to use pesticides in eastern India: Application of an extended version of the planned behavior theory. *Sustainable Production and Consumption*, *26*, 814–823. DOI: 10.1016/j.spc.2020.12.036

Grovermann, C., Schreinemachers, P., Riwthong, S., & Berger, T. (2017). 'Smart' policies to reduce pesticide use and avoid income trade-offs: An agent-based model applied to Thai agriculture. *Ecological Economics*, *132*, 91–103. DOI: 10.1016/j.ecolecon.2016.09.031

Gunstone, T., Cornelisse, T., Klein, K., Dubey, A., & Donley, N. (2021). Pesticides and soil invertebrates: A hazard assessment. *Frontiers in Environmental Science*, *9*, 122. DOI: 10.3389/fenvs.2021.643847

He, J., Zhou, W., Qing, C., & Xu, D. (2023). Learning from parents and friends: The influence of intergenerational effect and peer effect on farmers' straw return. *Journal of Cleaner Production*, *393*, 136143. DOI: 10.1016/j.jclepro.2023.136143

Houser, M., Marquart-Pyatt, S. T., Denny, R. C., Reimer, A., & Stuart, D. (2019). Farmers, information, and nutrient management in the US Midwest. *Journal of Soil and Water Conservation*, *74*(3), 269–280. DOI: 10.2489/jswc.74.3.269

Jørs, E., Konradsen, F., Huici, O., Morant, R. C., Volk, J., & Lander, F. (2016). Impact of training Bolivian farmers on integrated pest management and diffusion of knowledge to neighboring farmers. *Journal of Agromedicine*, *21*(2), 200–208. DOI: 10.1080/1059924X.2016.1143428 PMID: 26790134

Li, Y., Qing, C., Guo, S., Deng, X., Song, J., & Xu, D. (2023). Will farmers follow their peers in adopting straw returning? Evidence from rural Sichuan Province, China. *Environmental Science and Pollution Research International*, *30*(8), 21169–21185. DOI: 10.1007/s11356-022-23648-8 PMID: 36264456

Liu, T., & Wu, G. (2020). Does agricultural cooperative membership help reduce the overuse of chemical fertilizers and pesticides? Evidence from rural China. *Environmental Science and Pollution Research International*, *29*, 1–12. PMID: 34480697

López de Mesa, Y. P. (2020). The decision-making process of synthetic pesticide use in agricultural communities in Colombia: A grounded theory approach. *Revista de la Facultad Nacional de Salud Pública*, *38*(2), 38. DOI: 10.17533/udea.rfnsp.e331277

Mahmood, I., Imadi, S. R., Shazadi, K., Gul, A., & Hakeem, K. R. (2016). Effects of pesticides on environment. *Plant Soil Microbes*, *1*, 253–269. DOI: 10.1007/978-3-319-27455-3_13

Manski, C. F. (1993). Identification of endogenous social effects: The reflection problem. *The Review of Economic Studies*, *60*(3), 531–542. DOI: 10.2307/2298123

Matyjaszczyk, E., & Sobczak, J. (2017). Common EU registration rules and their effects on the availability of diverse plant protection products: A case study from oilseed rape and potato in five Member States. *Crop Protection (Guildford, Surrey)*, *100*, 73–76. DOI: 10.1016/j.cropro.2017.06.006

Moghadam, T. M., Raheli, H., Zariffian, S., & Yazdanpanah, M. (2021). The discovery of the barriers of agricultural water conservation behavior and related factors: Using a grounded theory approach (the case of Neyshabur Plain in Khorasan Razavi Province). *Majallah-i Tahqiqat-i Iqtisad va Tusiah-i Kishavarzi-i Iran*, *52*, 287–308.

Mubushar, M., Aldosari, F. O., Baig, M. B., Alotaibi, B. M., & Khan, A. Q. (2019). Assessment of farmers on their knowledge regarding pesticide usage and biosafety. *Saudi Journal of Biological Sciences*, *26*(8), 1903–1910. DOI: 10.1016/j.sjbs.2019.03.001 PMID: 31762673

Nguyen, N., & Drakou, E. G. (2021). Farmers' intention to adopt sustainable agriculture hinges on climate awareness: The case of Vietnamese coffee. *Journal of Cleaner Production*, *303*, 126828. DOI: 10.1016/j.jclepro.2021.126828

Pan, D., Kong, F., Zhang, N., & Ying, R. (2017). Knowledge training and the change of fertilizer use intensity: Evidence from wheat farmers in China. *Journal of Environmental Management*, *197*, 130–139. DOI: 10.1016/j.jenvman.2017.03.069 PMID: 28363140

Rahaman, M. M., Islam, K. S., & Jahan, M. (2018). Rice farmers' knowledge of the risks of pesticide use in Bangladesh. *Journal of Health & Pollution*, *8*(20), 181203. DOI: 10.5696/2156-9614-8.20.181203 PMID: 30560002

Ren, Z., & Jiang, H. (2022). Risk cognition, agricultural cooperatives training, and farmers' pesticide overuse: Evidence from Shandong Province, China. *Frontiers in Public Health*, *10*, 1032862. DOI: 10.3389/fpubh.2022.1032862 PMID: 36478731

Sarma, P. (2022). Farmer behavior towards pesticide use for reduction production risk: A Theory of Planned Behavior. *Cleaner Circular Bioeconomy*, *1*, 100002. DOI: 10.1016/j.clcb.2021.100002

Sharifzadeh, M. S., Abdollahzadeh, G., Damalas, C. A., Rezaei, R., & Ahmadyousefi, M. (2019). Determinants of pesticide safety behavior among Iranian rice farmers. *The Science of the Total Environment*, *651*, 2953–2960. DOI: 10.1016/j.scitotenv.2018.10.179 PMID: 30463146

Sharma, A., Kumar, V., Shahzad, B., Tanveer, M., Sidhu, G. P. S., Handa, N., Kohli, S. K., Yadav, P., Bali, A. S., Parihar, R. D., Dar, O. I., Singh, K., Jasrotia, S., Bakshi, P., Ramakrishnan, M., Kumar, S., Bhardwaj, R., & Thukral, A. K. (2019). Worldwide pesticide usage and its impacts on ecosystem. *SN Applied Sciences*, *1*(11), 1446. DOI: 10.1007/s42452-019-1485-1

Taghdisi, M. H., Besheli, B. A., Dehdari, T., & Khalili, F. (2019). Knowledge and practices of safe use of pesticides among a group of farmers in northern Iran. *The International Journal of Occupational and Environmental Medicine*, *10*(2), 66–72. DOI: 10.15171/ijoem.2019.1479 PMID: 31041923

Takahashi, K., Muraoka, R., & Otsuka, K. (2020). Technology adoption, impact, and extension in developing countries' agriculture: A review of the recent literature. *Agricultural Economics*, *51*(1), 31–45. DOI: 10.1111/agec.12539

Teng, Y., Chen, X., Jin, Y., Yu, Z., & Guo, X. (2022). Influencing factors of and driving strategies for vegetable farmers' green pesticide application behavior. *Frontiers in Public Health*, *10*, 907788. DOI: 10.3389/fpubh.2022.907788 PMID: 36159273

Tian, Q., Yu, Y., Xiang, Z., & Li, C. (2021). Agricultural technical education, interpersonal trust, and pesticide use by vegetable farmers in China. *Journal of Agricultural Education and Extension*, *27*(2), 211–227. DOI: 10.1080/1389224X.2020.1844769

Tu, Z. G., Cheng, L., & Zhang, M. (2022). Study on the regional difference and characteristic of temporal-spatial evolution of Chinese business environment. *Research on Quantitative Economics and Technological Economics*, *39*, 3–25.

Xia, H., Li, C., Zhou, D., Zhang, Y., & Xu, J. (2020). Peasant households' land use decision-making analysis using social network analysis: A case of Tantou Village, China. *Journal of Rural Studies*, *80*, 452–468. DOI: 10.1016/j.jrurstud.2020.10.023

Xu, Z., Zhang, K., Zhou, L., & Ying, R. (2022). Mutual proximity and heterogeneity in peer effects of farmers' technology adoption: Evidence from China's soil testing and formulated fertilization program. *China Agricultural Economic Review*, *14*(2), 395–415. DOI: 10.1108/CAER-10-2020-0250

Yang, W., Qi, J., Arif, M., Liu, M., & Lu, Y. (2021). Impact of information acquisition on farmers' willingness to recycle plastic mulch film residues in China. *Journal of Cleaner Production*, *297*, 126656. DOI: 10.1016/j.jclepro.2021.126656

Zhang, S., Kang, B., & Zhang, Z. (2020). Evaluation of business environment in Chinese provinces: Indicator system and quantitative analysis. *Economic Management*, *42*, 5–19.

Zhao, L., Wang, C., Gu, H., & Yue, C. (2018). Market incentive, government regulation and the behavior of pesticide application of vegetable farmers in China. *Food Control*, *85*, 308–317. DOI: 10.1016/j.foodcont.2017.09.016

Zheng, W., Luo, B., & Hu, X. (2020). The determinants of farmers' fertilizers and pesticides use behavior in China: An explanation based on label effect. *Journal of Cleaner Production*, *272*, 123054. DOI: 10.1016/j.jclepro.2020.123054

Zhou, L., Zhang, F., Zhou, S., & Turvey, C. G. (2020). The peer effect of training on farmers' pesticides application: A spatial econometric approach. *China Agricultural Economic Review*, *12*(3), 481–505. DOI: 10.1108/CAER-01-2019-0003

Chapter 14
A Qualitative Study on Consumer Coping Strategies to Deal With a Pandemic:
The Case of an Emerging Country

Tutku Eker İşcioğlu
https://orcid.org/0000-0002-4794-6368
Piri Reis University, Turkey

M. G. Serap Atakan
https://orcid.org/0000-0001-8786-233X
Istanbul Bilgi University, Turkey

Özge Kirezli
Yeditepe University, Turkey

ABSTRACT

COVID-19 pandemic is an important disease of recent decades, affecting the health of the population, as well as the global economies, social interaction patterns, and human psychology. To cope with this uncertain and stressful situation people devised different strategies. This study aims to reveal the coping strategies adopted by the consumers of an emerging country, namely Türkiye. It is a qualitative study aiming to understand the way Turkish consumers coped with the COVID-19 pandemic. Data was collected through conducting 60 in depth interviews and content analysis was applied using QDA Miner 6 software package. The findings reveal that mostly preferred coping strategies are avoidance, active coping, emotional support and positive thinking whereas rational thinking, emotional venting, and instrumental

DOI: 10.4018/979-8-3373-4432-4.ch014

support are the least preferred coping strategies of the Turkish sample. The findings contribute to the literature by adapting the multidimensional model of consumer coping developed by Duhachek (2005) and proposing coping strategies that are specific to a pandemic.

1 INTRODUCTION

The global crisis of the Coronavirus disease (COVID-19) pandemic damaged the world economy and healthcare system generating fear, panic, and uncertainty among billions of people globally. COVID-19 virus first seen in China in December 2019, spread to other countries of the world at an extraordinary speed. The disease was not limited to a specific geographical region and seriously affected the whole world.

People were put into quarantine, and most businesses were locked down. During the lockdowns, people demonstrated fear of uncertainty about their health and safety which led to stress and anxiety. In that manner, COVID-19 is a period of uncertainty and a stressful situation leading to symptoms of anxiety, depression, and sleep problems. People faced with these stressful situations need to solve problems and regulate their emotions by using some coping strategies that help to reduce the adverse impacts of the pandemic. In 2005, Duhachek proposed a model to understand how consumers cope with stressful situations, such as active coping (active coping, rational thinking, positive thinking), expressive support seeking (emotional venting, instrumental support, emotional support), and avoidance (avoidance, denial). COVID-19 being a major stressful situation of recent times, how well this model would fit consumers in an emerging market during a pandemic is the major motivation for this chapter. Thus, this chapter aims to reveal the coping strategies adopted by Turkish consumers. To the knowledge of the authors this study is the first attempt to understand the effect of coping mechanisms during a pandemic in an emerging country and fills the gap in the literature by providing implications for marketing managers and policy makers.

The chapter provides a literature review on consumer coping strategies. An emerging country, namely Türkiye, is taken as the case and Turkish consumers' coping strategies adopted during the pandemic are presented. A qualitative research design was used to explore the various coping strategies. The sample is comprised of 60 participants selected with convenience sampling and in-depth interviews were made to collect data. All the transcribed recordings were analyzed via QDA Miner 6 software to identify the emerging codes, categories and themes. The findings describe each coping strategy in detail. The chapter ends with a comprehensive discussion, including implications, limitations, and suggestions for future research.

2 CONSUMER COPING STRATEGIES

COVID-19 is a highly infectious, fast-spreading, and lethal virus. It causes a threat to human health. It is a threat not only to the health of the population but also to the economy, society, and businesses of the countries globally. The stay-at-home and physical distancing orders led consumers to lessen their social interactions, being a threat to socialization (Li *et al.* 2021). Economically it also created threats to consumers' income and quality of life, where economic hardships and job losses appeared.

Pandemics are highly stressful situations that can easily induce negative emotions and adverse mental health states (Mishra *et al.*, 2022; Kemp *et al.*, 2021). Besides stress, consumers experienced feelings of fear, anxiety, and depression which in turn affected their behavior (Ross, 2023; Di Crosta *et al.*, 2021; Sundaray *et al.*, 2021; Bae, 2023; Jain *et.al.*, 2023). Coping with COVID-19 disease is traumatic, it increased stress and anxiety levels, due to fear of contracting the virus and uncertainty about how the outbreak will affect individuals socially and financially. People experienced increased emotions of anxiety, sadness, disappointment, and confusion. This pandemic changed every aspect of daily life, leading to isolation and social distancing, financial problems, fear of the virus, feelings of helplessness, and hopelessness about how life will continue.

Coping strategies are methods developed by individuals to counteract the effects that cause tension, stress, and anxiety. These cognitive, emotional, and behavioral efforts aim to manage stressful situations (Folkman *et al.* 1986). Lazarus (1991) defines coping strategies as the consciously formed strategies of individuals against coping with stress and the negative life events that they experience.

Considered to be the pioneer in this research area, Lazarus and Folkman's (1984) classification of coping strategies is the most influential typology of coping strategies with stress, problem-focused and emotion-focused coping strategies. Problem-focused strategies are related to resolve the stressful event or alter the source of the problem. These strategies aim to vary one's environment by trying to change a stressful situation and engaging in activities directed to self - such as learning new skills and procedures and changing one's aspirations. On the other hand, emotion-focused coping strategies aim to manage the negative emotions associated with a stressful event. These strategies include thoughts and actions directed at regulating one's emotions such as avoiding a situation and engaging in activities to get one's mind off a problem or a stressful situation.

Stone *et al.* (1988) reviewed and classified the methods of coping with stress into the following categories: problem-solving (rational decision-making), avoidance (cognitive or behavioral), tension reduction behaviors (i.e., use of alcohol), social support (from the professional community), situation redefinition (seeing the situation differently and diminishing its perceived severity) and religion (praying) as cited in Mathur *et al.* (1999).

Carver et al. (1989) developed the Coping Orientation to Problems Experienced Inventory (COPE) by identifying 13 dimensions of coping: five problem-focused dimensions (active coping, planning, suppression of competing activities, restraint coping, seeking social support for instrumental reasons), five emotion focused dimensions (seeking social support for emotional reasons, positive reinterpretation and growth, acceptance, denial, turning to religion) and less useful three dimensions (focus on and venting of emotions, behavioral disengagement, mental disengagement).

In 2005, Duhachek studied various ways consumers coped with stress and negative emotions. He developed a multi-dimensional model of consumer coping strategies. In this model, Duhachek (2005) identified emotional (feelings such as threat, anger, sadness, challenge, stress, etc.) and cognitive (situational efficacy and situational control) antecedents to coping. He then suggested eight coping dimensions/strategies, namely active coping, rational thinking, positive thinking, emotional support, instrumental support, emotional venting, avoidance, and denial. Duhachek (2005, p. 44-46) defines each coping strategy as follows:

- Active Coping – "direct, objective attempts to manage a source of stress."
- Rational Thinking – "deliberate attempts to prevent subjective emotions from directing behavior."
- Positive Thinking – "attempts to psychologically reconstrue a source of stress to make it more tolerable."
- Emotional Support – "attempts to marshal social resources to improve one's emotional and/or mental state."
- Instrumental Support – "attempts to marshal social resources to take action towards ameliorating a stressor."
- Emotional Venting – "attempts to recognize and express one's emotions."
- Avoidance – "attempts to create psychic or physical distance between oneself and a stressor."
- Denial - "attempts to completely close off oneself mentally from a source of stress."

This model was utilized by many researchers like Sundaray *et al.* (2021), Ertz *et al.* (2022), Bae (2023) and Jain *et al.* (2023).

Coping strategies during the COVID-19 pandemic are examined in both psychology and consumer behavior-oriented research especially. A summary table of previous research in consumer behavior related resources is presented in Table 1. Accordingly, there are various studies held globally, in which quantitative research designs were more prevalent than qualitative or mixed methods. The coping strategies that were identified in these studies are problem-focused coping (e.g., self-justification, mental disengagement), social support, resilience (e.g., optimism, passive and active resilience), avoidance (e.g., avoidance of public places, avoidance of consumption), shopping-related coping (e.g., therapeutic shopping, conformity in consumption, hygienic behavior, do-it-yourself behaviors), religiosity and emotional venting. Among these, resilience and social support appear to be common themes identified in many studies (e.g. Ha *et al.*, 2021; Ertz *et al.*, 2022; Im *et al.*, 2022; Nath *et al.*, 2022).

3 CONSUMER COPING STRATEGIES DURING PANDEMIC IN A SELECTED COUNTRY – THE CASE OF TÜRKİYE

This qualitative study aims to understand how consumers coped with the COVID-19 pandemic in a developing country. The qualitative research is exploratory and inductive and is conducted in situations where not much is known about a particular phenomenon or concept as suggested by Thomas (2021).

This case examines the coping strategies applied by Turkish consumers by basing on the study of Duhachek's (2005) multidimensional model of consumer coping. This is a theoretical framework for investigating the dynamic effects of coping in order to gain insight on the coping mechanisms and strategies used during the pandemic, the most and the least preferred coping strategies, as well as whether there was any coping strategies or mechanisms specific to the COVID-19 pandemic.

The data is collected through in-depth interviews. This data collection method reveals experiences, opinions, and impressions of respondents (Thomas, 2021) and helps researchers to gain a deeper understanding of the consumer coping behavior. Other studies that examined consumer coping behavior before the COVID-19 pandemic and after (e.g. Nath *et al.*, 2022; Liu *et al.*, 2022; McKinlay *et al.*, 2021), also used in-depth interviews to understand this issue.

In-depth interviews were held in March-April 2022. The interviews were made online via Zoom to watch social distancing. The duration of the interviews ranged between 30 and 73 minutes, with an average of 40 minutes. The video recording of each interview was saved with the consent of the interviewees to be used as a reference for transcription and data analysis. Each interview was transcribed by the researchers. Semi-structured interviews were conducted with an interview guide as a data collection instrument. These semi-structured interviews make use of an interview guide with some broad open-ended questions or issues, that are to be discussed for possible investigation during an interview (Thomas, 2021). Therefore, this method was preferred over structured or unstructured interviews. The interview guide started with a short description of the research in a way that would not direct participants' answers. Then opening questions were asked to warm up the participants, such as whether they or their family members were infected of COVID-19, if so, how severe it was, whether they had to stay in the hospital, or whether there were any deaths in their family etc. After participants were set at ease, questions related to the research questions such as "How did you deal with the uncertain and stressful situation of COVID-19 pandemic?" "What coping mechanisms did you apply to feel better and less stressful?" were asked. First, the respondents were allowed to describe their own experiences of how they coped with the pandemic. More targeted questions were also asked about the predetermined categories presented in Duhachek's (2005) consumer coping model.

Using convenience sampling method sixty participants were interviewed in total. This sampling method is a good fit when researchers plan to gain insight on people's perceptions and attitudes with certain inclusion criteria (Golzar and Tajik, 2022). By utilizing convenience sampling method, the researchers selected respondents being conveniently available at the moment and who met the criterion of openly describing how they went through and coped with the pandemic which was a sensitive topic. At the point when theoretical saturation was achieved, the inclusion of new participants was stopped (Guest *et.al.*, 2006). Each participant's gender, age, education, occupation, marital status, and number of children are presented in Table 2. The list is sorted based on their age, from teenagers to adults in order to make it easier to interpret the results. Concerning participants' demographics, females account for 66,6% of the sample. Participants above 51 years old correspond to 31,6%. The majority of the sample has higher education level, where high school graduates consist 16,6%, undergraduates 45% and graduates 38,3% of the sample. While 63,4% of the sample is employed in various occupations, 36,6% of the sample is unemployed 20% students and 16,6% retired. As for marital status, there are an equal number of single and married participants. Finally, almost half of the sample has no child (48,3%); 36,8% have one child; 13,3% have two children; no participant has three children and 1,6% have four children.

To thematize participants' experiences with the COVID-19 pandemic and understand the coping strategies they applied, content analysis was used. This study utilized a directed approach as suggested by Hsieh and Shannon (2005), which aims to validate or extend conceptually a theoretical framework or theory, thus helping to determine the initial coding scheme or relationships between codes.

Taking Duhachek's (2005) multidimensional model of consumer coping as a theoretical framework, the researchers first identified the coping strategies described by Duhachek (2005) as the initial codes and the relevant categories. After transcribing all the interviews, the three researchers read the transcriptions repeatedly and carefully to detect and classify all instances of coping mechanisms related to the COVID-19 pandemic. Any text that could not be categorized according to initial coding scheme was then given a new code and category, as suggested by Hsieh and Shannon (2005). The codes and the relevant categories of coping strategies are presented in Figure 1 and Table 3. The categories that are shown in italic are the ones that Duhachek (2005) already mentioned, whereas the remaining ones are peculiar to this study.

Hsieh and Shannon (2005) recommend presenting the findings by showing codes with exemplars and by offering descriptive evidence. Thus, the findings of this study are reported together with the quotations of participants with the related codes as well as the rank order of the codes' frequencies. To calculate frequencies, QDA Miner 6 was used. This qualitative data analysis software is used for organizing, coding, retrieving, and analyzing collections of transcripts of interviews.

Validity and reliability issues specific to content analysis were also carried out. Since a directed approach to content analysis requires the use of a theoretical framework to design the coding scheme, it forms the basis for face validity (Potter and Levine-Donnerstein, 2009). For reliability, inter-coder reliability was attained as the three researchers of the study were the coders. As discussed by Potter and Levine-Donnerstein (2009), the researchers decided to have all the coders independently code the same material. After finishing the coding, researchers came together and went over each participant's transcription one by one to see whether the identified codes overlapped. If there was any disparity between the codings of the researchers, the codes were re-examined until all researchers reached consensus.

After identifying the codes, then grouping them into categories, the researchers tried to match the themes presented in Figure 1. The categories shown in italics are the ones that appear in Duhachek's (2005) study, whereas the remaining categories emerged in this study and are specifically related to the COVID-19 pandemic. The coping strategies of the respondents grouped as Duhachek's (2005) classification are presented below.

Figure 1. Categories of emotional support codes.

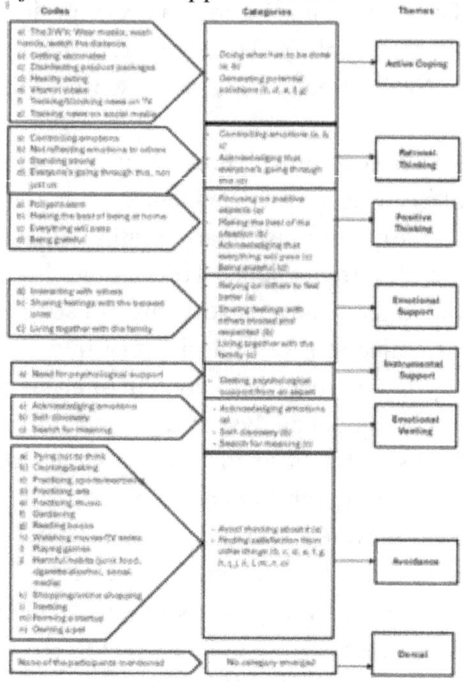

3.1 Active Coping

The majority of the participants stated that they applied the 3 W's rule – washing hands, wearing masks, and watching distance – and got vaccinated as these precautions were highly recommended by the government and the media. Therefore, these findings were categorized under the "Doing what has to be done" statement as proposed by Duhachek (2005).

Some of the participants stated that they disinfected the product packages after each shopping, especially at the beginning of the pandemic.

"When I bought alcohol to drink at home, I even disinfected the beer bottle. My disinfectant and cologne expenditure has increased tremendously during the pandemic." (Participant ID 13).

"Keeping the grocery bags on the balcony, taking them inside the house after washing them... Taking such precautions made me feel safe." (Participant ID 36).

For some participants, eating healthy food and taking vitamins were other ways of actively coping with the stress caused by the pandemic.

"To lessen our worry and anxiety of being infected from COVID-19, we ate healthy food and took vitamins to protect and strengthen our health." (Participant ID 18).

When it comes to coping with the pandemic by following news, some respondents preferred watching the news on TV, some others especially blocked TV news arguing that the truth was disguised, and preferred following social media accounts of the experts they trusted.

"At the beginning, I tracked the news quite a lot. I was following the daily number of infected people. After a few months, I stopped tracking because the numbers were getting worse every day." (Participant ID 4).

"At first I got stuck with the numbers, then question marks started to appear about the truth. Then I started following the news only on Twitter, switching from online sites or TV." (Participant ID 27).

"My family and I followed the doctors we trusted on Twitter. We also trusted the infection numbers announced by the Medical Association." (Participant ID 48).

As can be seen, participants took different measures and applied various precautions to actively cope with COVID-19 pandemic. Therefore, disinfecting product packages, healthy eating, vitamin intake, tracking or blocking news on TV, and tracking news on social media were all categorized under the "Generating potential solutions" as suggested by Duhachek (2005).

3.2 Rational Thinking

A few respondents stated that controlling their emotions, not reflecting their emotions to others, and standing strong were some of the ways they coped with the pandemic.

"Being a mother, I turned things into fun and stood strong in the eyes of my kid's. I controlled my emotions believing I could get out of this trouble." (Participant ID 29).

"I did not communicate my negative feelings to my spouse and children not to upset them, and reflect my fears." (Participant ID 31).

"When my parents were diagnosed with COVID-19, I had to stand strong and somehow I felt that power to take care of them." (Participant ID 23).

Since all of these attempts are related to how respondents tried to control themselves and the situation felt a lot easier to manage, they are categorized under "Controlling emotions" heading of Duhachek (2005).

Another code classified under rational thinking is acknowledging that everyone's going through tough times during the pandemic. Considering that this is something the whole world has to face, some respondents expressed relief that they are not alone in this difficult situation. Therefore, "Acknowledging that everyone's going through this" appears to be a new category that was not mentioned by Duhachek (2005).

"This was not just a situation that happened to me, everyone was going through this. I convinced myself that I am not the only one suffering from this pandemic." (Participant ID 6).

"Although there were times when I thought I lost control of my life, then I thought it is not just me alone, it is not a disease affecting only me and my country. I tried not to be desperate." (Participant ID 11).

3.3 Positive Thinking

With respect to positive thinking, some respondents stated they were relieved by telling themselves that everything will pass.

" Most of the time, I found myself comforted that everything will pass and this will end. People faced other similar diseases in the past, such as Spanish flu, bird flu and all past. I always thought that as the world coped with the other diseases with less medical development and technological opportunities, we can also cope with this pandemic." (Participant ID 13).

"I'm an optimistic person. I usually believe that difficult situations do not last forever, they pass sooner or later. I also told myself that this too will pass, especially during the times when I was feeling extra stressed." (Participant ID 35).

Putting two of these findings together, everything will pass and being optimistic are categorized as "Focusing on positive aspects", as suggested by Duhachek (2005). Additionally, some participants claimed they coped with the pandemic by making the best of being at home and this is categorized as "Making the best of the situation" considering Duhachek's model (2005).

"It was a situation I had to adapt to. I tried to enjoy the time I was staying at home. I had more time to play music with my reed flute. I started doing Chinese Qigong exercises more often. I had more time to spend with my daughter. I was reading a lot of books. I enriched myself spiritually. This lock down period was like heaven for me." (Participant ID 34).

"As a family we engaged in activities to entertain ourselves, we played card games, scrabble, etc. We tried to make life fun at home. I kept myself busy at the kitchen, I started a ketogenic diet and cooked new recipes." (Participant ID 16).

"The kitchen was a therapy for me. I simply gamified staying at home. Kids and my husband ordered what they wanted as if they were in a restaurant, and I cooked their orders for them." (Participant ID 20).

With respect to positive thinking, some respondents also stated they were relieved by telling themselves that everything will pass. "Acknowledging that everything will pass" can be considered as a new item to the Duhacheck's (2005) scale.

"I'm usually an optimistic person. I believe that difficult situations do not last forever. I told myself that this too will pass, especially when I was extra stressed." (Participant ID 35).

Some respondents stated that, even if the pandemic had negative consequences in their lives, they were able to cope with it by being grateful that they are still alive and had a chance to do things that they couldn't do before or wouldn't do otherwise. Therefore "Being grateful" is a new category which was not revealed by Duhachek (2005).

"Of course, it was a bad situation on one hand, but on the other hand, I'm grateful because I was able to take care of my baby, breastfeed him properly, eat healthy food, and take care of the house." (Participant ID 15).

"Instead of putting myself into a crisis, I was always grateful that my family and I were healthy. Some people got sick, even children. I felt very lucky that this didn't happen to us" (Participant ID 37).

3.4 Emotional Support

Some participants expressed how it was important to spend days together with their family members during the lockdown. Some of them especially emphasized that they wouldn't cope with their negative emotions without the support of their family.

"I was studying in a boarding high school and was back at my family house during the quarantine. It would have been very difficult without my family." (Participant ID 3).

"I was at home with my brother. If I were alone, I probably would have 'going nuts!'. I was happy having someone by my side." (Participant ID 19).

"What kept me standing still was my family. My wife put me at ease. Being together, eating together, taking care of the children together...We all felt great." (Participant ID 21).

Interacting with others was another coping mechanism for some participants. Even though most of them could not meet and see each other for a long time, online meetings and celebrations made them feel better. This type of coping is similar to "Relying on others to feel better" in Duhachek's (2005) model.

"It was really good to have online conversations with my classmates from whom I was separated. Zoom made communication easier." (Participant ID 4).

"Video conferencing has certainly increased a lot. We celebrated birthdays. We brought the family together on Zoom meetings and had a great time." (Participant ID 36).

"I talked to friends I hadn't seen for a while. We had online dinners, celebrations and toasted with friends. Somehow we tried to 'make life more interesting and exciting." (Participant ID 52).

Finally, sharing feelings with beloved ones appeared to be another aspect of emotional support. This was mentioned in Duhachek's (2005) model as "Sharing feelings with others trusted and respected".

"If I didn't have my girlfriend and work permit, I would have gotten through this period so hard. I was devastated. My permit allowed me to go out and thus I could see my girlfriend. It was good to see her and talk to her, even if it was for a limited time." Participant ID 13).

3.5 Instrumental Support

Some respondents needed and took psychological support to deal with the emotions they experienced during the pandemic. While some of them didn't openly talk about the main reasons for such a necessity, some felt this need because of living alone, and some others because of their family members (spouses, kids, etc.). Although Duhachek (2005) used "trying to get advice from someone about what to do" as a scale item, participants of this study emphasized the importance of getting psychological support from an expert and thus it should be used as a new item.

"I had intense emotions than usual. I realized that I couldn't deal with those feelings on my own and had to take psychological support. There were other reasons as well, like living alone, but I think the main trigger was the pandemic." (Participant ID 22).

"I preferred psychological support. Frankly, I had a little trouble." (Participant ID 32).

"I had problems with my husband since we spent so much time at home. We were overwhelmed with the kids on top of it. Towards the end, I really needed the support of a relationship counselor." (Participant ID 36).

3.6 Emotional Venting

Several participants stated that having so much time by themselves during the pandemic led them to face or acknowledge their emotions and so that they could feel better. This was stated by Duhachek (2005) as "Acknowledging emotions."

" I had troubles from time to time, but once I confronted my emotions, I felt good. I was able to throw away negative thoughts out of my mind." (Participant ID 9).

Some participants also indicated that during the pandemic they were able to discover their feelings, and were able to deal with the negative aspects of the pandemic. Therefore, this new type of coping was named as "Self-discovery" by the authors since it was not mentioned in Duhachek's (2005) study.

"I turned into myself and discovered who I was. COVID-19 pandemic was a good developmental period for me." (Participant ID 1).

"When you have more free time, you can find different emotions of your own. For instance, I took time for myself and discovered that I had different tastes and research interests. I started listening to and watching things that I was not interested in before. I watched a lot of documentaries about politics during the pandemic, which did not interest me before, and I learned a lot about it. I am very satisfied with the knowledge that I gained during the pandemic." (Participant ID 28).

"I questioned a lot about life, where I do stand, where I want to go, what kind of human and family relationships I want to have in the future. I started to discover things about myself that I didn't normally know much about. In that respect, the pandemic period was a process where I discovered the person I want to be in the future." (Participant ID 22).

"Search for meaning", which is another new category, had been practiced by one participant only as a way of coping with the pandemic, even though she claimed it didn't help a lot.

"I am a person focusing on work, and finding meaning in it, but it is completely gone. I almost lost everything that motivated me, I can't find any meaning in anything. Existential questions such as "what is the meaning of life" arose. I searched for the meaning of life." (Participant ID 22).

3.7 Avoidance

Avoiding the consequences resulting from the pandemic appears to be the coping strategy that participants applied most commonly. They either tried not to think about COVID or found satisfaction from other activities and practices. A few participants stated that they could deal with the pandemic by simply not thinking about its negative effects.

"At the beginning, I was thinking a lot that these were my best years and I didn't want to stay at home doing nothing. Then I stopped thinking about it and convinced myself that these days will pass too." (Participant ID 2).

On the other hand, the majority of the participants tried to find satisfaction from other activities such as cooking/baking, practicing sports/exercising, practicing arts, practicing music, gardening, reading books, watching movies/TV series, playing games, harmful habits (too much junk food, alcohol and cigarette consumption, increasing time spent on social media), shopping/online shopping, travelling, owning a pet and reorganizing home.

3.7.1 Cooking

Some participants engaged in cooking to enjoy their time, some others because they were obliged to. No matter what, it was a way to escape from reality.

"I cooked a lot. Cooking in my kitchen became like therapy" (Participant ID 20).

"In the past, I wasn't cooking for nine years. Our helper cooked all the time but when she had to quit because of the pandemic, I had to take her role and cooked a lot. I enjoyed making desserts and trying new recipies." (Participant ID 29).

3.7.2 Engaging in sports/arts/music.

Similar to cooking, doing sports was either an entertaining activity or something to be done out of necessity during the pandemic. Some participants practiced sports to spend their time effectively or to be in a good mood, some others engaged in it to lose the weight they gained because of the pandemic and get into shape.

"During this period, I took up new hobbies to clear my mind. That's how I saved the day. For example, I biked a lot." (Participant ID 4).

"I was really in a bad mood, then I started exercising regularly with the help and support of my brother." (Participant ID 7).

"In the beginning of the pandemic, my husband and I baked pizzas, pastries, cookies at home. So we gained a lot of weight of course. I started to exercise at home to lose weight. I followed Zumba classes from Youtube videos. I also did home walking exercises with Leslie. I enjoyed it so much." (Participant ID 37).

Related to arts, some participants engaged in drawing/painting, and some others visited online museums/exhibitions.

"Drawing relaxed me a lot during the quarantine" (Participant ID 5).

"I painted the flowerpots of my balcony for a while. I bought the pots and painted them. It was a lot of fun." (Participant ID 43).

"It was the most amazing time for me. I almost feared that the pandemic would end. I was happy making ceramics, visiting museums online, etc. during the pandemic." (Participant ID 47).

As for music, participants either listened to music, sang songs, or played musical instrument to cope with the pandemic.

"Playing guitar helped me to feel good. I used to play with tabs, I didn't know exactly where to press. I've made a lot of progress by self-training during the pandemic. I believe I improved my skills." (Participant ID 7).

"Music and sports are important to me. I listened to a lot of music while doing sports. Listening to music puts me in a positive mood." (Participant ID 11).

3.7.3 Gardening

A few participants got interested in or spent more time gardening or growing plants at home.

"I grew flowers and rearranged my balcony. These were the activities that I always wanted to do but couldn't find time earlier." (Participant ID 28).

"I wasn't interested in plants before, so I tried to take the time to grow plants at home. Buying flowers for home made me very happy. Seeing them bud and bloom gave me hope, it motivated me a lot." (Participant ID 30).

3.7.4 Reading books

Some of the participants revealed that they read books and some others listened to audio books from apps such as Storytell.

"In general, instead of trying something new during this period, I tried to do things that I could not find time for. There were a lot of books I wanted to read, so I read them. I also was interested in watching movies and read movie reviews as well." (Participant ID 3).

"There were so many books in my library that I bought but couldn't read, so I started reading them. I've also listened to a lot of books during the pandemic through Storytell app. As I did daily chores like ironing, cleaning, cooking, etc. I listened to too many audio books and it helped me to have fun while dealing with household chores". (Participant ID 28).

3.7.5 Watching movies/series

Too many participants stated that watching movies or series made them relaxed and took them away from the reality of COVID-19 pandemic.

"During this period, I started to watch almost every movie I saw. It was like I was running away from the pandemic while watching them and diving into other places than home. It's been very good for me." (Participant ID 2).

"At nights, after everyone had gone to sleep, I watched Netflix until 3 am. I needed this 'me time' to relax and empty my busy mind" (Participant ID 36).

3.7.6 Self-training

As another way of coping with the pandemic, some participants wanted to self-train themselves either in terms of their professional life or in the areas they are talented.

"I invested a lot in my professional development. I took online training and completed an international certification program." (Participant ID 17).

"During my summer holidays, I attended sculpture courses in Florence, Italy. Before the pandemic, I was planning to attend a painter's course in Rome. When COVID-19 pandemic outbursted, he offered online courses and I started taking his classes in November 2020. We were 5 students in total. So, the pandemic created an opportunity. It's been very good for me. I even had an online exhibition of my artwork." (Participant ID 23).

"I colored the old black and white family photos with apps and I was very surprised to see old family photos in color. I worked on it for 3 months, and it positively changed my mood and created positive emotions." (Participant ID 40).

3.7.7 Playing games

During the pandemic participants played various games, such as computer games, puzzles, board games, etc. Although most of them stated they had fun, some of them revealed that it started to be boring after some time.

"I started doing jigsaw puzzles during this period. I put all the pieces on the table and placed 10-15 pieces every day. I liked doing the 1000-piece puzzle." (Participant ID 2).

"I played a lot of computer games. I've played quite a variety of games on the computer, and I got a lot of fun from it. However, after a while, it started to get boring, it was not fun any more." (Participant ID 8).

"We did a lot of puzzles with the kids. We glued everything we did together, but I didn't have time to have them framed yet." (Participant ID 37).

"We did a lot of puzzles and played a lot of board games. The whole family was spending time together, we had a lot of fun" (Participant ID 48).

4.7.8 Harmful habits

Most of the participants claimed they sought pleasure by eating junk food, drinking alcohol, or smoking and a few of them considered spending an enormous time on social media as examples of harmful habit.

"Unfortunately, spending so much time at home during lock down period, I ate lots of junk food whenever I got bored." (Participant ID 3).

"In the first year of the pandemic, I gained 3-4 kg. Looking back, I realized that I had finished a total of 1.5 kg of Nutella jar. I couldn't stop. I didn't want salty chips, but I craved a lot for sugary foods which I normally never eat. My alcohol consumption had skyrocketed during this period as well. I can say I was drinking five days a week. From time to time, I could skip two or three evenings, but

there were many weeks when I drank for 7 days. It was terrible. Ok, it improves the mood up to a certain point, but it can also worsen it after a certain level of continuous alcohol consumption. I enjoyed one or two glasses, but with the third glass, I started to weep. If you can't regulate it, you wake up bad the next morning. It also may trigger depression." (Participant ID 22).

"I got very addicted to social media during the pandemic. That's why I had to delete apps, so I could get away from it." (Participant ID 4).

"As we watched movies until the morning, we ate a lot of junk food that we did not consume before. For example, I like dried mulberry very much. I ate packages of dried mulberry. That psychology was bad. There was nothing to do, you either watch something on TV, eat something, lie down, get up, sleep and nothing else. That's why my consumption of junk food has increased. Plus cigarettes. Because I started to work from home, I started to smoke more in my own house without any smoking restrictions of the workplace." (Participant ID 33).

3.7.9 Shopping

Participants revealed they either tried to find satisfaction from shopping or had to shop because they wanted to increase their comfort at home.

"I started to buy things for my house. Objects that would look good at home, glassware, etc. I bought things that I would be happy to see. I bought a bookcase and a table set. I bought a lot." (Participant ID 19).

"I realized that I needed to be a little more open to alternatives that I rejected. Because life is going on, you can die from COVID-19 the next day, but I said to myself not to go there without a floral dress. I grew up as a rocker and I usually don't like dresses with flowers, lace, scallops, etc. During the pandemic, I bought dresses with flowers, I even bought false eyelashes. I tried them on because I wanted to feel good." (Participant ID 22).

"We had more time surfing on the Internet. Since we couldn't spend time or money at a restaurant, we wanted to bring home the joy. So, we started to search for exclusive cheese or wine and ordered them online." (Participant ID 44).

"Normally I wouldn't but I had to do online shopping. I started buying everything online, household items, clothes, food, etc. That was convenient." (Participant ID 58).

3.7.10 Travelling

Although not many, a few participants stated that travelling and seeing new places during the end of the pandemic made them feel good.

"We started to travel without giving up the precautions. We haven't been able to go abroad yet, but we travelled domestically. After all, it's a matter of survival. Normally, we wouldn't rent a house, but we did that summer and we had a great time." (Participant ID 45).

"We went to Barcelona and Budapest. It was very good. At that time, we were in the summer period, so we took our measures. Budapest was completely open. The vaccination rate was high. I was very comfortable. It felt good not wearing mask outside. It was nice to see faces without masks. It was kind of a stress reliever. We enjoyed our trips." (Participant ID 49).

"I am normally addicted to travelling. I save my money to see new places. My biggest limitation during the pandemic was that I couldn't travel, it affected me a lot. I at least did two domestic short-distance flights but avoided long-distance flights." (Participant ID 57).

3.7.11 Owning a pet

Two participants owned a new pet during the pandemic.

"We bought a dog. I used to say I would never get an animal inside the house but gave up thinking so because it really was good for all of us." (Participant ID 20).

"We adopted a cat from the street. In the winter of 2021, we hosted a cat from outside, it kept me busy for two weeks. Since our cat didn't want the new one, I stayed on the balcony day and night for two weeks with our new guest. " (Participant ID 35).

3.7.12. Reorganizing home

Some participants tried to find satisfaction by organizing their houses, sorting out clothes, toys, or things that are not used, tidying drawers, etc.

"The first thing I did in the first weeks of the lockdowns was to organize my closets. I took everything out and replaced them. The first week passed tidying my room, it distracted me from reality." (Participant ID 25).

"I did things like arranging closets, throwing away excess items, sorting out my clothes, giving unnecessary or excess ones to those in need. I dealt with such things and felt relieved." (Participant ID 38).

In Table 3 below, every code, with its corresponding category and theme, is rank ordered according to their frequencies. Not to be surprised, avoidance from the COVID-19 pandemic has the highest frequency (mentioned 393 times in total), followed by active coping (mentioned 369 times in total). The remaining coping strategies do not seem to appear as much as the first two. Emotional support and positive thinking were mentioned 83 and 61 times respectively. The coping strategies that are least applied are rational thinking (frequency 16), emotional venting (frequency 15), and instrumental support (frequency 11).

The word cloud of the mostly preferred coping mechanisms is presented in Figure 2. Accordingly, 3W's rule, engaging in sports, getting vaccinated, vitamin intake, healthy eating, disinfecting product packages, interacting with others got highest frequencies.

Figure 2. Word cloud of coping mechanisms.

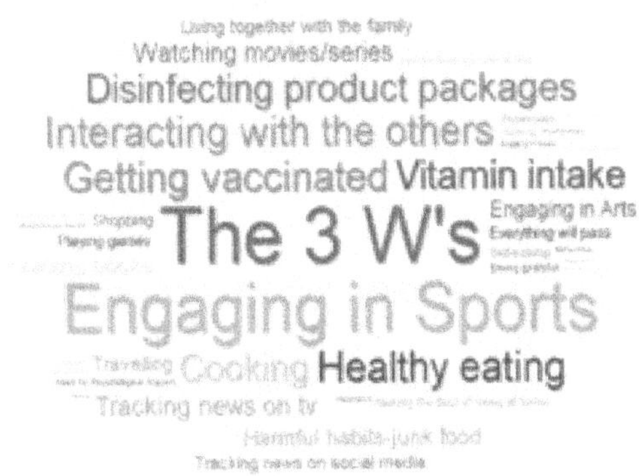

4. RECOMMENDATIONS AND CONCLUSION

Coping is a complex psychological process, embedded in a network of cognitive, attitudinal, and behavioral correlates. To develop a better understanding of coping strategies in an emerging country, during the pandemic, this study took Duhachek's (2005) multi-dimensional model of consumer coping as a theoretical background and aimed to answer how Turkish consumers coped during the pandemic.

The study identified all coping strategies proposed by Duhachek (2005) except denial. None of the participants claimed s/he coped with the COVID-19 pandemic by denying it. Obviously, for a disastrous pandemic with too many unknowns and uncertainties, denying it may not be a coping strategy. Avoidance and active coping were found to be the most preferred strategies. One of the aspects of avoidance, finding satisfaction from other things, resulted in too many areas that participants tried to find satisfaction in and had not been deeply investigated in other academic studies. These results may especially be important for consumer researchers because purchasing products and services was mentioned mostly as a way of finding satisfaction from other things. More specifically, engaging in sports, arts, music, reading, cooking, etc. required them to buy related products or they simply relied on shopping as an escape mechanism. Adopting active coping also necessitated shopping for some essential products such as disinfectants, colognes, soaps, masks, vitamins, healthy food, etc. Similar findings were also reported by Guthrie *et al.* (2021) who examined French consumers and reported that they purchased products during the pandemic to make life less stressful and well-being goods to take their minds away from the problem. Also, Jain *et al.* (2023) found that Indian consumers used more cleaning agents and sanitizers as preventive solutions against COVID-19. Consumers purchased face masks, gloves, protective equipment, disinfectants, sanitizers, soaps, hydro-alcoholic gels, organic and natural personal care items, and veterinary products. In this study rational thinking, emotional venting, and instrumental support were found to be the least preferred coping strategies.

Another contribution of this study is the identification of newly emerged coping mechanisms during the pandemic. Acknowledging that everyone's going through this, self-discovery, and search for meaning were not mentioned in previous studies and they appear to be the coping mechanisms specifically adopted during the COVID-19 pandemic. On the other hand, being grateful was not mentioned by Duhachek (2005) but they were identified by other COVID-19 related studies. For instance, religiosity was found to be a source of coping with COVID-19 led stress for subsistence consumers in India and Bangladesh (Nath *et al.*, 2022). Even though what is detected in this study as being grateful is not directly related to religion, both mechanisms can be considered as spiritual comforting. Living together with the family is another mechanism that was not mentioned by Duhachek (2005) but in some of the COVID-19 focused studies it was expressed as social support, ie. access to and use of individuals, groups, or organizations in dealing with life's ups and downs (Song *et al.*, 2011). For example, Nath *et al.* (2022) found that social support with family members in times of grief and hardship is beneficial for the overall well-being of subsistence consumers during the pandemic. In Mahapatra and Mishra's (2022) study, social support was stated as a significant coping response, while it existed in maximizing focus of consumers more than those with a satisfying focus for coping. Finally, Im *et al.* (2021) predicted that people from a collectivist culture (i.e., Koreans) find social coping resources more effective whereas those from an individualist culture (i.e., Americans) find personal coping resources more effective in protecting themselves from experiencing negative emotions (e.g., anxiety, loneliness), however, social support was found to reduce loneliness equally in both groups. The authors explain that having some constant face-to-face contact with a significant other or family members and receiving remote emotional support were sufficient to protect oneself from feeling lonely. However, in this study living together with family and interacting with others emerged as different coping mechanisms. Some participants highlighted that they would not cope with COVID-19 if they lived away from their families. Living together with family requires a higher level of physical touch such as hugging, kissing, or any other physical contact, and as indicated by Thomas and Kim (2021) theory and evidence suggest that physical touch may buffer stress.

Managerial implications of the study are targeted at policymakers and marketing managers. Even though the devastating effects of the COVID-19 pandemic seem to be over, governmental officials and policymakers should take lessons from it for future possible disasters. The findings of this study support the view that in case of a pandemic, most people would want to actively cope with it and thus government officials should consistently highlight the importance of imposing the 3W's rule and provide necessary resources to limit the spread of the virus. Since the usage of emotional support, positive thinking, rational thinking, emotional venting, and instrumental support appears to be less employed, the importance of such coping strategies can also be communicated to people through marketing communications tools. On the other hand, marketing managers may benefit from coping strategies described in this study and they may develop a better understanding of their consumers and plan product decisions and brand communication in a time of pandemic accordingly. Apparently, the major coping strategy marketing managers need to watch out for is avoidance, because a high number of consumers wanted to find satisfaction through shopping or engage in activities that require the purchase of related products and services. Therefore, examination of how consumers tried to find satisfaction may help marketing managers design related products and services, as well as design promotional messages to be targeted to consumers.

This study provides a thorough examination of consumer coping strategies in an emerging market. While most of the previous studies were quantitative, applying a qualitative research method allowed researchers to gain a deeper exploration of coping strategies and thus unraveling coping mechanisms specific to a pandemic. This method enabled researchers the identification of both previously documented coping strategies and mechanisms, such as the ones proposed by Duhachek (2005) and other researchers listed in Table 1, as well as the detection of novel coping mechanisms reported by the Turkish consumers.

As in many qualitative studies, this study also has its limitations, and they are twofold. Firstly, due to the selection of a nonprobability sampling method findings are pertinent to the consumers in Türkiye and thus may not be generalized to the whole world. Secondly, the interviews were made in Spring 2022, at a time when lockdowns were already finished and people got back to normal to a certain extent. Even though they honestly expressed how they went through the pandemic, it might have been better to interview them earlier when they were at lock down and actually coping with the pandemic.

This study reveals that Duhachek's (2005) multi-dimensional model of consumer coping can be applied to a situation of a worldwide pandemic, future research may attempt to understand the interrelations between the constructs. First of all, the effect of the demographic characteristics in explaining coping strategies can be investigated. Additionally, how shopping was used as a coping mechanism during the pandemic requires more delineation, attention to describe whether consumers had to change their shopping preferences, whether they had to encounter difficulties, and how they coped with the stress created by the shopping process during COVID-19 pandemic.

Ethics Approval: This research received ethics approval from Piri Reis University Ethics Committee with the approval number of 2023/5.

REFERENCES

Abdulmuhsin, A. A., Değirmenci, B., Efendİoğlu, İ. H., & Durmaz, Y. (2022). The perception of COVID-19 and avoidance behavior in Turkey: The role of income level, gender and education. *International Journal of Emerging Markets, 17*(10), 2558–2583. DOI: 10.1108/IJOEM-11-2020-1308

Bae, M. (2023). Coping strategies initiated by COVID-19 related stress, individuals' motives for social media use and perceived stress reduction. *Internet Research, 33*(1), 125–151. DOI: 10.1108/INTR-05-2021-0269

Carver, C. S., Scheier, M. F., & Weintraub, J. K. (1989). Assessing coping strategies: A theoretically based approach. *Journal of Personality and Social Psychology, 56*(2), 267–283. DOI: 10.1037/0022-3514.56.2.267 PMID: 2926629

Compas, B. E., Connor-Smith, J. K., Saltzman, H., Thomsen, A. H., & Wadsworth, M. E. (2001). Coping with stress during childhood and adolescence: Problems, progress, and potential in theory and research. *Psychological Bulletin, 127*(1), 87–127. DOI: 10.1037/0033-2909.127.1.87 PMID: 11271757

Di Crosta, A., Ceccato, I., Marchetti, D., La Malva, P., Maiella, R., Cannito, L., Cipi, M., Mammarella, N., Palumbo, R., Verrocchio, M. C., Palumbo, R., & Di Domenico, A. (2021). Psychological factors and consumer behavior during the COVID-19 pandemic. *PLoS One, 16*(8), e0256095. Advance online publication. DOI: 10.1371/journal.pone.0256095 PMID: 34398916

Duhachek, A. (2005). Coping: A multidimensional, hierarchical framework of responses to stressful consumption episodes. *The Journal of Consumer Research, 32*(1), 41–53. DOI: 10.1086/426612

Ertz, M., Tandon, U., Quenum, G. G. Y., Salem, M., & Sun, S. (2022). Consumers' coping strategies when they feel negative emotions in the face of forced deconsumption during the COVID-19 pandemic lockdowns. *Frontiers in Psychology, 13*, 13. DOI: 10.3389/fpsyg.2022.1018290 PMID: 36524184

Folkman, S., Lazarus, R. S., Gruen, R. J., & DeLongis, A. (1986). Appraisal, coping, health status, and psychological symptoms. *Journal of Personality and Social Psychology, 50*(3), 571–579. DOI: 10.1037/0022-3514.50.3.571 PMID: 3701593

Golzar, J., & Tajik, O. (2022). Convenience sampling. *International Journal of Education and Language Studies, 1*(2).

Guest, G., Bunce, A., & Johnson, L. (2006). How many interviews are enough? An experiment with data saturation and variability. *Field Methods*, *18*(1), 59–82. DOI: 10.1177/1525822X05279903

Guthrie, C., Fosso-Wamba, S., & Arnaud, J. B. (2021). Online consumer resilience during a pandemic: An exploratory study of e-commerce behavior before, during, and after a COVID-19 lockdown. *Journal of Retailing and Consumer Services*, *61*, 102570. DOI: 10.1016/j.jretconser.2021.102570

Ha, S., Childs, M., Sneed, C. T., & Berry, A. (2021). Consumer sustainable shopping practices for small business during COVID-19. *Sustainability (Basel)*, *13*(22), 12451. DOI: 10.3390/su132212451

Hsieh, H. F., & Shannon, S. E. (2005). Three approaches to qualitative content analysis. *Qualitative Health Research*, *15*(9), 1277–1288. DOI: 10.1177/1049732305276687 PMID: 16204405

Im, H., Kim, N. L., & Lee, H. K. (2022). Why did (some) consumers buy toilet papers? A cross-cultural examination of panic buying as a maladaptive coping response to COVID-19. *The Journal of Consumer Affairs*, *56*(1), 391–413. DOI: 10.1111/joca.12426

Jain, A., Dash, S., & Malhotra, N. K. (2023). Consumption coping to deal with pandemic stress: Impact on subjective well-being and shifts in consumer behavior. *European Journal of Marketing*, *57*(5), 1467–1501. DOI: 10.1108/EJM-11-2021-0864

Kemp, E., Bui, M., & Porter, M.III. (2021). Preparing for a crisis: Examining the influence of fear and anxiety on consumption and compliance. *Journal of Consumer Marketing*, *38*(3), 282–292. DOI: 10.1108/JCM-05-2020-3841

Kim, J., Yang, K., Min, J., & White, B. (2021). Hope, fear, and consumer behavioral change amid COVID-19: Application of protection motivation theory. *International Journal of Consumer Studies*, *46*(5), 558–574. PMID: 34220343

Kirk, C. P., & Rifkin, L. S. (2020). I'll trade you diamonds for toilet paper: Consumer reacting, coping, and adapting behaviors in the COVID-19 pandemic. *Journal of Business Research*, *117*, 124–131. DOI: 10.1016/j.jbusres.2020.05.028 PMID: 32834208

Lazarus, R. S. (1991). Progress on a cognitive-motivational-relational theory of emotion. *The American Psychologist*, *46*(8), 819–834. DOI: 10.1037/0003-066X.46.8.819 PMID: 1928936

Lazarus, R. S. (1996). The role of coping in the emotions and how coping changes over the life course. In Magai, C., & McFadden, S. H. (Eds.), *Handbook of emotions across the lifespan* (pp. 289–306). Academic Press., DOI: 10.1016/B978-012464995-8/50017-0

Lazarus, R. S., & Folkman, S. (1984). *Stress, appraisal, and coping*. Springer Publishing Company.

Li, J., Jin, X., Zhao, T., & Ma, T. (2021). Conformity consumer behavior and external threats: An empirical analysis in China during the COVID-19 pandemic. *SAGE Open*, *11*(3), 21582440211032152. Advance online publication. DOI: 10.1177/21582440211032152

Liu, C., Xia, S., & Lang, C. (2022). Coping with COVID-19: How resilient consumers adjust clothing consumption during the pandemic. *International Textile and Apparel Association Annual Conference Proceedings*, 78(1). DOI: 10.31274/itaa.13561

Mahapatra, S., & Mishra, A. (2022). Crying over spilt milk? Effect of post-consumption dissonance on coping behaviour for online purchases. *International Journal of Consumer Studies*, *46*(1), 1035–1054. DOI: 10.1111/ijcs.12744

Mathur, A., Moschis, G. P., & Lee, E. (1999). Stress and consumer behavior: Coping strategies of older adults. *Journal of Marketing Practice: Applied Marketing Science*, 5(6-7-8), 233-247. https://doi.org/DOI: 10.1108/EUM0000000004572

McKinlay, A. R., Fancourt, D., & Burton, A. (2022). Factors affecting the mental health of pregnant women using UK maternity services during the COVID-19 pandemic: A qualitative interview study. *BMC Pregnancy and Childbirth*, *22*(1), 1–15. DOI: 10.1186/s12884-022-04602-5 PMID: 35413807

Mishra, A., Indwar, R. G., & Samu, S. (2022). If tomorrow never comes: The impact of anxiety, materialism and attitude on consumption behavior in a pandemic. *Journal of Consumer Marketing*, *39*(3), 294–305. DOI: 10.1108/JCM-02-2021-4439

Nath, S. D., Jamshed, K. M., & Shaikh, J. M. (2022). The impact of the COVID-19 pandemic on subsistence consumers' well-being and coping strategies: Insights from India and Bangladesh. *The Journal of Consumer Affairs*, *56*(1), 180–210. DOI: 10.1111/joca.12440 PMID: 35603322

Potter, W. J., & Levine-Donnerstein, D. (2009). Rethinking validity and reliability in content analysis. *Journal of Applied Communication Research*, *27*(3), 258–274. DOI: 10.1080/00909889909365539

Ross, G. M. (2023). Now that everyone else has panicked, well… I have to panic: Regulatory focus predicts panic buying during COVID-19. *Journal of Consumer Marketing*, *40*(4), 521–534. DOI: 10.1108/JCM-06-2022-5409

Skinner, E. A., Edge, K., Altman, J., & Sherwood, H. (2003). Searching for the structure of coping: A review and critique of category systems for classifying ways of coping. *Psychological Bulletin*, *129*(2), 216–269. DOI: 10.1037/0033-2909.129.2.216 PMID: 12696840

Song, L., Son, J., & Lin, N. (2011). Social support. In Scott, J., & Carrington, P. J. (Eds.), *The SAGE handbook of social network analyses* (pp. 116–128). SAGE Publications Ltd.

Stone, A., Helder, L., & Schneider, M. (1988). Coping with stressful events: Coping dimensions and issues. In Cohen, L. (Ed.), *Life events and psychological functioning: Theoretical and methodological issues* (pp. 182–210). Sage.

Sundaray, B. K., Sarangi, P., & Patra, S. K. (2021). Psychological vulnerability and coping among management students during COVID-19 pandemic. *The Journal of Mental Health Training, Education and Practice*, *16*(6), 454–467. DOI: 10.1108/JMHTEP-05-2021-0041

Szymkowiak, A., Gaczek, P., Jeganathan, K., & Kulawik, P. (2021). The impact of emotions on shopping behavior during epidemic: What a business can do to protect customers. *Journal of Consumer Behaviour*, *20*(1), 48–60. DOI: 10.1002/cb.1853 PMID: 38607848

Thomas, C. G. (2021). *Research methodology and scientific writing*. Springer. DOI: 10.1007/978-3-030-64865-7

Thomas, P. A., & Kim, S. (2021). Lost touch? Implications of physical touch for physical health. *The Journals of Gerontology. Series B, Psychological Sciences and Social Sciences*, *76*(3), e111–e115. DOI: 10.1093/geronb/gbaa134 PMID: 32845008

Yang, Y., Li, O., Peng, X., & Wang, L. (2020). Consumption trends during the COVID-19 crisis: How awe, coping, and social norms drive utilitarian purchases. *Frontiers in Psychology*, *11*, 588580. DOI: 10.3389/fpsyg.2020.588580 PMID: 33192921

Yap, S. F., Xu, Y., & Tan, L. (2021). Coping with crisis: The paradox of technology and consumer vulnerability. *International Journal of Consumer Studies*, *45*(6), 1239–1257. DOI: 10.1111/ijcs.12724

Zulauf, K., & Wagner, R. (2022). Online shopping therapy: If you want to be happy, shop around. *Journal of International Consumer Marketing*, *34*(3), 332–345. DOI: 10.1080/08961530.2021.1955425

APPENDIX

Table 1. Summary table of the studies of consumer coping strategies during the Covid-19 Pandemic

Author/s	Year	Country	Coping Strategies	Type of the Study
Kirk and Rifkin	2020	Global study	Coping (maintaining social connectedness, do-it-yourself behaviors, changing views of the brand)	Commentary
Szymkowiak *et al.*	2020	Global study	Shopping time	Quantitative
Yang *et al.*	2020	China	Problem-focused coping	Quantitative
Guthrie *et al.*	2021	France	Shopping-related coping (growth in sales of face masks, gloves, protective equipment, disinfectants and sanitizers and soaps and hydro alcoholic gels; growth in sales of organic and natural, hair, body and face care, and makeup and perfume categories, growth in sales of veterinary products.)	Qualitative
Di Crosta *et al.*	2021	Italy	Self-justification for purchasing	Quantitative
Mahapatra and Mishra	2021	India and Bangladesh	Mental disengagement, seeking social support and product return.	Quantitative
Yap *et al.*	2021	Global study	Technology-mediated consumption as a coping strategy, social media as a resource for consumer resilience, innovations.	Review paper
Li *et al.*	2021	China	Conformity consumer behavior- informative conformity consumer behaviors, normative conformity consumer behaviors	Quantitative
Ha *et al.*	2021	USA	Optimism, passive resilience, and active resilience	Quantitative
Im *et al.*	2022	USA and Korean	Resilience, social support, and panic buying	Quantitative
Kim *et al.*	2021	USA	Hygienic behavior supports local and conscious consumption	Quantitative
Sundaray *et al.*	2021	India	Protective and avoidance coping strategies	Quantitative
Liu *et al.*	2022	USA	Coping self-efficacy	Mixed method
Zulauf and Wagner	2022	Europe and Latin America	Online therapeutic shopping motivation, crisis coping shopping, and therapeutic shopping purchases intention	Quantitative
Nath et al.	2022	India and Bangladesh	Religiosity and social support	Qualitative

continued on following page

Table 1. Continued

Author/s	Year	Country	Coping Strategies	Type of the Study
Ertz *et al.*	2022	North America	Confrontive coping, Mental disengagement, Seeking social support, Acceptance, Positive reinterpretation, Behavioral disengagement, Self- control, Planful problem-solving	Quantitative
Abdulmuhsin *et al.*	2022	Türkiye	Avoidance behavior- cognitive avoidance, avoiding public places, avoiding personal contact	Quantitative
Jain *et al.*	2023	India	Approach consumption coping (Reducing touchpoints, Immunity boosting measures, Preventive coping, Adopting technology, financial prudence) Avoidance consumption coping (Emotion regulation consumption, Distraction, Impulsive buying)	Mixed method
Bae, M.	2023		Active coping, expressive support, emotional venting, avoidance coping strategies	Quantitative

Table 2. Sample demographics

PARTICIPANT ID	GENDER	AGE	EDUCATION	OCCUPATION	MARITAL STATUS	NUMBER OF CHILDREN
1	F	18	High School	Student	Single	0
2	F	18	High School	Student	Single	0
3	F	18	High School	Student	Single	0
4	F	18	High School	Student	Single	0
5	F	18	High School	Student	Single	0
6	M	18	High School	Student	Single	0
7	M	18	High School	Student	Single	0
8	M	18	High School	Student	Single	0
9	M	18	High School	Student	Single	0
10	F	22	University	Student	Single	0
11	M	22	University	Student	Single	0
12	M	22	Master	Academician	Single	0
13	M	28	University	Engineer	Single	0
14	F	28	Master	Academician	Single	0
15	F	30	University	Engineer	Married	1
16	F	31	PhD	Student	Married	0
17	F	32	Master	Pedagogue	Single	0
18	M	33	Master	Manager	Married	1
19	F	35	PhD	Academician	Single	0
20	F	37	PhD	Academician	Married	2
21	M	38	University	Engineer	Married	2
22	F	38	PhD	Academician	Single	0
23	F	38	Master	Researcher	Single	0
24	F	39	PhD	Academician	Married	1
25	M	40	PhD	Academician	Married	1
26	M	40	University	Manager	Married	1
27	M	40	University	Freelancer	Single	1
28	F	41	University	Finance Specialist	Single	1
29	F	41	University	Lawyer	Married	1
30	F	41	Master	Engineer	Married	1
31	F	41	University	Architect	Married	2
32	M	42	PhD	Academician	Married	0
33	F	42	University	Manager	Single	0
34	F	42	University	Nurse	Single	1
35	M	42	PhD	Academician	Single	0
36	F	45	PhD	Academician	Married	2
37	F	45	University	Retired	Married	2
38	F	48	Secondary School	Cleaning Lady	Married	4
39	M	49	PhD	Academician	Married	1
40	M	50	University	Manager	Single	0
41	F	50	University	Retired	Married	1
42	F	51	PhD	Academician	Married	1
43	F	53	University	Architect	Single	1
44	F	54	University	Retired	Married	0
45	M	54	Master	Manager	Married	1
46	F	54	Master	Consultant	Married	2
47	F	55	Master	Retired	Single	1
48	F	55	University	Manager	Married	2
49	F	55	University	Retired	Married	2
50	F	56	PhD	Pedagogue	Single	0
51	M	56	University	Freelancer	Married	1
52	F	57	University	Retired	Married	1
53	F	60	University	Physical Therapist	Married	0
54	M	62	University	Engineer	Married	1
55	F	64	University	Retired	Married	0
56	F	66	Master	Retired	Single	1
57	F	55+	University	Retired	Single	0
58	F	60+	Master	Business Owner	Married	1
59	M	60+	University	Designer	Married	0
60	F	65+	University	Retired	Single	1

Table 3. Coping mechanisms & strategies

Code	Frequency	Cases	% of Cases	Category	Category Frequency	Theme	Theme Frequency
Engaging in Sports	88	42	70,00%	Finding satisfaction from other things	386	Avoidance	393
Cooking	48	34	56,70%				
Watching movies/series	33	26	43,30%				
Reading books	31	21	35,00%				
Harmful habits-junk food	29	27	45,00%				
Engaging in Arts	27	19	31,70%				
Traveling	26	15	25,00%				
Playing games	20	18	30,00%				
Shopping	20	18	30,00%				
Self-training	15	13	21,70%				
Harmful habit-cigarette alcohol	13	13	21,70%				
Engaging in music	9	9	15,00%				
Gardening	9	6	10,00%				
Re-organizing home	8	8	11,70%				
Acknowledging emotions	5	3	5,00%				
Harmful habit-social media	3	3	5,00%				
Pet ownership	2	2	3,30%				
Trying not to think	7	6	10,00%	Avoid thinking about it	7		
Healthy eating	54	44	73,30%	Generating potential solutions	214	Active Coping	369
Vitamin intake	51	42	70,00%				
Disinfecting product packages	51	36	60,00%				
Tracking news on tv	35	31	51,70%				
Tracking news on social media	23	23	38,30%				
The 3 W's	100	52	86,70%	Doing what has to be done	155		
Getting vaccinated	55	44	73,30%				

continued on following page

Table 3. Continued

Code	Frequency	Cases	% of Cases	Category	Category Frequency	Theme	Theme Frequency
Interacting with the others	54	42	70,00%	Relying on others to feel better	54	Emotional Support	83
Living together with the family	25	21	35,00%	Living together with the family	25		
Sharing feelings with the beloved ones	4	4	6,70%	Sharing feelings with others	4		
Everything will pass	19	16	26,70%	Focusing on positive aspects	31	Positive Thinking	61
Polyannaism	12	10	16,70%				
Making the best of being at home	16	12	20,00%	Making the best of the situation	30		
Being grateful	14	11	18,30%	Being grateful			
Standing strong	7	5	8,30%	Controlling emotions	11	Rational Thinking	16
Controlling emotions	2	2	3,30%				
Not reflecting emotions to others	2	2	3,30%				
Everyone's going through this, not just us	5	3	5,00%	Acknowledging that everyone's going through this	5		
Self-discovery	8	7	11,70%	Self-discovery	8	Emotional Venting	15
Acknowledging emotions	5	3	5,00%	Acknowledging emotions	5		
Search for Meaning	2	2	3,30%	Search for Meaning	2		
Need for Psychological Support	11	8	13,30%	Getting advice from someone about what to do	11	Instrumental Support	11

Chapter 15
Exploring Tourism Studies:
Key Insights From Strategic Stakeholder Matrices – A Case Study in Fez and Meknes

Aarab Hamza
https://orcid.org/0009-0005-4686-9822
Sidi Mohamed Ben Abdellah University, Morocco

Boukil Mehdi
https://orcid.org/0009-0007-2439-3039
Sidi Mohamed Ben Abdellah University, Morocco

ABSTRACT

Field interviews and qualitative studies play a crucial role in tourism research, offering in-depth insights into the experiences and perspectives of key stakeholders. This study explores the tourism landscape of Fez, Morocco, using a multi-method qualitative approach. In-depth interviews were conducted with tourism professionals, including hoteliers, tour guides, and restaurant owners, as well as government officials. These interviews provided valuable perspectives on local tourism challenges, opportunities, and the role of technology in smart tourism management. Additionally, stakeholder matrices were employed to systematically analyze the roles, relationships, and influences of different actors in the tourism sector. By integrating these qualitative methodologies, this research offers a comprehensive understanding of tourism dynamics in Fez, contributing to strategic planning and sustainable tourism development.

DOI: 10.4018/979-8-3373-4432-4.ch015

INTRODUCTION

Field interviews and qualitative studies are fundamental methodologies in tourism research, offering deep insights into the experiences and perspectives of various stakeholders. This chapter examines the tourism landscapes of Fez and Meknes, Morocco, through a comprehensive qualitative approach, aiming to uncover key opportunities and challenges in these historic cities.

To achieve this, we conducted in-depth interviews with a diverse range of stakeholders, including hoteliers, tour guides, restaurant owners, and government officials. These discussions provided valuable perspectives on local tourism practices, management strategies, and the role of technology in enhancing industry efficiency. Each stakeholder group operates with distinct priorities: hoteliers seek profitability and customer satisfaction, tour guides focus on curating enriching visitor experiences, restaurant owners aim to attract customers through quality service, and government officials prioritize economic growth, cultural heritage preservation, and regulatory frameworks. Tourists, on the other hand, value immersive cultural experiences, service quality, and affordability.

To analyze the complex relationships and power dynamics among these stakeholders, we employ the **MACTOR method** (Method of Actor Analysis and Their Relations). This framework helps identify key players—such as travel agencies, local governments, hotel businesses, local communities, and tourists—while mapping their objectives, potential collaborations, and conflicts. By highlighting stakeholder interactions, the MACTOR method facilitates the development of balanced strategies that promote sustainable tourism and address the diverse interests of all parties involved.

Through this chapter, we aim to advance both theoretical and practical understandings of qualitative methodologies in tourism research. By applying rigorous analytical tools, we contribute to the enhancement of research quality while offering actionable insights for industry practitioners and policymakers seeking to navigate the evolving tourism landscapes of Fez and Meknes.

I- THE ROLE OF INTERVIEWS AS A QUALITATIVE RESEARCH TOOL IN TOURISM

While the interview is a formalisation of a method that is already used in everyday life and is part of common sense, it can make even more sense in the development of tourism knowledge when it is systematised and accountable (**Picken, Felicity. 2018**)

Interviews serve as a vital means of collecting research data, conducted in a dialogue format within an interactive framework between the researcher and the interviewee.

The primary goal is to seek answers relevant to the researcher's analysis and the development of the research path and problem. In the context of tourism studies, interviews provide invaluable insights into the experiences and perspectives of various stakeholders.

One of the primary advantages of field interviews is the ability to capture responses in a natural context. This spontaneous interaction creates a relaxed and open environment that fosters authentic communication. By allowing interviewees to express themselves freely, researchers can gain deeper insights into the nuances of their experiences and viewpoints, which might not emerge in more structured settings.

Field interviews promote spontaneity and genuine expression of ideas, leading to a wealth of diverse and rich information. This dynamic environment encourages participants to share thoughts and experiences that may be more challenging to elicit through structured formats like surveys or questionnaires. The result is a broader spectrum of insights, making qualitative research more robust and reflective of real-world complexities.

To highlight this methodology, we conducted several semi-directed field interviews with various professionals and stakeholders in the tourism sector in Fez and Meknes, particularly those with extensive experience in the field. Among them, tour guides are considered the main actors, as they are closest to the tourism process and can provide valuable evaluations and insights into the local tourism landscape. Their perspectives are crucial for understanding both the challenges and opportunities within the tourism sector, enriching our overall analysis and contributing to a more comprehensive understanding of the dynamics at play in Fez. (Darbi, W., & Hall, C. 2014) provides a review of some of the major issues involved in the conduct of elite interviews and highlights some of the tactics that researchers may use in the interview process.

Subsequently, a qualitative study was conducted on the outputs of the multiple field interviews, followed by the selection of relevant information to address the research problem. This effort required significant time and effort, involving many days of meetings, information gathering, and establishing complex relationships with sector stakeholders.

II- UNDERSTANDING STAKEHOLDERS IN THE TOURISM SECTOR THROUGH STAKEHOLDER MATRICES

The **methodology** for identifying and analyzing stakeholders in the tourism sector involves several key steps. First, we will compile an exhaustive list of stakeholders, which includes travel agencies, hotels, tour guides, local authorities, transportation companies, and tourist associations. Next, we will create separate matrices for each aspect of our research; for instance, one matrix will focus on stakeholders involved in tourism development while another will address those related to cultural heritage preservation. In these matrices, we will document stakeholders' roles, interests, power, level of influence, and their relationships with other stakeholders. The stakeholder analysis will populate these matrices by identifying relevant stakeholders for each specific area of study, utilizing both qualitative and quantitative information gathered through documentary research, interviews with industry professionals, and on-site observations. We will then map out the relationships between different stakeholders using the information from the matrices, identifying interactions, partnerships, potential conflicts, and collaboration opportunities. Visual tools such as graphs, diagrams, or concept maps will be employed to clearly represent these relationships. By following this methodology, we aim to gain a comprehensive understanding of the stakeholders involved in the tourism sector, as well as their interactions, which will facilitate a deeper understanding of the sector's dynamics and inform effective and sustainable tourism management recommendations.

3. RESULTS AND DISCUSSION

3.1 Analyzing the Roles of Tourism Actors in the Cities of Fez and Meknes through the "Mactor" Methodology

In our observation and analysis of the relationships between tourism professionals in the cities of Fez and Meknes, we focused on the results of our field surveys. This was done to track the nature of the relationships that connect a significant network of actors representing both the public and private sectors, in their administrative and professional branches. We applied a vertical approach linked to the central administration, represented by the Ministry of Tourism, and a horizontal approach where local tourism structures converge within the framework of professional integration among them.

Initially, we assigned a specific code to each tourism actor for use through the "Mactor" technique (Méthode Acteurs, Objectifs, Rapports de force).

We then identified the dominant tourism actors involved in marketing the tourism product in both cities and selected 15 actors. Typically, between 10 and 20 actors are used in studies associated with this statistical tool, and their names are abbreviated using Latin letters for technical purposes, as follows:

Table 1. Distribution of the Study Sample of Tourism Actors

Abbreviation	Tourism Actor	Number
M.T	Ministry of Tourism	1
O.N.M.T	Moroccan National Tourism Office	2
D.T	Tourism Delegation	3
C.T	Tourism Council	4
H.C	Classified Hotels	5
M.H	Guest Houses	6
R.T	Tourist Restaurants	7
A.V	Travel Agencies	8
T.T	Tourist Transport	9
G.T	Tour Guides	10
B	Bazaars	11
A.L.V	Car Rental Agencies	12
B.C	Currency Exchange Offices	13
T.E	Foreign Tourists	14
T.M	Moroccan Tourists	15

Source: Fieldwork, 2021-2022

The process of using the Mactor technique required the completion of processing the fieldwork data, particularly those related to the relationships between the actors. Accordingly, a technical card was created detailing the roles they play in producing the tourism product in the cities of Fez and Meknes. This served as an initial step to help analyze the network of relationships that connect them. Their strategic roles are represented in the following table:

Table 2. Distribution of the Roles of Actors in the Production of the Tourism Product in the Cities of Fez and Meknes

Tourism Actor	Key Role in Building the Tourism Product
Ministry of Tourism	- Establishing legal regulations governing tourism professions. - Announcing programs to develop tourism activities. - Conducting studies, creating future scenarios, and ensuring strategic competitiveness for the tourism sector.
Moroccan National Tourism Office	Implementing advertising policies for tourist destinations nationally and internationally.
Tourism Delegation	- Implementing Ministry of Tourism strategies for tourism product development locally and regionally. - Monitoring tourism projects.

continued on following page

Table 2. Continued

Tourism Actor	Key Role in Building the Tourism Product
Tourism Council	- Implementing the Moroccan National Tourism Office's advertising policies regionally. - Supporting professional representations.
Classified Hotels	Providing accommodation rooms, with some offering dining and entertainment services.
Guest Houses	Providing accommodation rooms, with some offering dining and entertainment services.
Tourist Restaurants	Offering traditional and modern meals, and serving all types of beverages.
Travel Agencies	- Selling tickets and booking rooms at accommodation establishments. - Organizing tourist tours in Fez and Meknes for foreign tourists.
Tourist Transport	- Transporting tourists from Fez-Saïss Airport to hotels. - Transporting tourists between hotels and the old cities of Fez and Meknes. - Transporting tourists between Fez and Meknes.
Tour Guides	- Welcoming tourists upon arrival at Fez-Saïss Airport. - Accompanying tourists throughout their tour.
Bazaars	Displaying and selling traditional crafts (copper, leather, carpets, etc.).
Car Rental Agencies	Renting private cars to tourists upon request.
Currency Exchange Offices	Exchanging currencies from the dirham to foreign currencies, or vice versa, upon request.
Foreign Tourists	The primary target group of the tourism product and the key element of tourism demand.
Moroccan Tourists	A group not primarily targeted by the tourism offerings in the cities of Fez and Meknes.

Source: Fieldwork, 2021-2022

3.2 Direct and Indirect Influences Between Tourism Actors in the Cities of Fez and Meknes

The results of our fieldwork on the relationships between tourism actors at the local level constituted an important step in deducing the type of mutual influences between them. We used our findings in the "Mactor" program for this purpose, through a matrix known as the Matrix of Direct Influences (MID). This process involves assessing the degree of influence each actor has on the other, using a scale from 0 to 4. Accordingly, the levels of influence in our sample were as follows:

- **0**: No influence
- **1**: Influence on operations
- **2**: Influence on projects
- **3**: Influence on tasks

- **4**: Influence on continuity

We were able to create a matrix that illustrates the degree of influence between the actors, as shown in the figure 1:

Figure 1. Distribution of the matrix of direct influences between the actors

MID	M.T	O.N.M.T	D.T	C.T	H.C	M.H	R.T	A.V	T.T	G.T	B	A.L.V	B.C	T.E	T.M
M.T	0	4	3	3	4	4	4	4	4	4	1	1	2	4	4
O.N.M.T	1	0	1	4	3	3	3	3	3	3	1	1	1	1	1
D.T	0	0	0	3	4	4	4	4	4	4	1	1	1	1	0
C.T	0	0	1	0	1	1	1	1	1	1	0	0	0	4	1
H.C	0	0	0	0	0	3	3	4	3	3	1	0	1	1	0
M.H	0	0	0	0	3	0	3	3	3	4	4	1	1	1	0
R.T	0	0	0	0	1	1	0	1	1	1	0	0	0	1	0
A.V	0	0	0	0	1	1	1	0	1	1	1	0	0	1	0
T.T	0	0	0	0	1	0	1	1	0	3	1	4	1	1	0
G.T	0	0	0	0	3	1	3	2	1	0	4	0	1	4	0
B	0	0	0	0	0	2	0	1	0	3	0	0	3	4	0
A.L.V	0	0	0	0	0	0	0	1	2	0	0	0	0	1	1
B.C	0	0	0	0	1	1	0	0	0	1	1	0	0	2	0
T.E	0	0	2	2	3	3	3	4	4	3	3	2	2	0	1
T.M	0	0	0	0	1	1	0	0	1	0	0	1	0	0	0

Source: Mactor Progra m Results, 2022

Source: *Mactor Program Results, 2022*

The previous table shows the diversity in the degree of influence between each actor, particularly between those representing the governmental body (Ministry of Tourism) along with its administrative structures (Moroccan National Tourism Office and Tourism Delegation), and the other local actors who organize tourism services regionally.

3.3 Analysis of the Degree of Direct and Indirect Influence and Dependency Between Tourism Actors in the Cities of Fez and Meknes

We sought to deepen our understanding of these influences, so we developed a Matrix of Direct and Indirect Influences (MIDI), which helps extract two main indicators:

- **(Li)**: This indicator, located on the far right of the table in blue, determines the degree of direct and indirect influences by calculating the horizontal sum of each actor's influence. The actor with the highest absolute sum of values is the most influential.

- **(Di)**: This indicator, located at the bottom of the table in blue, determines the degree of direct and indirect dependency for each actor. It is calculated by adding up the vertical values of each actor's dependency, with the actor having the highest absolute sum being the most dependent.

Our data from this program resulted in a matrix of direct and indirect influences, which we explain in figure 2:

Figure 2. Distribution of the matrix of direct and indirect impacts among tourism stakeholders

MIDI	M.T	O.N.M.T	D.T	C.T	H.C	M.H	R.T	A.V	T.T	G.T	B	A.L.V	B.C	T.E	T.M	Ii
M.T	1	4	7	12	25	23	25	28	26	28	18	11	11	22	8	248
O.N.M.T	1	1	4	7	18	16	18	19	18	20	14	9	9	18	5	176
D.T	0	0	2	4	16	14	17	19	16	20	14	7	7	16	3	153
C.T	0	0	3	3	11	10	10	11	12	10	9	6	7	11	2	102
H.C	0	0	1	1	11	9	12	13	10	14	11	5	6	10	1	93
M.H	0	0	1	1	11	10	12	13	11	17	13	5	8	15	2	109
R.T	0	0	1	1	6	5	6	6	6	6	6	3	5	6	1	52
A.V	0	0	1	1	6	6	6	7	6	7	6	3	6	7	1	56
T.T	0	0	1	1	8	7	7	8	7	9	8	5	5	10	2	71
G.T	0	0	2	2	11	12	12	13	11	14	12	4	9	14	1	103
B	0	0	2	2	10	8	9	9	8	10	10	3	7	11	1	80
A.L.V	0	0	1	1	4	3	3	3	5	4	3	4	2	3	2	34
B.C	0	0	2	2	5	6	5	6	5	6	6	3	6	6	1	53
T.E	0	0	3	4	17	16	17	19	19	21	14	9	10	17	3	152
T.M	0	0	0	0	3	2	3	4	4	3	3	3	3	4	1	32
Di	1	4	29	39	151	137	156	171	157	175	137	76	95	153	33	1514

From the previous table, we can conclude that the most influential actor (**Li**) in this system of actors is the "Ministry of Tourism" (**Li (M.T) = 248**), through the role of its central offices in devising and proposing strategies for developing and marketing the tourism product. Following it is the "Moroccan National Tourism Office" (**Li (O.N.M.T) = 176**), owing to its position as a public institution with financial independence, which enables it to set its agenda for local tourism marketing.

The Tourism Delegation ranks as the third most influential actor on other professionals in the two cities (**Li (D.T) = 153**), representing the Ministry of Tourism at the regional and local levels. Its various departments serve as the tool for managing the tourism sector in Fez and Meknes.

Foreign tourists come in fourth place (**Li (T.E) = 152**). Their strong role in tourism demand exerts significant influence on potential changes in various tourism services. Guest houses rank fifth (**Li (M.H) = 109**), as the first tourism service to directly influence other professionals, highlighting their prominent control over the operation of tourism services in the two cities.

Two notable observations:

- The first point highlights that travel agencies rank eleventh among the most influential actors, reflecting their marginal role in crafting local tourism products and attracting foreign tourists to visit Fez and Meknes. According to Aârab and Serrhini (2020), only 13% of travel agencies in Fez specialize in inbound tourism.
- The second is the last-place position of Moroccan tourists on this list, reflecting the neglect and marginalization of plans to develop domestic tourism in relation to the cultural tourism offerings of the old cities. Thus, Recognizing the significance of domestic tourism alongside international markets is essential for fostering a more inclusive and resilient tourism industry.

Regarding the analysis of the degree of direct and indirect dependency (**Di**) for each tourism actor, we find that tour guides (**Di (G.T) = 175**) are the most dependent group among the actors. Despite the importance of the tour guide profession, it remains closely tied to other components of the tourism system, such as the Tourism Delegation, Tourism Council, travel agencies, guest houses, and tourist transport. Most of the time, tour guides rely on other professionals to provide tourists for guided tours, making their professional activity highly dependent on the efforts of other tourism actors in Fez and Meknes.

This matrix revealed some key insights, including the second-place ranking of travel agencies in terms of their dependency on other actors (**Di (A.V) = 171**), indicating that their impact on tourism activity in Fez and Meknes remains weak. As for the last three ranks, which represent the actors least dependent on others, they are the same actors with the most influence in the tourism system: the Tourism Delegation (**Di (D.T) = 29**), the Moroccan National Tourism Office (**Di (ONMT) = 4**), and the Ministry of Tourism (**Di (M.T) = 1**).

Based on these results and the Matrix of Direct Influences, we sought to extract the "Plan of Influences and Dependencies Between Actors" by dividing the actors into four groups as highlighted in the figure 3:

Figure 3. Plan of influences and dependencies between tourism actors in the cities of Fez and Meknes

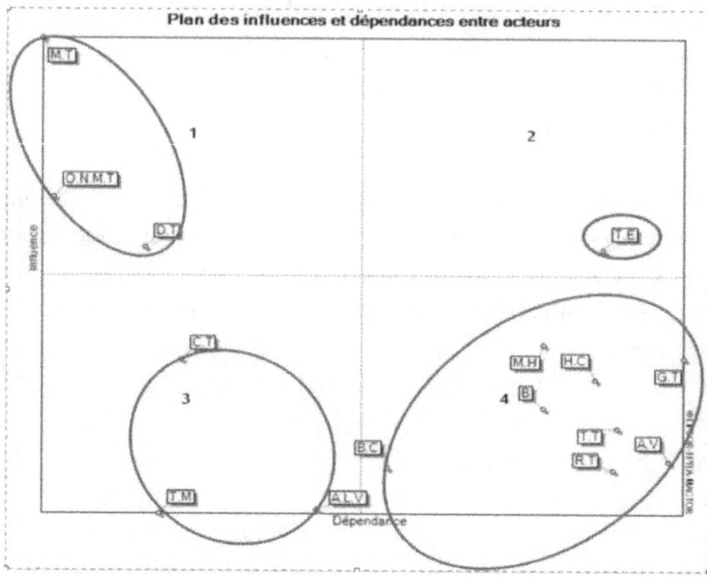

Source: Mactor Program Results, 2022

Source: *Mactor Program Results, 2022*

Based on the previous diagram, we can classify the influence and dependency of tourism actors in developing the tourism product in the cities of Fez and Meknes into four groups as follows:

- **Dominant Actors (Zone 1)**: These include the Ministry of Tourism (M.T), the Moroccan National Tourism Office (O.N.M.T), and the Tourism Delegation (D.T). They are institutions formed by the higher authority and have the ability to prioritize, establish, or prevent projects, distribute tasks, and even manage the daily operations of the sector.
- **Relay Actors (Zone 2)**: This zone includes one actor, which is foreign tourists (T.E). They are among the actors with significant influence and dependency, but their status as the target group of the tourism product underscores their important role within the actor system.
- **Autonomous Actors (Zone 3)**: This group consists of the Tourism Council (C.T), Moroccan tourists (T.M), and car rental agencies (A.L.V). These actors enjoy a certain degree of independence, but their influence on the actor system remains weak due to their limited roles in the tourism product in the cities of Fez and Meknes.
- **Dependent Actors (Zone 4)**: This group includes eight actors represented by guest houses (M.H), tour guides (G.T), travel agencies (A.V), tourist trans-

port (T.T), tourist restaurants (R.T), classified hotels (H.C), and bazaars (B). This group is the weakest in terms of influence and most dependent on other actors. The degree of their influence and dependency varies from one actor to another, and it is contingent upon the strategies and choices of the dominant and relay actors. This means that this group is governed by decisions from the Ministry of Tourism, represented at the regional and local levels by the Tourism Delegation, and the marketing strategies for the tourism product by the Moroccan National Tourism Office guide the nature of the work of various professionals in the two cities. Foreign tourists, as "relay actors," have their suggestions and demands taken seriously by professionals.

4 POWER BALANCE DIAGRAM BETWEEN TOURISM ACTORS IN THE CITIES OF FEZ AND MEKNES

The power balance coefficient for an actor is calculated from the matrix of maximum direct influences and dependencies, where absolute values are assigned to the power balances and ranked from strongest to weakest, or vice versa" (Jaziri & Cherif, 2005), denoted by the symbol Qi. The Power Balance Diagram (MMIDI) between tourism actors in the cities of Fez and Meknes can be presented as shown in Figure 4.

Figure 4. Power balance scheme among tourism actors in the cities of Fez and Meknes

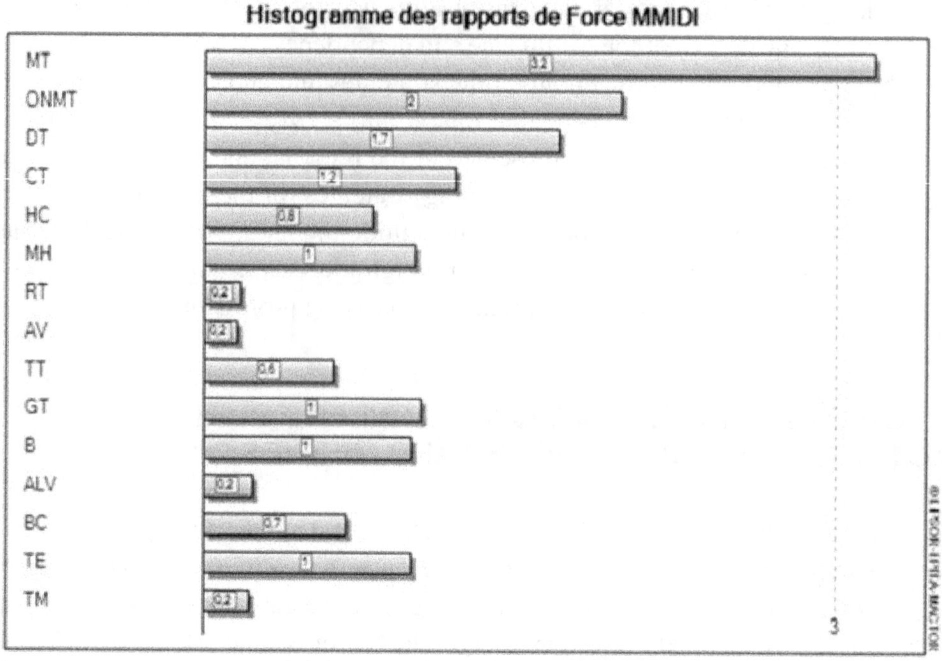

Source*: Mactor Program Results, 2022*

Based on the previous chart, we can interpret the absolute values for each actor, which we will clarify in figure 5:

Figure 5. Distribution of absolute values for power balances

	Qi
M.T	3,2
O.N.M.T	2,0
D.T	1,7
C.T	1,2
H.C	0,8
M.H	1,0
R.T	0,2
A.V	0,2
T.T	0,6
G.T	1,0
B	1,0
A.L.V	0,2
B.C	0,7
T.E	1,0
T.M	0,2

Source: *Mactor Program Results, 2022*

Based on the previous table, we can classify the power balances into three main categories, as follows:

- **Category One ($Qi \geq 2$)**: Represented by the Ministry of Tourism and the Moroccan National Tourism Office, which exert a clear influence on the power dynamics in the tourism system of the cities of Fez and Meknes.
- **Category Two ($2 < Qi < 1$)**: This includes, in order, the Tourism Delegation of the two cities, the Tourism Council, foreign tourists, guest houses, bazaars, and tour guides. This category remains directly dependent on Category One.
- **Category Three ($Qi \leq 1$)**: This includes, in order, classified hotels, currency exchange shops, tourist transport, car rental agencies, tourist restaurants, travel agencies, and Moroccan tourists. This is the weakest category in terms of power balances among tourism actors in the cities of Fez and Meknes.

3.4 Recommendations for a Balanced Tourism Development in Fez and Meknes

The findings of this research highlight the key role of central institutions, such as the Ministry of Tourism, ONMT, and the Tourism Delegation, in shaping the tourism system of Fez and Meknes. Local professionals, with their diversity, also contribute to the tourism product, and there is an opportunity for greater collaboration to enhance the product's development. This collaborative potential could foster a more complementary and sustainable tourism framework, strengthening the cities' competitiveness in the tourism market. To foster better synergy among stakeholders and strengthen the sector's resilience, we propose the following recommendations:

3.4.1. Strengthening the Autonomy of Local Actors

The autonomy of dependent actors can be enhanced through targeted initiatives aimed at reducing their structural vulnerability:

- **Diversification of Commercialization Channels:** Encourage the use of specialized digital platforms, such as Online Travel Agencies (OTAs), mobile applications, and AI-driven tools, to expand access for tour guides, travel agencies, and guesthouses to a broader clientele. These platforms enable local actors to bypass reliance on central institutions and directly engage with a global market. Furthermore, AI technologies, including machine learning algorithms for personalized marketing, dynamic pricing models, and recommendation systems, enhance service delivery by tailoring experiences and improving customer targeting. Research shows that integrating AI into digital platforms can significantly increase efficiency, customer satisfaction, and profitability by allowing local actors to better understand traveler preferences and anticipate market trends (Xiang et al., 2022; Zeng et al., 2023).

3.4.2. Improving Inter-Actor Cooperation

Better coordination among different levels of actors would optimize the integration of tourism into the local economy:

- **Establishment of a Collaborative Network**: Create a regional platform allowing guides, agencies, and service providers to co-develop offerings and pool resources. This platform should function as an inclusive space where all actors—public and private—can contribute to decision-making and benefit from collective actions. The creation of a Destination Management Organization (DMO) is critical in this context. Based on this work using the MACTOR tool to analyze the stakeholders in Fez and Meknes, the involvement of a DMO would help align the interests of various actors, reduce conflicts, and ensure the effective management of the destination. The DMO can also play a crucial role in coordinating marketing efforts and guiding the development of sustainable tourism strategies that benefit all stakeholders.
- **Dialogue and Consultation:** Organize periodic meetings between tourism professionals and institutional representatives to foster a more inclusive governance framework. Special attention should be given to integrating voices from marginalized actors, ensuring that all stakeholders contribute to shaping the destination's future. This process supports collaborative governance, which has been shown to enhance decision-making and foster a sense of shared responsibility among actors (Bichler & Lösch, 2019).

3.3.3. Structuring and Enhancing the Tourism Product

The tourism offerings in Fez and Meknes would benefit from diversification and better structuring to meet the evolving expectations of travelers while ensuring the active involvement of all key actors:

- **Segmentation and Adaptation of Tourism Products:** Develop specific offerings (e.g., spiritual tourism, immersive historical and handicrafts circuits) based on an experiential approach. The process should be participatory, ensuring that local artisans, guides, and travel agents or travel planners are actively involved in designing and refining these experiences, thereby promoting authenticity and local ownership. This collaborative approach not only enriches the visitor experience but also strengthens the role of local stakeholders in the tourism value chain, contributing to sustainable and inclusive tourism development.

- **Labeling and Certification**: By developing certification programs that focus on quality standards, local actors—such as guides, accommodations, and service providers—can differentiate themselves from competitors and attract more customers (Vellas & Becherel, 2020). Furthermore, labeling encourages continuous improvement among local service providers, as it fosters a culture of excellence and accountability. As a result, the overall tourism offering in a destination is enhanced, contributing to both economic growth and sustainable development.
- **Territorial Marketing Strategy**: Strengthen the destination's visibility through targeted digital campaigns and promotional actions in key markets. Local actors should be empowered to participate in promotional efforts, ensuring that their authentic narratives and experiences shape the destination's image. Research indicates that including local stakeholders in the marketing process enhances destination branding and fosters a sense of ownership, leading to more sustainable and appealing tourism promotions (Hernández-Mogollón et al., 2021).

These recommendations align with an integrated territorial development approach and the capacity-building of local actors. Their implementation would require a participatory approach involving all stakeholders to ensure more balanced governance and a more evenly distributed economic impact across the territory.

CONCLUSION

This study contributes to tourism research by employing a qualitative approach to examine the intricate relationships between various tourism actors in the cities of Fez and Meknes. By using stakeholder interviews as a primary data collection method, we gained in-depth insights into the roles, perceptions, and interactions within the tourism ecosystem. These first-hand accounts enriched our analysis, offering a nuanced understanding of power dynamics and dependencies.

The application of the Mactor methodology enabled a systematic classification of actors based on their influence and dependency, revealing a clear hierarchy within the system. The Ministry of Tourism, the Moroccan National Tourism Office, and the Tourism Delegation emerged as dominant actors, shaping strategies and policies that directly impact local tourism dynamics. The qualitative data provided further context, illustrating how these entities navigate challenges and opportunities in marketing and tourism product development.

Our analysis of direct and indirect influences highlighted the complex dependencies within the system. Tour guides and travel agencies, despite playing essential roles in service delivery, remain highly reliant on the decisions of dominant actors. This interdependence underscores the need for stronger collaboration and communication among all stakeholders to enhance service quality and better meet the evolving needs of tourists.

In conclusion, this research deepens our understanding of power structures and stakeholder relationships within the tourism sector of Fez and Meknes. It also reinforces the value of qualitative methods, particularly stakeholder interviews, in uncovering the complexities of tourism governance. By integrating these findings into future tourism strategies, policymakers and industry professionals can develop more sustainable, competitive, and inclusive tourism models that benefit all actors involved.

REFERENCES

Aârab, H., & Serrhini, S. (2020). L'impact des agences de voyages sur la destination touristique de Fès (Maroc). *EL-Manhel Economique, 3*(2), 453–468.

Ben Massou, S. (2011). *Quel tourisme pour le Maroc dans le cadre d'une société des réseaux? Le cas de la région de Marrakech-Tensift-Al Haouz* (Doctoral thesis, Ecole Doctorale: DEGEST, FLSH de Marrakech).

Bichler, B. F., & Lösch, M. (2019). Collaborative Governance in Tourism: Empirical Insights into a Community-Oriented Destination. *Sustainability (Basel), 11*(23), 6673. DOI: 10.3390/su11236673

Darbi, W. P. K., & Hall, C. M. (2014). Elite interviews: Critical practice and tourism. *Current Issues in Tourism, 17*(9), 832–848. DOI: 10.1080/13683500.2014.887663

Ethier, G.. (2012). *Epistémologie des études touristiques.* Presses Universitaires du Québec.

Hernández-Mogollón, J. M., González-González, I., & Manso, J. P. (2021). The role of local actors in the destination branding process. *Journal of Destination Marketing & Management, 19*, 100460.

Jaziri, R., & Cherif, M. (2005). Projet de contractualisation des universités tunisiennes: Analyse du jeu d'acteurs par la méthode MACTOR. In *Actes du colloque international sur: La connaissance: un nouveau levier et défis de la compétitivité pour les entreprises et les économies.* Faculté des Sciences Economiques et de Management de Biskra.

Picken, F. (2018). The interview in tourism research. In Hillman, W., & Radel, K. (Eds.), *Qualitative Methods in Tourism Research: Theory and Practice* (pp. 200–223). Channel View Publications., DOI: 10.21832/9781845416416-014

Presenza, A., & Cipollina, M. (2010). Analysing tourism stakeholders networks. *Tourism Review, 65*(4), 17–30. DOI: 10.1108/16605371011093845

Riadh, H. (2021). Intelligent tourism system using prospective techniques and the MACTOR methodology: A case study of Tunisian tourism. *Current Issues in Tourism, 25*(9), 1376–1398. DOI: 10.1080/13683500.2021.1937072

Vellas, F., & Becherel, L. (2020). *Tourism Marketing: A Strategic Approach.* Routledge.

Xiang, Z., Du, Q., Ma, Y., & Fan, W. (2022). Research agenda for artificial intelligence in tourism: A case study of data-driven tourism service innovation. *Tourism Management, 88*, 104403.

Zeng, Q., Liu, Y., & Li, L. (2023). AI-enabled smart tourism: Impacts and opportunities for destination management and marketing. *Journal of Destination Marketing & Management, 24*, 100704.

Compilation of References

Aârab, H., & Serrhini, S. (2020). L'impact des agences de voyages sur la destination touristique de Fès (Maroc). *EL-Manhel Economique*, *3*(2), 453–468.

Abdulmuhsin, A. A., Değirmenci, B., Efendioğlu, İ. H., & Durmaz, Y. (2022). The perception of COVID-19 and avoidance behavior in Turkey: The role of income level, gender and education. *International Journal of Emerging Markets*, *17*(10), 2558–2583. DOI: 10.1108/IJOEM-11-2020-1308

Achard, P., Davallon, J., Durand, J.-L., Pêcheux, M., & Orlandi, E. P. (1999). *The Role of Memory*. Pontes Publishers.

Adhikari, E. (2023). *Ethnography as a research methods: A critique*. Kathmandu University.

Agbo, J. (2018). Heidegger through Husserl to phenomenological ontology. *An African Journal Philosophy VOL*, *18*(2), 1–13.

Aguinis, H., & Solarino, A. M. (2019). Transparency and replicability in qualitative research: The case of interviews with elite informants. *Strategic Management Journal*, *40*(8), 1291–1315. DOI: 10.1002/smj.3015

Agustian, A., Malik, J., & Yuwono, U. (2020). Content analysis approach in critical discourse analysis of newspapers' news. [IJHI]. *International Journal of Humanities and Innovation*, *3*(1), 1–6.

Ahmed, S. (2024). The pillars of trustworthiness in qualitative research. *Journal of Medicine, Surgery, and Public Health*, *2*, 1–4. DOI: 10.1016/j.glmedi.2024.100051

Ajegbile, M. D., Olaboye, J. A., Maha, C. C., & Tamunobarafiri, G. (2024). Integrating business analytics in healthcare: Enhancing patient outcomes through data-driven decision making. *World J Biol Pharm Health Sci*, *19*(1), 243–250. DOI: 10.30574/wjbphs.2024.19.1.0436

Akinyode, B. F., & Khan, T. H. (2018). Step by step approach for qualitative data analysis. *International Journal of built environment and sustainability, 5*(3).

Akter, S., Bandara, R., Hani, U., Wamba, S. F., Foropon, C., & Papadopoulos, T. (2019). Analytics-based decision-making for service systems: A qualitative study and agenda for future research. *International Journal of Information Management, 48*, 85–95. DOI: 10.1016/j.ijinfomgt.2019.01.020

Al Umma, K. (2015). Ferdinand De Saussure: Structuralism and his role in modern lingustics. *Jurnal Lisanu ad-Dhad Vol. 02, No. 01*, 357-376.

Alam, M. K. (2021). A systematic qualitative case study: Questions, data collection, NVivo analysis and saturation. *Qualitative Research in Organizations and Management, 16*(1), 1–31. DOI: 10.1108/QROM-09-2019-1825

Alase, A. (2017). The interpretative phenomenological analysis (IPA): A guide to a good qualitative research approach. *International Journal of Education and Literacy Studies, 5*(2), 9–19. DOI: 10.7575/aiac.ijels.v.5n.2p.9

Alberti, V. (n.d.). *História oral: A experiência do CPDOC*. Fundação Getúlio Vargas.

Alcadipani, R., & Hassard, J. (2010). Actor-Network Theory, organizations and critique: Towards a politics of organizing. *Organization, 17*(4), 419–435. DOI: 10.1177/1350508410364441

Alcadipani, R., & Tureta, C. (2009). Actor-Network Theory and organizational analysis: Contributions and research possibilities in Brazil. *Organizações & Sociedade, 16*, 647–664. DOI: 10.1590/S1984-92302009000400003

Alhazmi, A., & Kaufmann, A. (2022). Phenomenological qualitative methods applied to the analysis of cross-cultural Experience in novel educational social contexts. *Frontiers in Psychology, 13*, 1–12. DOI: 10.3389/fpsyg.2022.785134 PMID: 35548502

Ali, M. P., Kabir, M. M. M., Haque, S. S., Qin, X., Nasrin, S., Landis, D., Holmquist, B., & Ahmed, N. (2020). Farmer's behavior in pesticide use: Insights study from smallholder and intensive agricultural farms in Bangladesh. *The Science of the Total Environment, 747*, 141160. DOI: 10.1016/j.scitotenv.2020.141160 PMID: 32781314

Allsop, D. B., Chelladurai, J. M., Kimball, E. R., Marks, L. D., & Hendricks, J. J. (2022). Qualitative methods with Nvivo software: A practical guide for analyzing qualitative data. *Psych, 4*(2), 142–159. DOI: 10.3390/psych4020013

Almeida, C. (2020). The Analytical Procedure Through the Discourse of the External Auditor. Available at https://teses.usp.br/teses/disponiveis/12/12136/tde-29062020-141256/pt-br.php

Almusallam, I. (2021). Using NVivo for Data Analysis in Qualitative Research. In *Paper presentation], Conference: Academic Improvement Committee at English Language & Literature Department*. King Saud University.

Alqudah, M. (2019). Deconstruction: A cornucopia of Esoteric meanings . *International Journal of English Literature and Social Sciences (IJELS) Vol-4, Issue-2*, 500-504.

Alsaigh, R., & Coyne, I. (2021). Doing a hermeneutic phenomenology Research Underpinned by Gadamer'sPhilosophy: A framework to facilitate data analysis. *International Journal of Qualitative Methods*, *20*, 1–10. DOI: 10.1177/16094069211047820

Althusser, L. (1980). *Ideology and Ideological State Apparatuses*. Editorial Presença.

Alyssa, E., & Anderman, L. (n.d.). Using interpretive phenomenological analysis to advance theory and research in educational psychology,. Retrieved 09 21, 2024, from DOI: 10.1080/00461520.2020.1787170

Amankwaa, L. (2016). Creating protocols for trustworthiness in qualitative research. *Journal of Cultural Diversity*, *23*(3). PMID: 29694754

Andersson, E., & Isgren, E. (2021). Gambling in the garden: Pesticide use and risk exposure in Ugandan smallholder farming. *Journal of Rural Studies*, *82*, 76–86. DOI: 10.1016/j.jrurstud.2021.01.013

Andrews, T. (2012). What is social constructionism? [online.]. *The Grounded Theory Review*, *11*(1), •••.

Andriani, G., Godoy, E., Koslovski, G., Obelheiro, R., & Pillon, M. (2019). An architecture for synchronising cloud file storage and organisation repositories. *International Journal of Parallel, Emergent and Distributed Systems : IJPEDS*, *34*(5), 538–555. DOI: 10.1080/17445760.2017.1422500

Annink, A. (2017). Using the research journal during qualitative data collection in a cross-cultural context. *Entrepreneurship Research Journal*, *7*(1), 20150063. DOI: 10.1515/erj-2015-0063

Archibald, M. M., Ambagtsheer, R. C., Casey, M. G., & Lawless, M. (2019). Using zoom videoconferencing for qualitative data collection: Perceptions and experiences of researchers and participants. *International Journal of Qualitative Methods*, *18*, 1609406919874596. DOI: 10.1177/1609406919874596

Ary, D., Jacobs, L., & Sorensen, C. (2010). *Introduction to research in education* (8th ed.). Wadsworth Cengage Learning.

Asmawi, A., & Alam, S. (2024). Qualitative research: Understanding its underlying philosophies. *Forum for Education Studies*, *2*(2), 1–14. DOI: 10.59400/fes.v2i2.1320

Asogwa, V., Hamisu, M., & Ede, M. (2023). Methodological triangulation in educational research: Pros and cons. *Propellers Journal of Education*, *2*(1), 79–87.

Ataei, P., Gholamrezai, S., Movahedi, R., & Aliabadi, V. (2021). An analysis of farmers' intention to use green pesticides: The application of the extended theory of planned behavior and health belief model. *Journal of Rural Studies*, *81*, 374–384. DOI: 10.1016/j.jrurstud.2020.11.003

Ayton, D. (2024, 10 25). *Chapter 6 Phenomenology*. Retrieved from Open Educational Resources Collective: https://oercollective.caul.edu.au/qualitative-research/chapter/__unknown__-6/

Ayton, D. (2023). Thematic analysis. In Ayton, D., Tsindos, T., & Berkovic, D. (Eds.), *Qualitative Research – a practical guide for health and social care researchers and practitioners* (pp. 199–209). Monash University.

Babchuk, W., & Boswell, E. (2023). Grounded theory. *International Encyclopedia of Education*, 107-122.

Bae, M. (2023). Coping strategies initiated by COVID-19 related stress, individuals' motives for social media use and perceived stress reduction. *Internet Research*, *33*(1), 125–151. DOI: 10.1108/INTR-05-2021-0269

Bakhitin, M. (1986). *Speech genre and other late essays*. Texas Press. DOI: 10.7560/720466

Bakhtin, M. (1997). *Marxism and the Philosophy of Language: Fundamental Problems of the Sociological Method in the Science of Language*. Hucitec Publishing.

Bakker, L., Sok, J., Van Der Werf, W., & Bianchi, F. (2021). Kicking the habit: What makes and breaks farmers' intentions to reduce pesticide use? *Ecological Economics*, *180*, 106868. DOI: 10.1016/j.ecolecon.2020.106868

Bardin, L. (2016). Content analysis. 3rd Reprint. *São Paulo: Edições, 70.*

Barendregt, B. (2021). Digital ethnography, or 'deep hanging out' in the age of big data. In Grasseni, C., & Barendregt, E. (Eds.), *F. de Maaker, D. Musso, A, M. Littlejohn, . . . M. Westmoreland, Audiovisual and digital ethnography. A practical guide* (pp. 168–190). Routledge. DOI: 10.4324/9781003132417-8

Bartesaghi, N. C. (2016). Intertexuality and interdiscursivity. In *Oxford Bibliographies in Anthropology* (pp. 1–44). OUP.

Bashir, M., Naqshbandi, M. M., & Farooq, R. (2020). Business model innovation: A systematic review and future research directions. *International Journal of Innovation Science, 12*(4), 457–476. DOI: 10.1108/IJIS-06-2020-0081

Basias, N., & Pollalis, Y. (2018). Quantitative and qualitative research in business & technology: Justifying a suitable research methodology. *Review of Integrative Business and Economics Research, 7*, 91–105.

Bassett, R. (2004). Qualitative data analysis software: Addressing the debates. *Journal of Management Systems, 16*(4), 33–39.

Bastos, L. C., & Santos, W. S. (2013). *The interview in qualitative research: Perspectives in narrative and interaction analysis*. Quartet Faperj.

Beauchamp, T. L., & Childress, J. F. (2013). *Principles of biomedical ethics* (7th ed.). Oxford University Press.

Becker, K. M. (2019). Beyond researcher as instrument: Researcher with instrument: musicking in qualitative data collection. *Qualitative Research Journal, 19*(4), 426–437. DOI: 10.1108/QRJ-02-2019-0021

Beer, S. (Undated). *Humanising Phenomenological analysis: using focus groups, food, and drink to collect data for Descriptive and Interpretative Phenomenological Analysis (IPA)*. Bournemouth University: Department of Tourism and Hospitality.

Bell, E., & Bryman, A. (2007). The ethics of management research: An exploratory content analysis. *British Journal of Management, 18*(1), 63–77. DOI: 10.1111/j.1467-8551.2006.00487.x

Ben Massou, S. (2011). *Quel tourisme pour le Maroc dans le cadre d'une société des réseaux? Le cas de la région de Marrakech-Tensift-Al Haouz* (Doctoral thesis, Ecole Doctorale: DEGEST, FLSH de Marrakech).

Benson, J. K. (1977). Organizations: A dialectical view. *Administrative Science Quarterly, 22*(1), 1–21. DOI: 10.2307/2391741

Berkovic, D. (2024, 10 6). *Qualitative Research: A practical guide to for health and Social Care Researchers and Practitioners*. Retrieved from Open Educational Resources Collective: https://oercollective.caul.edu.au/qualitative-research/chapter/__unknown__-21/

Berni, I., Menouni, A., El, I. G., Duca, R.-C., Kestemont, M.-P., Godderis, L., & Jaafari, S. E. (2021). Understanding farmers' safety behavior regarding pesticide use in Morocco. *Sustainable Production and Consumption, 25*, 471–483. DOI: 10.1016/j.spc.2020.11.019

Bhardwaj, P. (2019). Types of sampling in research . *Journal of the Practice of Cardiovascular Sciences ¦ Volume 5 ¦ Issue 3 ¦*, 157-163.

Bichler, B. F., & Lösch, M. (2019). Collaborative Governance in Tourism: Empirical Insights into a Community-Oriented Destination. *Sustainability (Basel), 11*(23), 6673. DOI: 10.3390/su11236673

Bingham, A. J., & Witkowsky, P. (2021). Deductive and inductive approaches to qualitative data analysis. *Analyzing and interpreting qualitative data: After the interview, 1*, 133-146.

Bingham, A. J. (2023). From data management to actionable findings: A five-phase process of qualitative data analysis. *International Journal of Qualitative Methods, 22*, 16094069231183620. DOI: 10.1177/16094069231183620

BIS. (2017a). *Guide to the BIS archives*. 45. www.bis.org

BIS. (2017b). *Basel III: Finalizing post-crisis reforms*.https://www.bis.org/BCBS/publ/d424.pdf

Blackledge, A. (2012). Discourse and power. In Gee, J., & Handford, M. (Eds.), *The Routledge handbook of discourse analysis* (pp. 616–627). Routledge.

Blandford, A. E. (2013). Semi-structured qualitative studies. In Soegard, M., & Dam, R. F. (Eds.), *The encyclopedia of human-computer interaction* (2nd ed.). The Interaction Design Foundation.

Bley, D. H. P., & Carvalho, A. B. G. (2019). Ciclos de codificação e o software Atlas ti: uma parceria criativa para análise de dados qualitativos em pesquisas sobre o uso das tecnologias digitais no campo da Educação. *In Teia | Revista De Educação Matemática E Tecnológica Iberoamericana, 10*(1). DOI: 10.36397/emteia.v10i1.240016

Blismas, N. G., & Dainty, A. R. (2003). Computer-aided qualitative data analysis: Panacea or paradox? *Building Research and Information, 31*(6), 455–463. DOI: 10.1080/0961321031000108816

Bloomfield, B. P., & Vurdubakis, T. (1999). The outer limits: Monsters, actor networks and the writing of displacement. *Organization, 4*(4), 625–647. DOI: 10.1177/135050849964004

Bloor, D. (1991). *Knowledge and social imagery*. University of Chicago Press.

Bogdan, R., & Biklen, S. (1994). *Qualitative research in education: An introduction to theory and methods*. Porto Editora.

Bourdon, S. (2002, May). The integration of qualitative data analysis software in research strategies: Resistances and possibilities. In *Forum Qualitative Sozialforschung/Forum: Qualitative. Social Research, 3*(2), .

Bradbury-Jones, C., Sambrook, S., & Irvine, F. (2009). The phenomenological focus group: An oxymoron. *Journal of Advanced Nursing, 65*(3), 663–671. DOI: 10.1111/j.1365-2648.2008.04922.x PMID: 19222664

Braun, V., & Clarke, V. (2012). Thematic analysis. In H. Cooper, P. Camic, L. Long, D. A. Panther, D. Rindskopf, & K. J. Sher, *APA handbook of research methods in psychology vol 2: Research designs: Quantitative,Qualitative neuropsychological and biological* (pp. 57-71). Washington: American Psychological Association.

Braun, V., & Clarke, V. (2006). Using thematic analysis in psychology. *Qualitative Research in Psychology, 3*(2), 77–101. DOI: 10.1191/1478088706qp063oa

Braun, V., & Clarke, V. (2021). Can I use TA? Should I use TA? Should I not use TA? Comparing reflexive thematic analysis and other pattern-based qualitative analytic approaches. *Counselling & Psychotherapy Research, 21*(1), 37–47. DOI: 10.1002/capr.12360

Braun, V., & Clarke, V. (2021). One size fits all? What counts as quality practice in (reflexive) thematic analysis? *Qualitative Research in Psychology, 18*(3), 328–352. DOI: 10.1080/14780887.2020.1769238

Braun, V., & Clarke, V. (2022). Toward good practice in thematic analysis: Voiding common problems and be(com)ing a knowing researcher. *International Journal of Transgender Health, 24*(1), 1–6. DOI: 10.1080/26895269.2022.2129597 PMID: 36713144

Bridges, D. (2017). *Philosophy in education al research: Epistemology, ethics and quality*. Springer. DOI: 10.1007/978-3-319-49212-4

Brinkmann, S. Jacobsen., & Kristiansen, S. (2014). Historical overview of qualitative research in the social sciences. In P. Leavy, *The Oxford Handbook of qualitative research* (pp. 17-42). Oxford: Oxford University Press.

Britannica, T. Editors of Encyclopaedia. (2023). Dasein. Encyclopedia Britannica.

Brodowicz, M. (2024, 07 2). *Strength and weaknesses of phenomenology*. Retrieved from Aithor: https://aithor.com/essay-examples/strengths-and-weaknesses-of-phenomenology

Brougham, D., & Haar, J. (2018). Smart technology, artificial intelligence, robotics, and algorithms (STARA): Employees' perceptions of our future workplace. *Journal of Management & Organization*, *24*(2), 239–257. DOI: 10.1017/jmo.2016.55

Brown, D. (2002, May). Going digital and staying qualitative: Some alternative strategies for digitizing the qualitative research process. In *Forum Qualitative Sozialforschung/Forum: Qualitative. Social Research*, *3*(2), .

Bryda, G., & Costa, A. P. (2023). Qualitative research in digital era: Innovations, methodologies and collaborations. *Social Sciences (Basel, Switzerland)*, *12*(10), 570. DOI: 10.3390/socsci12100570

Bryman, A., & Burgess, R. G. (Eds.). (1994). *Analyzing qualitative data* (Vol. 11). Routledge.

Burns, J., & Scapens, R. W. (2000). Conceptualizing management accounting change: An institutional framework. *Management Accounting Research*, *11*(1), 3–25. DOI: 10.1006/mare.1999.0119

Butcher, L. M., & Holbrook, A. (1996). Uses of qualitative data analysis software in educational research: The literature, the hard questions and some specific research applications. *Australian Educational Researcher*, *23*(3), 55–80. DOI: 10.1007/BF03219631

Butler, J. (1997). *The psychic life of power: Theories in subjection*. Stanford University Press. DOI: 10.1515/9781503616295

Cabantous, L., & Sergi, V. (2018). Seeing the potentialities at the intersection: A reflection on performativity and processuality mindsets. *M@n@gement*, 21(4), 1229–1243.

Cadena, D. L. (2019). The constitution of objectivities in consciousness in ideas I and ideas II. *Rev. Filos., Aurora, Curitiba, v. 31, n. 53*, 503-519.

Callon, M. (1984). Some elements of a sociology of translation: Domestication of the scallops and the fishermen of St Brieuc Bay. *The Sociological Review*, *32*(1, suppl), 196–233. DOI: 10.1111/j.1467-954X.1984.tb00113.x

Callon, M. (1987). Society in the making: The study of technology as a tool for sociological analysis. In Bijker, W. E., Hughes, T. P., & Pinch, T. J. (Eds.), *The social construction of technological systems: New directions in the sociology and history of technology* (pp. 83–103). MIT Press.

Callon, M. (1998). *The laws of the markets*. Sociological Review Monographs Series. Blackwell Publishers.

Callon, M. (1999). Actor-network theory—The market test. *The Sociological Review*, *47*(1, suppl), 181–195. DOI: 10.1111/j.1467-954X.1999.tb03488.x

Callon, M. (2010). Performativity, misfires, and politics. *Journal of Cultural Economics*, *3*(2), 163–169. DOI: 10.1080/17530350.2010.494119

Callon, M., & Latour, B. (1981). They were unscrewing the big Leviathan: How actors macro-structure reality and how sociologists help them. In Knorr-Cetina, K., & Cicourel, A. V. (Eds.), *Advances in social theory and methodology: Toward integrating micro-and macro-sociologies* (pp. 277–303). Routledge & Kegan Paul.

Callon, M., & Rip, A. (1992). Humans, non-humans: Morality of coexistence. In *La Terre Outragée. Les experts sont formels* (pp. 97–112). Autrement.

Camillis, P. K. D., & Antonello, C. S. (2016). From translation to enactment: Contributions of Actor-Network Theory to the processual approach to organizations. *Cadernos EBAPE.BR*, *14*(1), 61–82. DOI: 10.1590/1679-395131412

Campbell, M. (2018). Postmodernism and Educational Research. *Open Journal of Social Sciences*, *6*(7), 67–73. DOI: 10.4236/jss.2018.67006

Campbell, R., Goodman-Williams, R., Fehler-Cabral, G., & Feeney, H. (2020). Assessing Triangulation Across Methodologies, Methods, and Stakeholder Groups: The Joys, Woes, and Politics of interpreting convergent and divergent data. *The American Journal of Evaluation*, *41*(1), 125–144. DOI: 10.1177/1098214018804195

Carriger, J. F., Rand, G. M., Gardinali, P. R., Perry, W. B., Tompkins, M. S., & Fernandez, A. M. (2006). Pesticides of potential ecological concern in sediment from South Florida canals: An ecological risk prioritization for aquatic arthropods. *Soil & Sediment Contamination*, *15*(1), 21–45. DOI: 10.1080/15320380500363095

Carter, S. M., Shih, P., Williams, J., Degeling, C., & Mooney-Somers, J. (2021). Conducting qualitative research online: Challenges and solutions. *Patient*, *14*(6), 711–718. DOI: 10.1007/s40271-021-00528-w PMID: 34114170

Carver, C. S., Scheier, M. F., & Weintraub, J. K. (1989). Assessing coping strategies: A theoretically based approach. *Journal of Personality and Social Psychology*, *56*(2), 267–283. DOI: 10.1037/0022-3514.56.2.267 PMID: 2926629

Catalano, T., & Waugh, L. (2020). *Critical Discourse Analysis, Critical Discourse Studies and Beyond*. Springer. DOI: 10.1007/978-3-030-49379-0

Cecez-Kecmanovic, D., Davison, R. M., Fernandez, W., Finnegan, P., Pan, S. L., & Sarker, S. (2020). Advancing qualitative IS research methodologies: Expanding horizons and seeking new paths. *Journal of the Association for Information Systems, 21*(1), 1. DOI: 10.17705/1jais.00599

Cerretto, C., & Domenico, S. M. R. D. (2016). Change and Actor-Network Theory: Humans and non-humans in controversies during the implementation of a shared services center. *Cadernos EBAPE.BR, 14*, 83–115. DOI: 10.1590/1679-395131385

Chakraborty, T., Sharada, V. S., & Sharma, J. (2024). Data-Driven Decisions for Employee Motivation: Strategies for Implementing Technology in Performance Reviews in a Tech Landscape. In Employee Performance Management for Improved Workplace Motivation (pp. 311-338). IGI Global.

Chandra, Y., Shang, L., Chandra, Y., & Shang, L. (2019). Computer-assisted qualitative research: An overview. *Qualitative Research Using R: A Systematic Approach*, 21-31.

Charmaz, K. (2006). Constructing grounded theory: A practical guide through qualitative analysis. *Sage (Atlanta, Ga.).*

Charmaz, K. (2014). *Constructing grounded theory* (2nd ed.). SAGE Publications.

Cheligeer, C., Yang, L., Nandi, T., Doktorchik, C., Quan, H., Zeng, Y., & Singh, S. (2023). Natural language processing (NLP) aided qualitative method in health research. *Journal of Integrated Design & Process Science*, *27*(1), 41–58. DOI: 10.3233/JID-220013

Cheng, L., Varshney, K. R., & Liu, H. (2021). Socially responsible ai algorithms: Issues, purposes, and challenges. *Journal of Artificial Intelligence Research, 71*, 1137–1181. DOI: 10.1613/jair.1.12814

Choukaier, D. (2024). Enhancing English As A Foreign Language (EFL) Instruction Through Digital Teaching Platforms: Analyzing The Impact Of Microsoft Teams, Zoom, And Google Meet On Communication And Participation. *Educational Administration: Theory and Practice*, *30*(6), 2404–2418.

Chouliaraki, L., & Fairclough, N. (1999). *Discourse in late modernity: rethinking critical discourse analysis*. Edinburgh University Press.

Chowdhury, A., & Shil, N. (2021). Thinking 'Qualitative' through a case study: Homework for a researcher. *American Journal of Qualitative Research Vol. 5 No. 2, p*, 190-210.

Chueke, G. V., & Lima, M. C. (2012). Qualitative research: Evolution and criteria. *Revista Espaço Acadêmico, 11*(128), 63–69.

Cinar, R., & Benneworth, P. (2021). Why do universities have little systemic impact with social innovation? An institutional logics perspective. *Growth and Change*, *52*(2), 751–769. DOI: 10.1111/grow.12367

Cohen, L., Manion, L., & Morrison, K. (2018). *Research methods in education*. Routledge.

Cole, R. (2023). Inter-rater reliability methods in qualitative case study research. *Sociological Methods & Research*, •••, 00491241231156971.

Compas, B. E., Connor-Smith, J. K., Saltzman, H., Thomsen, A. H., & Wadsworth, M. E. (2001). Coping with stress during childhood and adolescence: Problems, progress, and potential in theory and research. *Psychological Bulletin*, *127*(1), 87–127. DOI: 10.1037/0033-2909.127.1.87 PMID: 11271757

Conley, T. G., & Udry, C. R. (2010). Learning about a new technology: Pineapple in Ghana. *The American Economic Review*, *100*(1), 35–69. DOI: 10.1257/aer.100.1.35

Conselho Monetário Nacional. (2017). *Resolução CMN n° 4.553 de 30/1/2017 – Estabelece a segmentação do conjunto das instituições financeiras e demais instituições autorizadas a funcionar pelo Banco Central do Brasil para fins de aplicação proporcional da regulação prudencial.* Resolução Do Concelho Monetário Nacional. https://www.CBB.gov.br/estabilidadefinanceira/exibenormativo?tipo=Resolução&numero=4553

Conselho Monetário Nacional. (2021). *Resolução CMN n° 4.958 de 21/10/2021 – Dispõe sobre os requerimentos mínimos de Patrimônio de Referência (PR), de Nível I e de Capital Principal e sobre o Adicional de Capital Principal (ACP).* Resolução Do Concelho Monetário Nacional. https://www.CBB.gov.br/estabilidadefinanceira/exibenormativo?tipo=Resolução CMN&numero=4958

Cooper, D. J., Ezzamel, M., & Robson, K. (2019). The multiplicity of performance management systems: Heterogeneity in multinational corporations and management sense-making. *Contemporary Accounting Research*, *36*(1), 451–485. DOI: 10.1111/1911-3846.12416

Cooper, K., & White, R. (2012). *Qualitative research in the post-modern era*. Springer. DOI: 10.1007/978-94-007-2339-9

Cresswell, K., Domínguez Hernández, A., Williams, R., & Sheikh, A. (2022). Key challenges and opportunities for cloud technology in health care: Semistructured interview study. *JMIR Human Factors*, *9*(1), e31246. DOI: 10.2196/31246 PMID: 34989688

Creswell, J. (2014). *Research design: Qualitative, quantitative and mixed approaches*. Sage.

Creswell, J. W. (1998). *Qualitative inquiry and research design: Choosing among five traditions*. Sage Publications.

Creswell, J. W. (2007). Qualitative inquiry and research design : Choosing among five traditions. *Sage (Atlanta, Ga.)*.

Creswell, J. W., & Creswell, J. D. (2017). *Research design: Qualitative, quantitative, and mixed methods approaches*. Sage publications.

Creswell, J. W., & Poth, C. N. (2018). *Qualitative inquiry and research design: Choosing among five approaches* (4th ed.). SAGE Publications.

Creswell, J., & Creswell, J. (2018). *Qualitative,quantitative, and mixed methods approaches*. Sage.

Creswell, J., & Creswell, J. D. (2018). *Research design qualitative, quantitative and mixed research*. Sage.

Crick, J. M. (2021). Qualitative research in marketing: What can academics do better? *Journal of Strategic Marketing*, *29*(5), 390–429. DOI: 10.1080/0965254X.2020.1743738

Cudjoe, E. (2022). Using diaries with interpretative phenomenological analysis: Guidelines from a study of children whose parents have mental illness. *International Journal of Qualitative Methods*, *21*, 1–9. DOI: 10.1177/16094069221084435

Cullen, M., & Brennan, N. (2021). Grounded Theory: Description, divergences and application. *Accounting. Finance & Governance Review*, *27*, 1–19. DOI: 10.52399/001c.22173

Cunliffe, A. L. (2010). Retelling tales of the field: In search of organizational ethnography 20 years on. *Organizational Research Methods*, *13*(2), 224–239. DOI: 10.1177/1094428109340041

Cypress, B. S. (2019). Data analysis software in qualitative research: Preconceptions, expectations, and adoption. *Dimensions of Critical Care Nursing*, *38*(4), 213–220. DOI: 10.1097/DCC.0000000000000363 PMID: 31145168

Czarniawska, B., & Hernes, T. (2005). *Actor-network theory and organizing*. Liber.

D'Angelo, L., & Marino, F. (2024). Multimodal analysis. In Kessler, M., & Polio, C. (Eds.), *Conducting genre-based research in applied linguistics: A methodological guide*. Routledge.

Dahlborg, E., Boman, A., Eriksson, H., & Tengelin, E. (2023). Encircling discourses—A guide to critical discourse analysis in caring science. *Scandinavian Journal of Caring Sciences*, *00*, 1–8. PMID: 37421160

Dalkin, S., Forster, N., Hodgson, P., Lhussier, M., & Carr, S. M. (2021). Using computer assisted qualitative data analysis software (CAQDAS; NVivo) to assist in the complex process of realist theory generation, refinement and testing. *International Journal of Social Research Methodology*, *24*(1), 123–134. DOI: 10.1080/13645579.2020.1803528

Darbi, W. P. K., & Hall, C. M. (2014). Elite interviews: Critical practice and tourism. *Current Issues in Tourism*, *17*(9), 832–848. DOI: 10.1080/13683500.2014.887663

David, A., Jeganathan, G. S., Sikandar, M. D., & Azam, S. A. (2023, May). Internet Users Top Concerns Ensuring Data Privacy, Security, and Protection. In *Arokiaraj David, Jeganathan Gomathi Sankar & Md Sikandar Azam, —Internet Users Top Concerns Ensuring Data Privacy, Security, and Protection , 10th International Conference on Multidisciplinary Research and Modern Education, ISBN* (No. 978-920, pp. 5-20241).

David, A., Sankar, J. G., Ganeshkumar, C., & Kashif, M. (2024). An Introduction to Generative AI: Crafting Worlds From Code. In *Generative AI and Implications for Ethics, Security, and Data Management* (pp. 1-16). IGI Global.

David, A., Ahmed, R. R., Ganesh Kumar, C., & Sankar, J. G. (2020). Consumer purchasing process of organic food product: An empirical analysis. *Journal of Management Systems. Quality - Access to Success*, *21*(177), 128–132.

David, A., Maheswari, M., Sankar, G., Christy, F., Karuppannan, A., & Banumathi, M. (2023). Mental illness of Pregnant Women determined by Covid-19: An Empirical Analysis of Pre-and- Post Delivery. *The Saybold Report Journal*, *18*(03), 1706–1721.

David, A., & Sankar, J. G. (2024). *Generative AI and Implications for Ethics, Security, and Data Management.* IGI Global Scientific Publishing.

David, A., & Sankar, J. G. (2024). *Generative AI for Transformational Management.* IGI Global Scientific Publishing.

Dawadi, S. (2020). Thematic analysis approach: A step by stepm guide for ELT research practitioners. *Journal of NELTA*, *25*(1-2), 62–71. DOI: 10.3126/nelta.v25i1-2.49731

Dawadi, S., Shrestha, S., & Giri, R. A. (2021). Mixed-methods research: A discussion on its types, challenges, and criticisms. *Journal of Practical Studies in Education*, *2*(2), 25–36. DOI: 10.46809/jpse.v2i2.20

de Holanda, L. A. (2003). *Formação e Institucionalização do Campo Organizacional do Turismo no Recife-PE* [Master's Dissertation. Postgraduate Program in Administration, Federal University of Pernambuco]. https://repositorio.ufpe.br/handle/123456789/1157

De Man, P., Bloom, H., & Miller, J. (1979). *Deconstruction and criticism*. Routledge and Kegan.

De Seta, G. (2020). Three lies of digital ethnography vol 2 No 1. *Journal of Digital Social Research*, *2*(1), 77–97. DOI: 10.33621/jdsr.v2i1.24

De Villiers, C., Farooq, M. B., & Molinari, M. (2022). Qualitative research interviews using online video technology–challenges and opportunities. *Meditari Accountancy Research*, *30*(6), 1764–1782. DOI: 10.1108/MEDAR-03-2021-1252

de Wildt, L. (2020). Qualitative data analysis software. In *The SAGE Encyclopedia of the Sociology of Religion* (pp. 643–644). SAGE Publications, Inc.

Deakin University Library. (2024, 10 5). *Qualitative study design*. Retrieved from Deakin University Library: https://deakin.libguides.com/qualitative-study-designs/narrative-inquiry#:~:text=Definition,ordered%20into%20a%20chronological%20narrative

Deane, P. (2018, 05 22). *A guide for interdisciplinary researchers: Adding axiology alongside ontology and epistemology*. Retrieved from A communiIntegration and Implementation Insights blog and repository of resources for improving research impact on complex real-world problems: https://i2insights.org/2018/05/22/axiology-and-interdisciplinarity/

Deep, B. (2020). Lived experience and the idea of the social in Alfred Schutz: A phenomenological study of contemporary relevance. *Journal of Indian Council of Philosophical Research*, *37*(3), 361–381. DOI: 10.1007/s40961-020-00211-9

Delve. (2024, 09 17). *Deductive and Inductive Coding*. Retrieved from Delve: https://delvetool.com/blog/deductiveinductive

Deng, L., Laghari, T., & Gao, X. (2021). A genre-based exploration of intertextuality and interdiscursivity in advertorial discourse. *English for Specific Purposes*, *62*, 30–42. DOI: 10.1016/j.esp.2020.11.003

Denzin, N. K., & Giardina, M. D. (2016). *Qualitative inquiry through a critical lens*. Routledge. DOI: 10.4324/9781315545943

Denzin, N. K., & Lincoln, Y. S. (2011). *The Sage handbook of qualitative research*. Sage.

Denzin, N., & Lincoln, Y. (2014). *The SAGE handbook of qualitative research*. Sage.

Denzin, N., & Lincoln, Y. (Eds.). (1994). *Handbook of qualitative research*. Sage.

Devakurubai, E. (2020). Quality in qualitative research. *Indian Journal of Continuing Nursing Education, 21*(1), 76–80. DOI: 10.4103/IJCN.IJCN_48_20

Dhakal, K. (2022). NVivo. *Journal of the Medical Library Association: JMLA, 110*(2), 270. DOI: 10.5195/jmla.2022.1271 PMID: 35440911

Di Crosta, A., Ceccato, I., Marchetti, D., La Malva, P., Maiella, R., Cannito, L., Cipi, M., Mammarella, N., Palumbo, R., Verrocchio, M. C., Palumbo, R., & Di Domenico, A. (2021). Psychological factors and consumer behavior during the COVID-19 pandemic. *PLoS One, 16*(8), e0256095. Advance online publication. DOI: 10.1371/journal.pone.0256095 PMID: 34398916

Dicks, B. (2019). Multimodal analysis. In Atkinson, P. D. (Ed.), *Research Methods Foundations*. SAGE.

Diemer, N., Staudacher, P., Atuhaire, A., Fuhrimann, S., & Inauen, J. (2020). Smallholder farmers' information behavior differs for organic versus conventional pest management strategies: A qualitative study in Uganda. *Journal of Cleaner Production, 257*, 120465. DOI: 10.1016/j.jclepro.2020.120465

DiMaggio, P. J., & Powell, W. W. (1983). The iron cage revisited: Institutional isomorphism and collective rationality in organizational fields. *American Sociological Review, 48*(2), 147–160. DOI: 10.2307/2095101

Dingwall, R. M., Watson, P., Greatbatch, D., & Parker, S. (1998). Cathing goldfish: Quality in quality research. *Journal of Health Services Research & Policy, 3*(3), 167–171. DOI: 10.1177/135581969800300308 PMID: 10185376

Dodgson, J. E. (2019). Reflexivity in qualitative research. *Journal of Human Lactation, 35*(2), 220–222. DOI: 10.1177/0890334419830990 PMID: 30849272

Dohan, D., & Sanchez-Jankowski, M. (1998). Using computers to analyze ethnographic field data: Theoretical and practical considerations. *Annual Review of Sociology, 24*(1), 477–498. DOI: 10.1146/annurev.soc.24.1.477

Donkoh, S., & Mensah, J. (2023). Application of triangulation in qualitative research. *Journal of Applied Biotechnology and Bioengineering, 10*(1), 6–10. DOI: 10.15406/jabb.2023.10.00319

Dorfler, V., & Stierand, M. (2020). Bracketing: A phenomenological theory applied through transpersonal reflexvity. *Journal of Organisational Management*, 1-22.

Dörfler, V., & Stierand, M. (2020). Bracketing: A phenomenological theory applied through transpersonal reflexivity. *Journal of Organizational Change Management*, •••, 1–22.

Doringer, S. (2020). "The problem-centred expert interview" Combining qualitative interviewing approaches for investigating implicit expert knowledge. *International Journal of Social Research Methodology vOL 24 Issue 3*, 265-278.

Dovetail Editorial Team. (2023, 02 06). *How to do thematic analysis*. Retrieved from Dovetail: https://dovetail.com/research/thematic-analysis/

Duarte, M. D. F., & Alcadipani, R. (2016). Contributions of organizing to organizational studies. *Organizações & Sociedade*, *23*, 57–72. DOI: 10.1590/1984-9230763

Duarte, R. (2004). (in Review). Interviews in qualitative research. *Educating*, (24), 213–225.

Duhachek, A. (2005). Coping: A multidimensional, hierarchical framework of responses to stressful consumption episodes. *The Journal of Consumer Research*, *32*(1), 41–53. DOI: 10.1086/426612

Dursun, B. (2023). A qualitative research technique: Interview. *The Journal of Interdsciplinary Qualitative Research*, *1*(13), 100–113.

Duzy, L. M., Campana, D. J., & Brain, R. (2023). Agroeconomic costs for meeting the Environmental Protection Agency's mitigation menu approach to pesticide regulation. *Agricultural & Environmental Letters*, *8*(2), e20119. DOI: 10.1002/ael2.20119

Ebrahim, S. A., Poshtan, J., Jamali, S. M., & Ebrahim, N. A. (2020). Quantitative and qualitative analysis of time-series classification using deep learning. *IEEE Access : Practical Innovations, Open Solutions*, *8*, 90202–90215. DOI: 10.1109/ACCESS.2020.2993538

EduBirdie. (2024, 09 7). *The evolution of discourse and development of discourse analysis*. Retrieved from EduBirdie: https://edubirdie.com/examples/the-evolution-and-development-of-discourse-analysis/

Elhami, A., & Khoshnevisan, B. (2022). Conducting an interview in qualitative research:The modus operandi. *MEXTESOL Journal*, *46*(1), 1–7. DOI: 10.61871/mj.v46n1-3

Ellett, J. (2011). Narrative and Phenomenology as methodology for understanding persistence in art teachers: A Reflective Journey. *Marilyn Zurmuehlen Working Papers in Art Education 2011*.

Elliott, J. (2022). The craft of using NVivo12 to analyze open-ended questions: An approach to mixed methods analysis. *The Qualitative Report*. Advance online publication. DOI: 10.46743/2160-3715/2022.5460

Elliott-Mainwaring, H. (2021). Exploring using NVivo software to facilitate inductive coding for thematic narrative synthesis. *British Journal of Midwifery*, *29*(11), 628–632. DOI: 10.12968/bjom.2021.29.11.628

Elliott, V. (2018). Thinking about the coding process in qualitative data analysis. *The Qualitative Report*, *23*(11). Advance online publication. DOI: 10.46743/2160-3715/2018.3560

Emanuel, E. J., Wendler, D., & Grady, C. (2000). What makes clinical research ethical? *Journal of the American Medical Association*, *283*(20), 2701–2711. DOI: 10.1001/jama.283.20.2701 PMID: 10819955

Embree, L. (2024, 10 28). *Routledge Encylopedia of Philosophy*. Retrieved from Overview phenomenological movement: https://www.rep.routledge.com/articles/overview/phenomenological-movement/v-1/sections/constitutive-phenomenology

Emiliussen, J., Engelsen, S., Christianse, R., & Klausen, S. (2021). We are all in it!: Phenomenological qualitative research and embeddedness. *International Journal of Qualitative Methods*, *20*, 1–6. DOI: 10.1177/1609406921995304

Emirbayer, M. (1997). Manifesto for a relational sociology. *American Journal of Sociology*, *103*(2), 281–317. DOI: 10.1086/231209

Erciyes, E. (2020). Reflections of a social constructivist on teaching methods. *European Journal of Educational Sciences*, *7*(4), 16–26. DOI: 10.19044/ejes.v7no4a2

Erlingsson, C., & Brysiewicz, P. (2017). A hands-on guide to doing content analysis. *African Journal of Emergency Medicine : Revue Africaine de la Médecine d'Urgence*, *7*(3), 93–99. DOI: 10.1016/j.afjem.2017.08.001 PMID: 30456117

Ertz, M., Tandon, U., Quenum, G. G. Y., Salem, M., & Sun, S. (2022). Consumers' coping strategies when they feel negative emotions in the face of forced deconsumption during the COVID-19 pandemic lockdowns. *Frontiers in Psychology*, *13*, 13. DOI: 10.3389/fpsyg.2022.1018290 PMID: 36524184

Estrada, L. M., & Koolen, M. (2018). Audiovisual media annotation using qualitative data analysis software: A comparative analysis. *The Qualitative Report*, *23*(13), 40–60.

Ethier, G.. (2012). *Epistémologie des études touristiques*. Presses Universitaires du Québec.

European Central Bank. (2017). *SSM thematic review on IFRS 9 Assessment of institutions' preparedness for the implementation of IFRS 9*. European Central Bank., DOI: 10.2866/732554

European Commission. (2018). *General Data Protection Regulation (GDPR)*. Retrieved from https://ec.europa.eu/info/law/law-topic/data-protection_en

Evers, J. C. (2018). Current issues in qualitative data analysis software (QDAS): A user and developer perspective. *The Qualitative Report*, *23*(13), 61–73. DOI: 10.46743/2160-3715/2018.3205

Fabbri, S., Silva, C., Hernandes, E., Octaviano, F., Di Thommazo, A., & Belgamo, A. (2016). Improvements in the StArt tool to better support the systematic review process. In *Proceedings of the 20th International Conference on Evaluation and Assessment in Software Engineering (EASE '16)*. Association for Computing Machinery, New York, NY, USA, Article 21, 1-5. https://doi.org/DOI: 10.1145/2915970.2916013

Fairclough, N. (2001). Critical discourse analysis as a method in social scientific research. In Wodak, R., & Meyer, M. (Eds.), *Methods of critical discourse analysis* (pp. 121–138). Sage.

Fairclough, N. (2003). *Analysing Discourse: Textual analysis for social research*. Routledge. DOI: 10.4324/9780203697078

Farrel, E. (2020). Researching lived experience in education: Misunderstood or missed opportunity? *International Journal of Qualitative Methods*, *19*, 1–8. DOI: 10.1177/1609406920942066

FEBRABAN. (2020). *2020 - LIVE: 10º International Risk Management Congress (1º day) - YouTube*. November 17, 2020. https://www.youtube.com/watch?v=GLJDypipDuU FEBRABAN (2021). *2021 - LIVE: 11º International Risk Management Congress – Day 14 (Portuguese)* October 14, 2021 - *YouTube*. https://www.youtube.com/watch?v=i59w9sO4Qz8

FEBRABAN. (2022a). *2022 - LIVE | 12º International Risk Management Congress (G- Risc) - Day 9 (Portuguese)* November 9, 2022 - *YouTube*.https://www.youtube.com/watch?app=desktop&v=eYFhpRGDouI

FEBRABAN. (2022b). *2022 - LIVE | 12º International Risk Management Congress (G-Risc) - Day 10 - (Portuguese) November 10, 2022 - YouTube*.https://www.youtube.com/watch?v=AnOde_eyS5g

FEBRABAN. (2022c). *Febraban - FEBRABAN*.https://portal.febraban.org.br/paginas/9/pt- br/institucional

Fernandes, C., & Sargentini, V. (2022). Saussure and Foucault; language and discourse. *Revista da Anpoll, Florianópolis*, *53*(2), 53–67. DOI: 10.18309/ranpoll.v53i2.1692

Fernando, A. N. (2021). Seeking the treated: The impact of mobile extension on farmer information exchange in India. *Journal of Development Economics*, *153*, 102713. DOI: 10.1016/j.jdeveco.2021.102713

Finlay, L. (2009). Debating phenomenological research methods. *Phenomenology & Practice*, *3*(1), 6–25. DOI: 10.29173/pandpr19818

Fiorin, J. L. (1990). Trends in Discourse Analysis. *Cadernos de Estudos Linguísticos*, *19*, 173–179.

Flick, U. (2022). *An introduction to qualitative research.*

Flick, U. (2022). *An introduction to qualitative research*. Sage.

Florêncio, A. M. G., Magalhães, B., Sobrinho, H. F. S., & Cavalcante, M. S. A. O. (2009). *Discourse Analysis: Foundations & Practice*. Federal University of Alagoas Publishing.

Floyd, G. (2020). he Sense of Phenomenology: A response to Tarek Dika. *The INPR Journal. Disputatio*, *I*, 163–168.

Folkman, S., Lazarus, R. S., Gruen, R. J., & DeLongis, A. (1986). Appraisal, coping, health status, and psychological symptoms. *Journal of Personality and Social Psychology*, *50*(3), 571–579. DOI: 10.1037/0022-3514.50.3.571 PMID: 3701593

Foucault, M. (1972). *The archaeology of knowledge and the discourse*. Phanteon Books.

Foucault, M. (2008). *The Archaeology of Knowledge*. Forense Universitária Publishing.

Fowler, R., Hodge, B., Kress, G., & Trew, T. (. (1979). Language and control. London: Boston and Henley.

Frechette, J., Bitzas, V., Aubry, M., Kilpatrick, K., & Lavole-Tremblay, L. (2020). Capturing lived experience: Methodological considerations for interpretive phenomenological inquiry. *International Journal of Qualitative Methods*, *19*, 1–12. DOI: 10.1177/1609406920907254

Friedland, R., & Alford, R. R. (1991). Bringing Society Back. In Powell, I. W. W., & DiMaggio, P. J. (Eds.), *Symbols, Practices and Institutional Contradictions* (pp. 232–263). The New Institutionalism in Organizational Analysis.

Friese, S. (2014). *Qualitative Data Analysis with ATLAS.ti* (2. Ed).

Friese, S. (2019). Atlas. ti 8 Mac-User Manual Updated for Program Version 8.4. *ATLAS. ti, Berlin.*

Fritz, R. L., & Vandermause, R. (2018). Data collection via in-depth email interviewing: Lessons from the field. *Qualitative Health Research, 28*(10), 1640–1649. DOI: 10.1177/1049732316689067 PMID: 29298576

Fuster, D. (2019). Qualitative Research Hermeneutical Phenomenological Method. *Propósitos y Representaciones, 7*(1), 201–229.

Gaber, M., Pastor, G. C., & Omer, A. (2020). Speech-to-Text technology as a documentation tool for interpreters: A new approach to compiling an ad hoc corpus and extracting terminology from video-recorded speeches. TRANS: revista de traductología, (24), 263-281.

Gad, C., & Bruun Jensen, C. (2010). On the consequences of post-ANT. *Science, Technology & Human Values, 35*(1), 55–80. DOI: 10.1177/0162243908329567

Ganeshkumar, C., Sankar, J. G., & David, A. (2023). Impact of Artificial Intelligence on Agriculture Value Chain Performance: Agritech Perspective. In *Blockchain, IoT, and AI Technologies for Supply Chain Management* (pp. 71-98). CRC Press.

Ganeshkumar, C., David, A., & Sankar, J. G. (2024). Blockchain Technology Acceptance in Agribusiness Industry. In Idrees, S. M., & Nowostawski, M. (Eds.), *Blockchain Transformations. Signals and Communication Technology*. Springer., DOI: 10.1007/978-3-031-49593-9_15

Ganeshkumar, C., David, A., Sankar, J. G., & Saginala, M. (2023). Application of Drone Technology in Agriculture: A Predictive Forecasting of Pest and Disease Incidence. In *Applying Drone Technologies and Robotics for Agricultural Sustainability* (pp. 50–81). IGI Global.

Ganeshkumar, C., Sankar, J. G., & David, A. (2023). Adoption of Big Data Analytics: Determinants and Performances Among Food Industries. [IJBIR]. *International Journal of Business Intelligence Research, 14*(1), 1–17. DOI: 10.4018/IJBIR.317419

Ganeshkumar, C., Sankar, J. G., & David, A. (2025). Impact of Artificial Intelligence on Organizational Performance of Agritech Firm. In *AI-Powered Business Intelligence for Modern Organizations* (pp. 265–282). IGI Global.

Gary, A., & Holmes, D. (2020). Researcher Positionality - A Consideration of Its influence and place in qualitative research - A New researcher guide. *Shanlax International Journal of Education, 8*(4), 1–10. DOI: 10.34293/education.v8i4.3232

Gąsior-Niemiec, A. (2008). Teoria dyskursu Laclau i Mouffe jako narzędzie analizy socjo logicznej – przypadek dyskursu „osiedli grodzonych. In Horolets, A. (Ed.), *Analiza dyskursu w socjologii i dla socjologii* (pp. 291–316). Wydawnictwo Adam Marszałek.

Gatti, B. A. (2015). Grupo Focal: fundamentos, perspectivas e procedimentos. In M. Tavares & R. J. Ruchardson. (Org.), *Metodologias Qualitativas: teoria e prática* – Editora CRV (231- 257). DOI: 10.24824/978854440558.1

Gephart, R. P., & Saylors, R. (2020). Qualitative designs and methodologies for business, management, and organizational research. In *Oxford research Encyclopedia of Business and management*. DOI: 10.1093/acrefore/9780190224851.013.230

Ghafouri, R., & Ofoghi, S. (2016). Trustworth and rigor in qualitative research. *International Journal of Advanced Biotechnology and Research*, *7*(4), 1914–1922.

Ghanbar, H., Cinaglia, C., Randez, R., & De Costa, P. I. (2024). A methodological synthesis of narrative inquiry research in applied linguistics: What's the story? *International Journal of Applied Linguistics*, *34*(4), 1–27. DOI: 10.1111/ijal.12591

Gilbert, L. S., Jackson, K., & Di Gregorio, S. (2014). Tools for analyzing qualitative data: The history and relevance of qualitative data analysis software. *Handbook of research on educational communications and technology*, 221-236.

Gilbert, L. S. (2002). Going the distance:'closeness' in qualitative data analysis software. *International Journal of Social Research Methodology*, *5*(3), 215–228. DOI: 10.1080/13645570210146276

Gill, P., & Baillie, J. (2018). Interviews and focus groups in qualitative research: An update for the digital age. *British Dental Journal*, *225*(7), 668–672. DOI: 10.1038/sj.bdj.2018.815 PMID: 30287965

Giorgi, A., & Giorgi, B. (2003). *Phenomenology*. Sage.

Girju, R. (2023, October). Understanding Lived Experience: Bridging Artificial Intelligence and Natural Language Processing with Humanities and Social Sciences. [). IOP Publishing.]. *IOP Conference Series. Materials Science and Engineering*, *1292*(1), 012020. DOI: 10.1088/1757-899X/1292/1/012020

Given, L. (2008). *The SAGE Encyclopedia of qualitative research methods*. Sage. DOI: 10.4135/9781412963909

Glaser, B., & Strauss, A. (1967). (*The discovery of Grounded Theory strategies for qualitative research*). CA Sociology Press.

Godoy, A. S. (1995). Introduction to qualitative research and its possibilities. *RAE - Journal of Business Administration, 35*(2), 57.

Goldsmith, L. (2021). Using Framework Analysis in Applied Qualitative Research. *The Qualitative Report, 26*, 2061–2076. DOI: 10.46743/2160-3715/2021.5011

Golzar, J., & Tajik, O. (2022). Convenience sampling. *International Journal of Education and Language Studies, 1*(2).

Gomes, R. D. P., Pimentel, V. P., Landim, A. B., & Pieroni, J. P. (2012). Clinical trials in Brazil: International competitiveness and challenges. *BNDES Sectorial, 36*, 45–84.

Goodman, L. A. (1961). Snowball sampling. The Annals of Mathematical Statistics. *Annals of Mathematical Statistics, 32*(1), 148–170. DOI: 10.1214/aoms/1177705148

Gowda, B., Sendhil, R., Adak, T., Raghu, S., Patil, N., Mahendiran, A., Rath, P. C., Kumar, G., & Damalas, C. A. (2021). Determinants of rice farmers' intention to use pesticides in eastern India: Application of an extended version of the planned behavior theory. *Sustainable Production and Consumption, 26*, 814–823. DOI: 10.1016/j.spc.2020.12.036

Grant, R. (2018). *Data visualization: Charts, maps, and interactive graphics*. Chapman and Hall/CRC. DOI: 10.1201/9781315201351

Green Jr, C. (2022). Industry 4.0 Technologies and Biotechnology Biosensors a Game Changer in US Supply Chain Food Safety: An Exploratory Study. Capella University.

Gregolin, M. R. (2005). Discursive Formation, Memory Networks, and Social Paths of Meaning: Media and Identity Production. Paper presented at the II Discourse Analysis Seminar (SEAD), at UFRGS, Porto Alegre, 2005. A more concise version was published on CD-ROM in the proceedings of the II SEAD.

Grochalska, M. (2020). Research practices in critical discourse analysis: Constraints and challenges. *Kultura i Edukacja, 2020*(2), 98–112. DOI: 10.15804/kie.2020.02.06

Grovermann, C., Schreinemachers, P., Riwthong, S., & Berger, T. (2017). 'Smart' policies to reduce pesticide use and avoid income trade-offs: An agent-based model applied to Thai agriculture. *Ecological Economics, 132*, 91–103. DOI: 10.1016/j.ecolecon.2016.09.031

Guba, E., & Lincoln, Y. (1989). Fourth generation evaluation. *Sage (Atlanta, Ga.).*

Guba, E., & Lincoln, Y. (2005). Paradigmatic controversies, contradictions & emerging confluences. In Denzin, N., & Lincoln, Y. (Eds.), *The sage handbook of qualitative research*. Sage.

Guest, G., Bunce, A., & Johnson, L. (2006). How Many Interviews Are Enough?: An Experiment with Data Saturation and Variability. *Field Methods*, *18*(1), 59–82. DOI: 10.1177/1525822X05279903

Guetterman, T. C., Chang, T., DeJonckheere, M., Basu, T., Scruggs, E., & Vydiswaran, V. V. (2018). Augmenting qualitative text analysis with natural language processing: Methodological study. *Journal of Medical Internet Research*, *20*(6), e231. DOI: 10.2196/jmir.9702 PMID: 29959110

Gulsia, O., & Yadav, S. (2023). Conceptualizing network approaches for a successful farm entrepreneurship using ATLAS. ti®. *Sustainable Technology and Entrepreneurship*, *2*(1), 100026. DOI: 10.1016/j.stae.2022.100026

Gundumogula, M. (2021). Importance of focus groups in qualitative research. *The International Journal of Humanities & Social Studies*, *8*(11), 299–302.

Gunstone, T., Cornelisse, T., Klein, K., Dubey, A., & Donley, N. (2021). Pesticides and soil invertebrates: A hazard assessment. *Frontiers in Environmental Science*, *9*, 122. DOI: 10.3389/fenvs.2021.643847

Gupta, A. (2023, January). A Qualitative Study of Management Practices & Influence on Middle Managers' Morale in Public Sector Banks Using ATLAS. ti. In *World Conference on Qualitative Research* (pp. 18-37). Cham: Springer International Publishing.

Gupta, D., & Rani, R. (2019). A study of big data evolution and research challenges. *Journal of Information Science*, *45*(3), 322–340. DOI: 10.1177/0165551518789880

Gupta, J., Mitra, B., & Garg, K. (2021). Technological disruptions reshaping management education: A qualitative study of stakeholder perspectives. *International Journal of Management Education*, *15*(2), 179–198.

Guraya, S., Harkin, D., Yusoff, M., & Guraya, S. (2023). Paradigms unfolded – developing, validating, and evaluating the Medical Education e-Professionalism framework from a philosophical perspective. *Frontiers in Medicine*, *10*, 1–13. DOI: 10.3389/fmed.2023.1230620

Guthrie, C., Fosso-Wamba, S., & Arnaud, J. B. (2021). Online consumer resilience during a pandemic: An exploratory study of e-commerce behavior before, during, and after a COVID-19 lockdown. *Journal of Retailing and Consumer Services*, *61*, 102570. DOI: 10.1016/j.jretconser.2021.102570

Habibullah, K., Mohammed, B. U., & Hamza, M. (2023). Conceptual framework in reflexive bracketing techniques in qualitative methodology. *Research Review Journal, 4*(2), 849–857.

Hammersley, M., & Atkinson, P. (2019). *Ethnography: Principles in practice* (4th ed.). Routledge. DOI: 10.4324/9781315146027

Haradhan, M. (2018). Qualitative Research Methodology in Social Sciences and Related Subjects. *Munich Personal RePEc Archive, 7*(01), 23–48.

Harman, G. (2009). *Prince of networks: Bruno Latour and metaphysics*. Re.press.

Harris, Z. (1952). Discourse analysis. *Language, 28*(1), 1–30. DOI: 10.2307/409987

Hartmann, J., Heitmann, M., Siebert, C., & Schamp, C. (2023). More than a feeling: Accuracy and application of sentiment analysis. *International Journal of Research in Marketing, 40*(1), 75–87. DOI: 10.1016/j.ijresmar.2022.05.005

Ha, S., Childs, M., Sneed, C. T., & Berry, A. (2021). Consumer sustainable shopping practices for small business during COVID-19. *Sustainability (Basel), 13*(22), 12451. DOI: 10.3390/su132212451

Hawkins, J. E. (2018). The practical utility and suitability of email interviews in qualitative research. *The Qualitative Report, 23*(2). Advance online publication. DOI: 10.46743/2160-3715/2018.3266

Heath, J., Williamson, H., Williams, L., & Harcourt, D. (2018). "It's just more personal": Using multiple methods of qualitative data collection to facilitate participation in research focusing on sensitive subjects. *Applied Nursing Research, 43*, 30–35. DOI: 10.1016/j.apnr.2018.06.015 PMID: 30220360

Heckathorn, D. D. (1997). Respondent-driven sampling-A new approach to the study of hidden populations. *Social Problems, 44*(2), 174–199. DOI: 10.2307/3096941

He, J., Zhou, W., Qing, C., & Xu, D. (2023). Learning from parents and friends: The influence of intergenerational effect and peer effect on farmers' straw return. *Journal of Cleaner Production, 393*, 136143. DOI: 10.1016/j.jclepro.2023.136143

Hendren, K., Newcomer, K., Pandey, S. K., Smith, M., & Sumner, N. (2023). How qualitative research methods can be leveraged to strengthen mixed methods research in public policy and public administration? *Public Administration Review, 83*(3), 468–485. DOI: 10.1111/puar.13528

Hendricks, G. P. (2016). Deconstruction the end of writing:'Everything is a text, there is nothing outside context. *Verbum et Ecclesia, 37*(1), 1–9. DOI: 10.4102/ve.v37i1.1509

Hennink, M., & Kaiser, B. (2022). Sample size for saturation in qualitative research: A systematic review of empirical literature. *Social Science and Medicine vol vol 292*, 1-10.

Hernández-Mogollón, J. M., González-González, I., & Manso, J. P. (2021). The role of local actors in the destination branding process. *Journal of Destination Marketing & Management, 19*, 100460.

Hesse, A., Glenna, L., Hinrichs, C., Chiles, R., & Sachs, C. (2019). Qualitative research ethics in the big data era. *The American Behavioral Scientist, 63*(5), 560–583. DOI: 10.1177/0002764218805806

Hesse-Biber, S. (2022). Mixed methods research designs in qualitatively driven research. *The SAGE Handbook of Qualitative Research Design*, 618-635.

Hesse-Biber, S., Dupuis, P., & Kinder, T. S. (1991). HyperRESEARCH: A computer program for the analysis of qualitative data with an emphasis on hypothesis testing and multimedia analysis. *Qualitative Sociology, 14*(4), 289–306. DOI: 10.1007/BF00989642

Hodge, B., & Kress, G. (1979). *Language as ideology*. Routledge.

Holloway, I., & Galvin, K. (2017). *Qualitative research in nursing and healthcare*. John Wiley., Retrieved from https://deakin.libguides.com/qualitative-study-designs/phenomenology

Hopf, T. (2004). Discourse and content analysis: Some fundamental incompatibilities. *Qualitative Methods*, (Spring), 31–33.

Hossain, S., Alam, K., & Ali, S. (2024). Phenomenological Approach in the Qualitative Study: Data Collection and Saturation. *ICRRD Journal, 5*(2), 148–172. DOI: 10.53272/icrrd.v5i2.4

Houser, M., Marquart-Pyatt, S. T., Denny, R. C., Reimer, A., & Stuart, D. (2019). Farmers, information, and nutrient management in the US Midwest. *Journal of Soil and Water Conservation, 74*(3), 269–280. DOI: 10.2489/jswc.74.3.269

Howlett, M. (2022). Looking at the 'field'through a Zoom lens: Methodological reflections on conducting online research during a global pandemic. *Qualitative Research, 22*(3), 387–402. DOI: 10.1177/1468794120985691 PMID: 35663097

Hsieh, H. F., & Shannon, S. E. (2005). Three approaches to qualitative content analysis. *Qualitative Health Research, 15*(9), 1277–1288. DOI: 10.1177/1049732305276687 PMID: 16204405

Huan, C., & Guan, X. (2020). Sketching landscapes in discourse analysis (1978–2018): A bibliometric study. *Discourse Studies*, *22*(6), 697–719. DOI: 10.1177/1461445620928814

Huarng, K. H., Rey-Martí, A., & Miquel-Romero, M. J. (2018). Quantitative and qualitative comparative analysis in business. *Journal of Business Research*, *89*, 171–174. DOI: 10.1016/j.jbusres.2018.02.032

Hussenot, A. (2008). Between structuration and translation: An approach to ICT appropriation. *Journal of Organizational Change Management*, *21*(3), 335–347. DOI: 10.1108/09534810810874813

Hussenot, A. (2014). Analyzing organization through disagreements: The concept of managerial controversy. *Journal of Organizational Change Management*, *27*(3), 373–390. DOI: 10.1108/JOCM-01-2012-0006

Hussenot, A., & Missonier, S. (2010). A deeper understanding of the evolution of the role of the object in the organizational process: The concept of "mediation object.". *Journal of Organizational Change Management*, *23*(3), 269–286. DOI: 10.1108/09534811011049608

Husserl, E. (1913.2013). *General introduction to pure phenomenology*. London: Routledge.

Im, H., Kim, N. L., & Lee, H. K. (2022). Why did (some) consumers buy toilet papers? A cross-cultural examination of panic buying as a maladaptive coping response to COVID-19. *The Journal of Consumer Affairs*, *56*(1), 391–413. DOI: 10.1111/joca.12426

Indursky, F. (2001). From the Heterogeneity of Discourse to the Heterogeneity of the Text and Its Implications in the Reading Process. In: *Reading and Writing as Discursive Practices*, Org: Ernst-Pereira, A. & Funck, S. B., Ed. EDUCAT: Pelotas.

Indursky, F. (2003). Lula There: Structure and Event. *Journal of the Institute of Letters of UFRGS*, *17*(35), 101–121.

Internet Encyclopedia of Philosophy (IEP). (2024, 10 24). *The Phenomenological reduction*. Retrieved from Internet Encyclopedia of Philosophy (IEP): https://iep.utm.edu/phen-red/

Internet Encyclopedia of Philosophy. (2024, 10 26). *Edmund Husserl: Intentionality and Intentional content*. Retrieved from Internet Encyclopedia of Philosophy: https://iep.utm.edu/huss-int/#:~:text=For%20Husserl%2C%20intentionality%20includes%20a,just%20to%20name%20a%20few

Ireland, E. (2020). Strategies Big Data Analytics Specialists Need to Improve the Analytical Methods Used for Classifying Structured and Unstructured Data (Doctoral dissertation, Colorado Technical University).

Irshad, M., & Chaudhry, A. (2023). Principle nature of ethnography: A systematic literature review of qualitative research methods in anthropology. *Pakistan Journal of Social Research*, *5*(4), 1–9. DOI: 10.52567/pjsr.v5i04.1326.

Jackall, R. (2015). Moral Mazes. *Wiley Encyclopedia of Management*, 1-2. https://doi.org/DOI: 10.1002/9781118785317.weom020150

Jain, A., Dash, S., & Malhotra, N. K. (2023). Consumption coping to deal with pandemic stress: Impact on subjective well-being and shifts in consumer behavior. *European Journal of Marketing*, *57*(5), 1467–1501. DOI: 10.1108/EJM-11-2021-0864

James Cook University. (2023). *Qualitative research methodologies*. Retrieved from James Cook university: https://jcu.pressbooks.pub/intro-res-methods-health/chapter/4-3-qualitative-research-methodologies/

James, I. (2023). Metaphor in Biosemiotics and Deconstruction. *Oxford Literary Review*, *45*(2), 229–250. DOI: 10.3366/olr.2023.0417

Jamieson, M., Govaart, G., & Pownall, M. (2023). Reflexivity in quantitative research: A rationale and beginner's guide. *Social and Personality Psychology Compass*, *17*(4), 1–15. DOI: 10.1111/spc3.12735

Jaspal, R. (2020). Content analysis, thematic analysis and discourse analysis. In Breakwell, G., Wright, D., & Barnett, J. (Eds.), *Research Methods in Pyschology* (pp. 1–52). Sage.

Javadian, G., Dobratz, C., Gupta, A., Gupta, V. K., & Martin, J. A. (2020). Qualitative research in entrepreneurship studies: A state-of-science. *The Journal of Entrepreneurship*, *29*(2), 223–258. DOI: 10.1177/0971355720930564

Jayathilaka, A. (2021). Ethnography and organisational ethnography: Research methodology. *Open Journal of Business and Management*, *9*(1), 91–102. DOI: 10.4236/ojbm.2021.91005

Jaziri, R., & Cherif, M. (2005). Projet de contractualisation des universités tunisiennes: Analyse du jeu d'acteurs par la méthode MACTOR. In *Actes du colloque international sur: La connaissance: un nouveau levier et défis de la compétitivité pour les entreprises et les économies*. Faculté des Sciences Economiques et de Management de Biskra.

Jeganathan, G. S., & Srinivasulu, Y. Impact of tourism on macro–economic indicators: A Comparative study of BRICS nations.

Jeganathan, G. S., & Srinivasulu, Y. The international journal of business & management.

Jeganathan, G. S., David, A., & Ilangovan, K. (2022). Determination of Hospitality Services Quality and Customer Satisfaction–a Holserv Approach. In *AU Virtual International Conference on" Entrepreneurship & Sustainability in Digital Era" under the theme of" Challenges of Organizational & Business Management in Dynamic Digital Dimension"* (Vol. 3, No. 1, pp. 325-334).

Jeganathan, G. S., & Srinivasulu, Y. (2014). Growth Drivers of Tourism Industry in India. *International Journal of Business and Management, 2*(11), 100.

Jeri, L. (2023). Strengthening the weakness of a generic qualitative research design. *The Qualitative Report, 28*(10), 1759–1768.

Jerolmack, C., & Khan, S. (2018). An introduction: An analytical approach ethnography. In Jerolmack, C., & Khan, S. (Eds.), *Approaches to ethnography analysis and representation in participant observation* (pp. 8–21). Oxford university Press.

Johansson, J., Malmström, M., Wincent, J., & Parida, V. (2021). How individual cognitions overshadow regulations and group norms: A study of government venture capital decisions. *Small Business Economics, 56*(2), 857–876. DOI: 10.1007/s11187-019-00273-3

Johnson, D. R., Scheitle, C. P., & Ecklund, E. H. (2021). Beyond the in-person interview? How interview quality varies across in-person, telephone, and Skype interviews. *Social Science Computer Review, 39*(6), 1142–1158. DOI: 10.1177/0894439319893612

Johnson, R., & Christensen, L. (2017). *Educational research,quantitative, qualitative and mixed approaches.* Sage.

Jørs, E., Konradsen, F., Huici, O., Morant, R. C., Volk, J., & Lander, F. (2016). Impact of training Bolivian farmers on integrated pest management and diffusion of knowledge to neighboring farmers. *Journal of Agromedicine, 21*(2), 200–208. DOI: 10.1080/1059924X.2016.1143428 PMID: 26790134

Kaimathiri, P. (2020). *A Narrative Inquir e Inquiry into the Daily Challenges t o the Daily Challenges to the Business o the Business Growth of W owth of Women's Microenterprises in Rur oenterprises in Rural Kenya.* A doctoral thesis submitted to the College of Management and Technology: Walden University.

Kakar, Z. U. H., Rasheed, R., Rashid, A., & Akhter, S. (2023). Criteria for assessing and ensuring the trustworthiness in qualitative research.

Kalpokaite, N., & Radivojevic, I. (2020). Teaching qualitative data analysis software online: A comparison of face-to-face and e-learning ATLAS. ti courses. *International Journal of Research & Method in Education*, *43*(3), 296–310. DOI: 10.1080/1743727X.2019.1687666

Kang, E., & Hwang, H. J. (2021). Ethical conducts in qualitative research methodology: Participant observation and interview process. *Journal of Research and Publication Ethics*, *2*(2), 5–10.

Kanger, L. (2017). Mapping 'the ANT multiple': A comparative, critical and reflexive analysis. *Journal for the Theory of Social Behaviour*, *47*(4), 435–462. DOI: 10.1111/jtsb.12141

Kant, I. (1999). *A critique of pure reason.* Big City Publishers. (Original work published 1781)

Karcher, K., McCuaig, J., & King-Hill, S. (2024). (Self-) Reflection / Reflexivity in Sensitive Qualitative Research: A Scoping Review. *International Journal of Qualitative Methods*, *23*, 1–15. DOI: 10.1177/16094069241261860

Kemp, E., Bui, M., & Porter, M.III. (2021). Preparing for a crisis: Examining the influence of fear and anxiety on consumption and compliance. *Journal of Consumer Marketing*, *38*(3), 282–292. DOI: 10.1108/JCM-05-2020-3841

Kessler, M. (2024). References. In Kessler, M. (Ed.), *Digital Multimodal Composing: Connecting Theory, Research and Practice in Second Language Acquisition* (pp. 133–147). Blue Ridge Summit Multilingual Matters. DOI: 10.21832/9781800416680-011

Khafaga, A. F. (2017). Discourse Interpretation: A Deconstructive, Reader-oriented Approach to Critical Discourse Analysis. *International Journal of Applied Linguistics and English Literature*, *6*(2), 138–146. DOI: 10.7575/aiac.ijalel.v.6n.2p.138

Khan, T. H., & MacEachen, E. (2022). An alternative method of interviewing: Critical reflections on videoconference interviews for qualitative data collection. *International Journal of Qualitative Methods*, *21*, 16094069221090063. DOI: 10.1177/16094069221090063

Kim, J., Yang, K., Min, J., & White, B. (2021). Hope, fear, and consumer behavioral change amid COVID-19: Application of protection motivation theory. *International Journal of Consumer Studies*, *46*(5), 558–574. PMID: 34220343

Kirk, C. P., & Rifkin, L. S. (2020). I'll trade you diamonds for toilet paper: Consumer reacting, coping, and adapting behaviors in the COVID-19 pandemic. *Journal of Business Research, 117*, 124–131. DOI: 10.1016/j.jbusres.2020.05.028 PMID: 32834208

Kitchenham, B. (2007). Guidelines for performing Systematic Literature Reviews in software engineering. Reviews in software engineering. EBSE Technical Report. 2007 (1). https://www.researchgate.net/publication/258968007_Kitchenham_B_Guidelines_for_performing_Systematic_Literature_Reviews_in_software_engineering_EBSE_Technical_ Report_EBSE-2007-01

Kivunja, C., & Kuyini, A. (2017). Understanding and applying research paradigms in educational contexts. *International Journal of Higher Education, 6*(5), 26–41. DOI: 10.5430/ijhe.v6n5p26

Klinke, M., & Fernandez, A. (2023). Taking phenomenology beyond the frst-person perspective: Conceptual grounding in the collectionand analysis of observational evidence. *Phenomenology and the Cognitive Sciences, 22*(1), 171–191. DOI: 10.1007/s11097-021-09796-1

Konecki, K. T. (2019). Creative thinking in qualitative research and analysis. *Qualitative Sociology Review, 15*(3), 6–25. DOI: 10.18778/1733-8077.15.3.01

Kozinets, R. (2019). Netnography: The essential guide to qualitative social media research.

Kraiwanit, T., Limna, P., & Siripipatthanakul, S. (2023). NVivo for social sciences and management studies: A systematic review. *Advance Knowledge for Executives, 2*(3), 1–11.

Kraus, S., Palmer, C., Kailer, N., Kallinger, F. L., & Spitzer, J. (2019). Digital entrepreneurship: A research agenda on new business models for the twenty-first century. *International Journal of Entrepreneurial Behaviour & Research, 25*(2), 353–375.

Kristeva, J. (1980). *Desire in language: A semiotic approach to literature and art.* Columbia University Press.

Kuckartz, U., Rädiker, S., Kuckartz, U., & Rädiker, S. (2019). Introduction: analyzing qualitative data with software. *Analyzing qualitative data with MAXQDA: text, Audio, and video*, 1-11.

Kuckartz, U., & Rädiker, S. (2022). Using Maxqda for integration in mixed methods research. In *The Routledge handbook for advancing integration in mixed methods research* (pp. 540–562). Routledge. DOI: 10.4324/9780429432828-41

Kuhn, T. (1962). *The structure of scientific revolutions*. International Encyclopedia of Unified Science.

Kumar, S., Kumar, S., Govindaraj, M., & Prabhu, N. (2020). Sampling framework for personal interviews in research . *Journal Of Archaeology Of Egypt/Egyptology 17(7)*, 7102-7114.

Kumar, P., Kumar, A., & Sankar, J. G. (2020). Service quality of health tourism in India: A structural equation modelling approach. *Solid State Technology*, *63*(2), 1187–1191.

Kumar, V., Ramachandran, D., & Kumar, B. (2021). Influence of new-age technologies on marketing: A research agenda. *Journal of Business Research*, *125*, 864–877. DOI: 10.1016/j.jbusres.2020.01.007

Kuzmina, J. (2015). Intertextuality and interdiscursivity in information technologies organisational discourse. *Baltic Journal of English Language. Literature and Culture*, *5*, 75–89.

Kuzmina, J. (2015). Intertexuality and interdiscursivity in information technologies organisational discourse. *Baltic Journal of English Language. Literature and Culture*, *5*, 75–89.

Kuzmina, J., Bartesaghi, M., & Noy, C. (2015). Interdiscursivity. In Karen, T., Ilie, C., & Sandel, T. (Eds.), *The International Encyclopedia of Language and Social Interaction* (pp. 1–6). JohnWiley & Sons.

Laffont, J. J., & Maskin, E. (1982). Theory of incentives: An overview (part 1). In *Advances in Economic Theory* (pp. 31–93). Cambridge University Press.

Langley, A. (1999). Strategies for theorizing from process data. *Academy of Management Review*, *24*(4), 691–710. DOI: 10.2307/259349

Langley, A., & Tsoukas, H. (Eds.). (2016). *The SAGE handbook of process organization studies*. SAGE. DOI: 10.4135/9781473957954

Langmead, B., & Nellore, A. (2018). Cloud computing for genomic data analysis and collaboration. *Nature Reviews. Genetics*, *19*(4), 208–219. DOI: 10.1038/nrg.2017.113 PMID: 29379135

LAPES - UFSCAR. (2021). *StArt - State of the Art through Systematic Review.* http://lapes.dc.ufscar.br/tools/start_tool

Latour, B. (1994). *We have never been modern*. Editora 34.

Latour, B. (1984). The powers of association. *The Sociological Review*, *32*(1, suppl), 264–280. DOI: 10.1111/j.1467-954X.1984.tb00115.x

Latour, B. (1993). *The pasteurization of France*. Harvard University Press.

Latour, B. (1996). On actor-network theory: A few clarifications. *Soziale Welt*, •••, 369–381.

Latour, B. (2000). *Science in action: How to follow scientists and engineers through society*. Unesp.

Latour, B. (2007). *Reassembling the social: An introduction to actor-network theory*. OUP Oxford.

Latour, B. (2010). The making of law: An ethnography of the Conseil d'Etat. *Polity*.

Latour, B. (2019). *Inquiry into modes of existence: An anthropology of the moderns*. Editora Vozes.

Law, J. (1992). Notes on actor-network theory: Ordering, strategy, and heterogeneity. *Systems Practice*, *5*(4), 379–393. DOI: 10.1007/BF01059830

Law, J. (1999). After ANT: Complexity, naming, and topology. *The Sociological Review*, *47*(1_suppl, S1), 1–14. DOI: 10.1111/j.1467-954X.1999.tb03479.x

Law, J. (2004). *After method: Mess in social science research*. Routledge. DOI: 10.4324/9780203481141

Law, J., & Hassard, J. (1999). *Actor-network theory and after*. Blackwell.

Lazarus, R. S. (1991). Progress on a cognitive-motivational-relational theory of emotion. *The American Psychologist*, *46*(8), 819–834. DOI: 10.1037/0003-066X.46.8.819 PMID: 1928936

Lazarus, R. S. (1996). The role of coping in the emotions and how coping changes over the life course. In Magai, C., & McFadden, S. H. (Eds.), *Handbook of emotions across the lifespan* (pp. 289–306). Academic Press., DOI: 10.1016/B978-012464995-8/50017-0

Lazarus, R. S., & Folkman, S. (1984). *Stress, appraisal, and coping*. Springer Publishing Company.

Leavy, P. (2017). *Quantitative,Qualitative, Mixed Methods Arts-Based, and Community-Based Participatory Research Approaches*. The Guildford Press.

Leavy, P. (2017). *Research design Qualitative, Mixed Methods,Arts-Based, and Community-Based Participatory Research Approaches*. The Guildford Press.

Leavy, P. (2017). *Research design, quantitative, qualitative and mixed methods, arts based and community based participatory research approaches*. The Guilford Press.

Leitão, C. (2021). The interview as an instrument of scientific research: Planning, execution and analysis. In *Methodology of scientific research in informatics in education: Qualitative research approach* (Vol. 3).

Lester, J. N., Cho, Y., & Lochmiller, C. R. (2020). Learning to do qualitative data analysis: A starting point. *Human Resource Development Review*, *19*(1), 94–106. DOI: 10.1177/1534484320903890

Leung, K.-H. (2022). Sartre and Marion on Intentionality and Phenomenality. *Theory, Culture & Society*, *39*(1), 41–60. DOI: 10.1177/02632764211024349

Lewis-Hickman, C. (2015). Strength and limitations of research. Walden: Walden University.

Lieber, E., Salmona, M., & Kaczynski, D. (2021). Introduction to Dedoose for mixed analysis. In *The Routledge Reviewer's Guide to Mixed Methods Analysis* (pp. 319–330). Routledge. DOI: 10.4324/9780203729434-27

Li, J., Jin, X., Zhao, T., & Ma, T. (2021). Conformity consumer behavior and external threats: An empirical analysis in China during the COVID-19 pandemic. *SAGE Open*, *11*(3), 21582440211032152. Advance online publication. DOI: 10.1177/21582440211032152

Lima, A. (2023). Understanding narrative inquiry through life story interviews with former prisoners. *Irish Educational Studies*, *42*(4), 775–786. DOI: 10.1080/03323315.2023.2257673

Limna, P. (2023). The impact of NVivo in qualitative research: Perspectives from graduate students. *Journal of Applied Learning and Teaching*, *6*(2).

Lim, W. M. (2024). What is qualitative research? An overview and guidelines. *Australasian Marketing Journal*, •••, 14413582241264619.

Lincoln, Y., & Guba, E. (1985). The naturalistic inquiry. Tousand Oaks: Sage. DOI: 10.1016/0147-1767(85)90062-8

Lincoln, Y., & Guba, E. (1985). Naturalistic inquiry. *Sage (Atlanta, Ga.)*.

Lindgren, B. M., Lundman, B., & Graneheim, U. H. (2020). Abstraction and interpretation during the qualitative content analysis process. *International Journal of Nursing Studies*, *108*, 103632. DOI: 10.1016/j.ijnurstu.2020.103632 PMID: 32505813

Linneberg, M. S., & Korsgaard, S. (2019). Coding qualitative data: A synthesis guiding the novice. *Qualitative Research Journal, 19*(3), 259–270. DOI: 10.1108/QRJ-12-2018-0012

Liu, J., & Eagan, J. (2021). ADQDA: a cross-device affinity diagramming tool for fluid and holistic qualitative data analysis. Proceedings of the ACM on Human-Computer Interaction, 5(ISS), 1-19. DOI: 10.1145/3488534

Liu, C., Xia, S., & Lang, C. (2022). Coping with COVID-19: How resilient consumers adjust clothing consumption during the pandemic. *International Textile and Apparel Association Annual Conference Proceedings*, 78(1). DOI: 10.31274/itaa.13561

Liu, N., Nikitas, A., & Parkinson, S. (2020). Exploring expert perceptions about the cyber security and privacy of Connected and Autonomous Vehicles: A thematic analysis approach. *Transportation Research Part F: Traffic Psychology and Behaviour, 75*, 66–86. DOI: 10.1016/j.trf.2020.09.019

Liu, T., & Wu, G. (2020). Does agricultural cooperative membership help reduce the overuse of chemical fertilizers and pesticides? Evidence from rural China. *Environmental Science and Pollution Research International, 29*, 1–12. PMID: 34480697

Li, Y., Qing, C., Guo, S., Deng, X., Song, J., & Xu, D. (2023). Will farmers follow their peers in adopting straw returning? Evidence from rural Sichuan Province, China. *Environmental Science and Pollution Research International, 30*(8), 21169–21185. DOI: 10.1007/s11356-022-23648-8 PMID: 36264456

Lobe, B., Morgan, D., & Hoffman, K. A. (2020). Qualitative data collection in an era of social distancing. *International Journal of Qualitative Methods, 19*, 1609406920937875. DOI: 10.1177/1609406920937875

Lochmiller, C. (2021). Conducting thematic analysis with qualitative data. *The Qualitative Report, 26*(6), 2029–2044.

Locke, K., Feldman, M., & Golden-Biddle, K. (2022). Coding practices and iterativity: Beyond templates for analyzing qualitative data. *Organizational Research Methods, 25*(2), 262–284. DOI: 10.1177/1094428120948600

Lodico, M., Spaulding, D., & Voegtle, K. (2010). *Methods in educational research* . USA: A Wiley Imprint Jossy-Bassey.

Longo, L. (2019, September). Empowering qualitative research methods in education with artificial intelligence. In *World conference on qualitative research* (pp. 1-21). Cham: Springer International Publishing.

López de Mesa, Y. P. (2020). The decision-making process of synthetic pesticide use in agricultural communities in Colombia: A grounded theory approach. *Revista de la Facultad Nacional de Salud Pública, 38*(2), 38. DOI: 10.17533/udea.rfnsp.e331277

Lounsbury, M. (2002). Institutional Transformation and Status Mobility: The Professionalization of the Field of Finance. *Academy of Management Journal, 45*(1), 255–266. DOI: 10.2307/3069295

Lovelace, S., Trudel, C., Dulude, C., & King, W. J. (2020, September). Cost vs. Benefit: what does NVivo video analysis of EMR simulations add to our understanding of user experience? In *Proceedings of the International Symposium on Human Factors and Ergonomics in Health Care* (Vol. 9, No. 1, pp. 24-32). Sage CA: Los Angeles, CA: SAGE Publications. DOI: 10.1177/2327857920091056

Lüdke, M., & André, M. E. D. A. (1986). *Research in education: Qualitative approaches*. EPU.

Lune, H., & Berg, B. (2017). *Qualitative research methods for the social sciences*. Pearson.

Mackenzie, U. P. (2019). Mackenzie launches unprecedented professional doctorate in controllership and finance Mackenzie. https://www.mackenzie.br/noticias/artigo/n/a/i/mackenzie-lanca-doutorado-profissional-inedito-em-controladoria-e-financas

MacMillan, K., & Koenig, T. (2004). The wow factor: Preconceptions and expectations for data analysis software in qualitative research. *Social Science Computer Review, 22*(2), 179–186. DOI: 10.1177/0894439303262625

Madison, D. S. (2020). *Critical ethnography: Method, ethics, and performance*. Sage. DOI: 10.4135/9781071878965

Madrigal, D., & McClain, B. (2024, 10 8). *Strength and weaknesses of qualitative Research*. Retrieved from UX Matters: https://www.uxmatters.com/mt/archives/2012/09/strengths-and-weaknesses-of-quantitative-and-qualitative-research.php

Magnani, J. G. C. (1986). Discourse and representation, or how the Baloma of Kiriwina can reincarnate in current research. In Cardoso, R. (Ed.), *The anthropological adventure: Theory and research* (pp. 127–140). Paz e Terra.

Mahapatra, S., & Mishra, A. (2022). Crying over spilt milk? Effect of post-consumption dissonance on coping behaviour for online purchases. *International Journal of Consumer Studies, 46*(1), 1035–1054. DOI: 10.1111/ijcs.12744

Maher, C., Hadfield, M., Hutchings, M., & De Eyto, A. (2018). Ensuring rigor in qualitative data analysis: A design research approach to coding combining NVivo with traditional material methods. *International Journal of Qualitative Methods, 17*(1), 1609406918786362. DOI: 10.1177/1609406918786362

Mahmood, I., Imadi, S. R., Shazadi, K., Gul, A., & Hakeem, K. R. (2016). Effects of pesticides on environment. *Plant Soil Microbes, 1*, 253–269. DOI: 10.1007/978-3-319-27455-3_13

Mahmood, Z., & Uddin, S. (2020). Institutional logics and practice variations in sustainability reporting: Evidence from an emerging field. *Accounting, Auditing & Accountability Journal, 34*(5), 1163–1189. DOI: 10.1108/AAAJ-07-2019-4086

Mahmud, M. (2017). Doing discourse analysis: An introduction. Kalangan: Phoenix Publisher.

Maia, I. (2024). How to overcome the secretiveness of a group: Opportunities of online interviews. *Sociological Research Online, 29*(1), 171–183. DOI: 10.1177/13607804231168253

Maibom, C., & Smith, P. (2016). Symbiosis across institutional logics in a social enterprise. *Social Enterprise Journal, 12*(3), 260–280. DOI: 10.1108/SEJ-02-2016-0002

Maingueneau, D. (1984). *Geneses of Discourse*. Mardaga.

Maingueneau, D. (2015). *Discourse and Discourse Analysis*. Parábola Editorial.

Manski, C. F. (1993). Identification of endogenous social effects: The reflection problem. *The Review of Economic Studies, 60*(3), 531–542. DOI: 10.2307/2298123

Manzano, A. (2024, 10 28). *Policy evaluation and approaches: Focus group*. Retrieved from esbc: https://scienceetbiencommun.pressbooks.pub/pubpolevaluation/chapter/focus-groups/

Mark, L. (2024, 06 23). *What is thematic analysis advantages and disadvantages*. Retrieved from Acadmemic Papers: https://www.theacademicpapers.co.uk/blog/2021/10/02/what-is-thematic-analysis-advantages-and-disadvantages/

Martucci, C. (2024). Qualitative evaluation of Power BI and its application in an Aerospace Industry cluster (Doctoral dissertation, Politecnico di Torino).

Mason, M. (2010). *Sample Size and Saturation in PhD Studies Using Qualitative Interviews*. https://www.qualitative-research.net/

Mason, M. (2010). Sample size and saturation in PhD studies using qualitative interviews. *Forum Qualitative Sozialforschung/Forum: Qualitative. Social Research, 11*(3).

Mathur, A., Moschis, G. P., & Lee, E. (1999). Stress and consumer behavior: Coping strategies of older adults. *Journal of Marketing Practice: Applied Marketing Science*, 5(6-7-8), 233-247. https://doi.org/DOI: 10.1108/EUM0000000004572

Mathysen, D., & Glorieux, I. (2021). Integrating virtual reality in qualitative research methods: Making a case for the VR-assisted interview. *Methodological Innovations, 14*(2), 20597991211030778. DOI: 10.1177/20597991211030778

Mattos, P. T. L. (2006). Decisional autonomy, administrative discretion and legitimacy of the State's regulatory function in Brazilian legal debate. In de Aragão, A. S. (Ed.), *The normative power of regulatory agencies* (pp. 333–364). Forense.

Matyjaszczyk, E., & Sobczak, J. (2017). Common EU registration rules and their effects on the availability of diverse plant protection products: A case study from oilseed rape and potato in five Member States. *Crop Protection (Guildford, Surrey), 100*, 73–76. DOI: 10.1016/j.cropro.2017.06.006

Maxwell, J. A. (2013). *Qualitative research design: An interactive approach* (3rd ed.). SAGE Publications.

Maykut, P., & Morehouse, R. (1995). *Beginning qualitative research: A philosophic and practical guide*. Falmer Press.

Mayring, P. (2019). Qualitative content analysis: Demarcation, varieties, developments. *Forum Qualitative Social Research, 20*(3), 1–14.

Mazzon, J. A. (1983). Avaliação do programa de alimentação do trabalhador. Subsídios à formulação de um plano de marketing social. *Revista ADM, 18*(4), 33–58. DOI: 10.1016/rausp.v18i4.167313

McCall, C., & Edwards, C. (2021). Theory new perspectives for implementing Grounded Theory. *Studies in Engineering Education, 1*(2), 93–107. DOI: 10.21061/see.49

McClelland, J. (1997). Knowing and being known: Parents' experiences with rural schools. *Journal of Research in Rural Education, 13*(2), 108–116.

McGrath, C., Palmgren, P. J., & Liljedahl, M. (2019). Twelve tips for conducting qualitative research interviews. *Medical Teacher, 41*(9), 1002–1006. DOI: 10.1080/0142159X.2018.1497149 PMID: 30261797

McKinlay, A. R., Fancourt, D., & Burton, A. (2022). Factors affecting the mental health of pregnant women using UK maternity services during the COVID-19 pandemic: A qualitative interview study. *BMC Pregnancy and Childbirth*, *22*(1), 1–15. DOI: 10.1186/s12884-022-04602-5 PMID: 35413807

Meißner, M., Pfeiffer, J., Peukert, C., Dietrich, H., & Pfeiffer, T. (2020). How virtual reality affects consumer choice. *Journal of Business Research*, *117*, 219–231. DOI: 10.1016/j.jbusres.2020.06.004

Meridian University. (2023, 10 2). *How positivism shaped our understanding of reality*. Retrieved from Meridian University: https://meridianuniversity.edu/content/how-positivism-shaped-our-understanding-of-reality

Merriam, S. B., & Tisdell, E. J. (2016). *Qualitative research: A guide to design and implementation* (4th ed.). Jossey-Bass.

Mertens, D. (2005). *Research evaluation in education and psychology: Integrating diversity with quantitative, qualitative, and mixed methods* (2nd ed.). Sage.

Mertens, D. M. (2014). *Research and evaluation in education and psychology: Integrating diversity with quantitative, qualitative, and mixed methods* (4th ed.). SAGE Publications.

Mertens, E., Heylighen, A., Declercq, A., Hannes, K., Truyen, F., Denier, Y., & Dierckx de Casterlé, B. (2017). QualiBuddy: An online tool to improve research skills in qualitative data analysis. *Qualitative Research Journal*, *17*(4), 306–318. DOI: 10.1108/QRJ-06-2016-0034

Mertova, P., & Webster, L. (2020). *Using narrative inquiry as a research method: An introduction to critical event narrative analysis in research teaching and professional practice*. Routledge.

Meyer, J. W., & Rowan, B. (1977). Institutionalized organizations: Formal structure as myth and ceremony. *American Journal of Sociology*, *83*(2), 340–363. DOI: 10.1086/226550

Milano, M., Vieira, F., & Carvalho, C. A. (2003). *Campos Organizacionais: de wallpaper à construção histórica do contexto de organizações culturais em Porto Alegre e em Recife*.

Miles, M. B., Huberman, A. M., & Saldaña, J. (1994). *Qualitative Data Analysis A Methods Sourcebook Edition*.

Miles, M., & Huberman, A. M. (1994). *Qualitative data analysis: An expanded sourcebook* (2nd ed.). Sage.

Miles, M., Huberman, A., & Saladana, J. (2014). *Qualitative data analysis*. Sage Publications.

Mills, K. A. (2018). What are the threats and potentials of big data for qualitative research? *Qualitative Research, 18*(6), 591–603. DOI: 10.1177/1468794117743465

Mirick, R. G., & Wladkowski, S. P. (2019). Skype in qualitative interviews: Participant and researcher perspectives. *The Qualitative Report, 24*(12), 3061–3072. DOI: 10.46743/2160-3715/2019.3632

Mishra, A., Indwar, R. G., & Samu, S. (2022). If tomorrow never comes: The impact of anxiety, materialism and attitude on consumption behavior in a pandemic. *Journal of Consumer Marketing, 39*(3), 294–305. DOI: 10.1108/JCM-02-2021-4439

Mishra, P., & Gupta, A. (2024). Exploring the Use of Diary Entries for Qualitative Researchers: Mitigating Challenges When Investigating Sensitive Topics on Indian Women. *The Qualitative Report, 29*(0), 5, 1279–1299. DOI: 10.46743/2160-3715/2024.6330

Mitev, N. (2009). In and out of actor-network theory: A necessary but insufficient journey. *Information Technology & People, 22*(1), 9–25. DOI: 10.1108/09593840910937463

Mocăna u, D. (2020). Determining sample size qualitative research. *International Multidisciplinary Scientific Conference on the Dialogue between Sciences & Arts, Religion & Education*, 181-187.

Mogaji, E., Jain, V., & Sharma, H. (2024). *Qualitative Research for a Digital World: A Practical Guide*. SAGE Publications Limited.

Moghadam, T. M., Raheli, H., Zariffian, S., & Yazdanpanah, M. (2021). The discovery of the barriers of agricultural water conservation behavior and related factors: Using a grounded theory approach (the case of Neyshabur Plain in Khorasan Razavi Province). *Majallah-i Tahqiqat-i Iqtisad va Tusiah-i Kishavarzi-i Iran, 52*, 287–308.

Mohajan, D., & Mohajan, H. (2022). Development of Grounded Theory in social sciences: A Qualitative Approach. *Munich Personal RePEc Archive (MPRA)*, 1-28.

Moisander, J., Närvänen, E., & Valtonen, A. (2020). Interpretive marketing research: Using ethnography in strategic market development. In *Marketing management* (pp. 237–253). Routledge. DOI: 10.4324/9780203710807-19

Mol, A. (1999). Ontological politics: A word and some questions. *The Sociological Review, 47*(1, suppl), 74–89. DOI: 10.1111/j.1467-954X.1999.tb03483.x

Mol, A. (2010). Actor-network theory: Sensitive terms and enduring tensions. *Kölner Zeitschrift für Soziologie und Sozialpsychologie*, *50*(1), 253–269.

Monios, J., & Ng, A. K. Y. (2021). Competing institutional logics and institutional erosion in environmental governance of maritime transport. *Journal of Transport Geography*, *94*, 103114. DOI: 10.1016/j.jtrangeo.2021.103114

Moody's Corporation. (2022). *About us*. https://www.moodys.com/web/en/us/about/who-we- are/about-us.html

Moon, M. D. (2020). Triangulation: A Method to Increase Validity, Reliability, and Legitimation in Clinical Research. *Journal of Emergency Nursing: JEN*, *45*(1), 103–105. DOI: 10.1016/j.jen.2018.11.004 PMID: 30616761

Moreira, M. A. (2018). Criteria and strategies to ensure rigor in qualitative research. *Brazilian Journal of Science and Technology Teaching*, *11*(1), 405–411.

Moreno, V. (2002). Validity Issues in phenomenological research: Bridging theory and practice in a study of IT driven radical organisational change. *Americas Conference on Information Systems (AMCIS)* (pp. 1-8). USA: AMCIS 2002.

Morgan, D. (2007). Paradigm lost and pragmatism regained: Methodological implications of combining qualitative and quantitative methods. *Journal of Mixed Methods Research*, *1*(1), 48–76. DOI: 10.1177/2345678906292462

Morison, M., & Moir, J. (1998). The role of computer software in the analysis of qualitative data: Efficient clerk, research assistant or Trojan horse? *Journal of Advanced Nursing*, *28*(1), 106–116. DOI: 10.1046/j.1365-2648.1998.00768.x PMID: 9687137

Morse, J. (2020). The changing face of qualitative inquiry. *International Journal of Qualitative Methods*, *19*, 1609406920909938. DOI: 10.1177/1609406920909938

Morse, J. M., & Richards, Lyn. (2002). Readme first for a user's guide to qualitative methods. *Sage (Atlanta, Ga.)*.

Mortelmans, D. (2019). Analyzing qualitative data using NVivo. The Palgrave handbook of methods for media policy research, 435-450. DOI: 10.1007/978-3-030-16065-4_25

Mubushar, M., Aldosari, F. O., Baig, M. B., Alotaibi, B. M., & Khan, A. Q. (2019). Assessment of farmers on their knowledge regarding pesticide usage and biosafety. *Saudi Journal of Biological Sciences*, *26*(8), 1903–1910. DOI: 10.1016/j.sjbs.2019.03.001 PMID: 31762673

Muijeen, K., Kongvattananon, P., & Somprasert, C. (2020). The key success factors in focus group discussions with the elderly for novice researchers: A review. *Warasan Wichai Witthayasat Kanphaet, 34*(4), 359–371. DOI: 10.1108/JHR-05-2019-0114

Mwita, K. (2022). Strengths and weaknesses of qualitative research in social science studies. *Research in Business & Social Science, 11*(6), 618–625. DOI: 10.20525/ijrbs.v11i6.1920

Myles, G. D. (1995). *Public economics*. Cambridge University Press. DOI: 10.1017/CBO9781139170949

Nadu, C. (2015). Text/Texts: Interrogating Julia Kristeva's Concept. *Research Journal of Humanities and Social Sciences, 3*, 77–80.

Naeem, M., Ozuem, W., Howell, K., & Ranfagni, S. (2023). A Step-by-Step Process of Thematic Analysis to Develop a Conceptual Model in Qualitative Research. *International Journal of Qualitative Methods, 22*, 1–18. DOI: 10.1177/16094069231205789

Naganathan, G., Srikanthan, S., Balachandran, A., Gladdy, A., & Shanmuganathan, V. (2022). Collaborative zoom coding—A novel approach to qualitative analysis. *International Journal of Qualitative Methods, 21*, 16094069221075862. DOI: 10.1177/16094069221075862

Nair, L. B. (2021). From 'whodunit'to 'how': Detective stories and auditability in qualitative business ethics research. *Journal of Business Ethics, 172*(2), 195–209. DOI: 10.1007/s10551-020-04479-4

Naping, M. (2020). Cleaning up the local environment: A study of virtual ethnography in the social media community of Makassah,Indonesia. *European Journal of Research in Social Sciences, 8*(2), 56–61.

Nardacchione, G., & Acevedo, M. H. (2013). Pragmatic-pragmatist sociologies tested in Latin America.

Nath, S. D., Jamshed, K. M., & Shaikh, J. M. (2022). The impact of the COVID-19 pandemic on subsistence consumers' well-being and coping strategies: Insights from India and Bangladesh. *The Journal of Consumer Affairs, 56*(1), 180–210. DOI: 10.1111/joca.12440 PMID: 35603322

Neubauer, B., Witkop, C. T., & Varpio, L. (2019). How phenomenology can help us learn from the experience of others. *Perspect Med Educ. 5;8(2)*, 90–97.

Nganga, C., Maroko, G., & Ong'onda, A. N. (2023). Intertextuality and interdiscursivity in postcolonial language policies in Kenya. *European Journal of Applied Linguistics Studies, 6*(2), 214–214. DOI: 10.46827/ejals.v6i2.480

Nguyen, N., & Drakou, E. G. (2021). Farmers' intention to adopt sustainable agriculture hinges on climate awareness: The case of Vietnamese coffee. *Journal of Cleaner Production*, *303*, 126828. DOI: 10.1016/j.jclepro.2021.126828

Nickerson, C. (2024, 02 13). *Interpretivism and research philosophy*. Retrieved from Simply Psychology: https://www.simplypsychology.org/interpretivism-paradigm.html

Nicmanis, M. (2024). Reflexive content analysis:An approach to data analysis, reduction, and description. *International Journal of Qualitative Methods*, *23*, 1–12. DOI: 10.1177/16094069241236603

Nicmanis, M., Chur-Hansen, A., & Linehan, K. (2023). The information needs and experiences of people living with cardiac implantable electronic devices: Qualitative content analysis of Reddit posts. *JMIR Cardio*, *7*, e46296. DOI: 10.2196/46296 PMID: 37766632

Nicolaou, C. (2021). Qualitative methods research through the Internet Applications and Services: The contribution of audiovisual media technology as technology-enhanced research. *International Research in Higher Education*, *6*(1), 1–14. DOI: 10.5430/irhe.v6n1p1

Niedbalski, J., & Ślęzak, I. (2022). Encounters with CAQDAS: Advice for beginner users of computer software for qualitative research. *The Qualitative Report*, *4*, 22–2022. DOI: 10.46743/2160-3715/2022.4770

Niedbalski, J., & Ślęzak, I. (2023, January). NVivo as a tool for supporting teamwork in the context of qualitative research conducted remotely-opportunities, limitations, and practical tips. In *World conference on qualitative research* (pp. 38-59). Cham: Springer International Publishing. DOI: 10.1007/978-3-031-31346-2_3

Nigar, N. (2020). Hermeneutic phenomenological narrative enquiry: A qualitative study design. *Theory and Practice in Language Studies*, *10*(1), 10–18. DOI: 10.17507/tpls.1001.02

Noble, H., & Smith, J. (2015). Issues of validity and reliability in qualitative research. *Evidence-Based Nursing*, *18*(2), 1–7. DOI: 10.1136/eb-2015-102054 PMID: 25653237

Nobre, J. C. D. A., & Pedro, R. M. L. R. (2010). Reflections on methodological possibilities of Actor-Network Theory. *Cadernos UniFOA*, *5*(14), 47–56. DOI: 10.47385/cadunifoa.v5.n14.1018

Novawan, A., Walker, S. A., & Ikeda, O. (2024). The new face of technology-enhanced language learning (TELL) with artificial intelligence (AI): Teacher perspectives, practices, and challenges. *Journal of English in Academic and Professional Communication, 10*(1), 1–18. DOI: 10.25047/jeapco.v9i1.3754

Ntsobi, M. P., Costa, A. P., Kasperiuniene, J., Brandão, C., & Ribeiro, J. (2024, January). Digital Tools and Techniques in Qualitative Research: Digital Skills and Research Optimisation. In World Conference on Qualitative Research (pp. 1-25). Cham: Springer Nature Switzerland.

O'Hagan, J., Saeghe, P., Gugenheimer, J., Medeiros, D., Marky, K., Khamis, M., & McGill, M. (2023). Privacy-enhancing technology and everyday augmented reality: Understanding bystanders' varying needs for awareness and consent. *Proceedings of the ACM on Interactive, Mobile, Wearable and Ubiquitous Technologies, 6*(4), 1–35. DOI: 10.1145/3569501

O'Halloran, K. L., Tan, S., Pham, D. S., Bateman, J., & Vande Moere, A. (2018). A digital mixed methods research design: Integrating multimodal analysis with data mining and information visualization for big data analytics. *Journal of Mixed Methods Research, 12*(1), 11–30. DOI: 10.1177/1558689816651015

Ocasio, W., & Radoynovska, N. (2016). Strategy and commitments to institutional logics: Organizational heterogeneity in business models and governance. *Strategic Organization, 14*(4), 287–309. DOI: 10.1177/1476127015625040

Odongo, I., & Ntara, C. (2024). Role of phenomenology in management research. *Int. J. Management Concepts and Philosophy, 17*(4), 427–441. DOI: 10.1504/IJMCP.2024.142324

Olapane, E. C. (2021). An in-depth exploration on the praxis of computer-assisted qualitative data analysis software (CAQDAS). *Journal of Humanities and Social Sciences Studies, 3*(11), 57–78. DOI: 10.32996/jhsss.2021.3.11.5

Oliveira, G. (2023). Developing a codebook for qualitative data analysis: Insights from a study on learning transfer between university and the workplace. *International Journal of Research & Method in Education, 46*(3), 300–312. DOI: 10.1080/1743727X.2022.2128745

Oliveira, M., Bitencourt, C., Teixeira, E., & Santos, A. C. (2013, July). Thematic content analysis: Is there a difference between the support provided by the MAXQDA® and NVivo® software packages. In *Proceedings of the 12th European Conference on Research Methods for Business and Management Studies* (pp. 304-314).

Olmos-Vega, F., Stalmeijer, R., Varpio, L., & Kahlke, R. (2023). A practical guide to reflexivity in qualitative research: AMEE Guide No. 149. *AMEE Guide No. 149. Medical Teacher*, *45*(3), 241–251. DOI: 10.1080/0142159X.2022.2057287 PMID: 35389310

Olorunfemi, D. (2023). Diary studies in research: More than a research method. *International Journal of Market Research*, *66*(4), 1–7.

Olson, R. E. (2023). Emotions in human research ethics guidelines: Beyond risk, harm and pathology. *Qualitative Research*, *23*(3), 526–544. DOI: 10.1177/14687941211039965

Opara, V., Spangsdorf, S., & Ryan, M. K. (2023). Reflecting on the use of Google Docs for online interviews: Innovation in qualitative data collection. *Qualitative Research*, *23*(3), 561–578. DOI: 10.1177/14687941211045192

Orb, A., Eisenhauer, L., & Wynaden, D. (2001). Ethics in qualitative research. *Journal of Nursing Scholarship*, *33*(1), 93–96. DOI: 10.1111/j.1547-5069.2001.00093.x PMID: 11253591

Orlikowski, W. J., & Scott, S. V. (2008). Sociomateriality: Challenging the separation of technology, work, and organization. *The Academy of Management Annals*, *2*(1), 433–474. DOI: 10.5465/19416520802211644

Osakinle, P. (2024). Discourse of Intertextuality in the language of political advertisements in selected Nigerian newspapers. *Elsya: Journal of English Language Studiesvol*, *6*(1), 1–19. DOI: 10.31849/elsya.v6i1.16121

Palinkas, L. A., Horwitz, S. M., Green, C. A., Wisdon, J. P., Duan, N., & Hoagwood, K. (2015). Purposeful sampling for qualitative data collection and analysis in mixed method implementation research. *Administration and Policy in Mental Health*, *42*(5), 533–544. DOI: 10.1007/s10488-013-0528-y PMID: 24193818

Pan, D., Kong, F., Zhang, N., & Ying, R. (2017). Knowledge training and the change of fertilizer use intensity: Evidence from wheat farmers in China. *Journal of Environmental Management*, *197*, 130–139. DOI: 10.1016/j.jenvman.2017.03.069 PMID: 28363140

Pandin, M., & Yanto, E. (2023). The what and how of existential phenomenological research. *The Qualitative Report*, *28*(3), 816–827. DOI: 10.46743/2160-3715/2023.6268

Panovski, A. (2024, 04 24). *5 Phases of phenomenology and the rise of existentialism*. Retrieved from The Collector: https://www.thecollector.com/phenomenology-rise-existentialism/

Park, G. (2019). The changing wind of data privacy law: A comparative study of the European Union's General Data Protection Regulation and the 2018 California Consumer Privacy Act. *UC Irvine L. Rev.*, *10*, 1455.

Patel, D. (2020). Scope and usage of discourse analysis as a research method in English studies. *International Journal for Innovative Research in Multidisciplinary Field*, *6*(3), 234–238.

Pathak, V. (2017). Phenomenological Research: A Study of Lived experiences. *IJARIIE*, *3*(1), 1719–1722.

Patricia, I., Ph, D., & Ness, L. R. (2015). *Are We There Yet? Data Saturation in Qualitative Research*. Walden Faculty and Staff Publications. https://scholarworks.waldenu.edu/facpubs/455

Patton, C., & Broward, J. (2023). The Giants and Forerunners of Phenomenology: Husserl, Heidegger, and their Predecessors. *American Journal of Qualitative Research*, *7*(4), 79–94. DOI: 10.29333/ajqr/13600

Patton, M. Q. (2015). *Qualitative research & evaluation methods: Integrating theory and practice* (4th ed.). SAGE Publications.

Paulus, T. M. (2023). Using qualitative data analysis software to support digital research workflows. *Human Resource Development Review*, *22*(1), 139–148. DOI: 10.1177/15344843221138381

Paulus, T. M., & Lester, J. N. (2020). Using software to support qualitative data analysis. In *Handbook of qualitative research in education* (pp. 420–429). Edward Elgar Publishing. DOI: 10.4337/9781788977159.00048

Paulus, T., Woods, M., Atkins, D. P., & Macklin, R. (2017). The discourse of QDAS: Reporting practices of ATLAS. ti and NVivo users with implications for best practices. *International Journal of Social Research Methodology*, *20*(1), 35–47. DOI: 10.1080/13645579.2015.1102454

Pearse, N. (2021). Guidelines for theory development using qualitative research approaches. *Electronic Journal of Business Research Methods*, *19*(2), 95–103. DOI: 10.34190/ejbrm.19.2.2512

Pêcheux, M. (1995). *Semantics and Discourse: A Critique of the Assertion of the Obvious*. UNICAMP Publishing.

Pêcheux, M. (1997a). Discourse Analysis: Three Periods (1983). In Gadet, F., & Hak, T. (Eds.), *Towards an Automatic Discourse Analysis: An Introduction to the Work of Michel Pêcheux* (pp. 311–319). UNICAMP Publishing.

Pêcheux, M. (2006). *Discourse: Structure or Event*. Pontes Publishing.

Pereira, A. D. (2011). *Auditing of Financial Statements: A Legal and Accounting Approach*. Atlas Publishing S.A.

Pereira, F. A. M. (2012). The Evolution of Institutional Theory in Organizational Studies: A Research Field to be Explored. *Journal of Organizations in Context*, *8*(16), 275–295.

Perrow, C. (1986). *Complex Organizations: A Critical Essay*. McGraw-Hill.

Pervin, N., & Mokhtar, M. (2022). The interpretivist research paradigm: A subjective notion of a social context. *International Journal of Academic Research in Progressive Education and Development*, *11*(2), 419–428. DOI: 10.6007/IJARPED/v11-i2/12938

Peters, V., & Wester, F. (2007). How qualitative data analysis software may support the qualitative analysis process. *Quality & Quantity*, *41*(5), 635–659. DOI: 10.1007/s11135-006-9016-8

Petty, N., Thomson, O., & Stew, G. (2012). Ready for a paradigm shift? Part 1: Introducing the philosophy of qualitative research. *National Institute of Health*, 1-8.

Pfeffer, J., & Salancik, G. R. (2003). *The External Control of Organizations: A Resource Dependence Perspective*. Stanford University Press.

Phillips, M. (2023). Towardsa social constructionist, criticalist Foucauldian informed qualitative research approach: Opportunties and challenges. *SN Soc Sci 3, 175*, 174-199. Retrieved 9 10, 2024, from DOI: 10.1007/s43545-023-00774-9

Phillips, M. (2023). Towards a social constructionist, criticalist, Foucaldian -informed qualitative research approaches:Opportunities and challenges. *SN Social Sciences*, *3*(175), •••.

Phillips, M., & Lu, J. (2018). A quick look at NVivo. *Journal of Electronic Resources Librarianship*, *30*(2), 104–106. DOI: 10.1080/1941126X.2018.1465535

Picken, F. (2018). The interview in tourism research. In Hillman, W., & Radel, K. (Eds.), *Qualitative Methods in Tourism Research: Theory and Practice* (pp. 200–223). Channel View Publications., DOI: 10.21832/9781845416416-014

Pietkiewicz, I., & Smith, J. (2012). Praktyczny przewodnik interpretacyjnej analizy fenomenologicznej w badaniach jakościowych w psychologii. *Czasopismo Psychologiczne*, *18*(2), 361–369.

Pillow, W. (2003). Confession, catharsis, or cure? Rethinking the uses of reflexivity as methodological power in qualitative research. *International Journal of Qualitative Studies in Education : QSE, 16*(2), 175–196. DOI: 10.1080/0951839032000060635

Pontes, E. S., et al. (2017). National Academic Production in Accounting: Analysis of Theses and Dissertations Produced Between 2007 and 2016. *REAd | Porto Alegre*, 23(1), 239–258.

Porte, M., Saur-Amaral, I., & Pinho, C. (2018). Research in Auditing: Main Topics. *Journal of Accounting & Finance*, *29*(76), 41–59.

Possenti, S. (n.d.). Introduction to Discourse Analysis.

Potter, W. J. (2017). Epistemology. In Matthes, S., & Potter, R. (Eds.), *The international encyclopedia of comunication research methods* (pp. 1–14). John Wiley.

Potter, W. J., & Levine-Donnerstein, D. (2009). Rethinking validity and reliability in content analysis. *Journal of Applied Communication Research*, *27*(3), 258–274. DOI: 10.1080/00909889909365539

Poucher, Z., Tamminena, K., Carona, J., & Sweet, S. (2019). hinking through and designing qualitative research studies: a focused mapping review of 30 years of qualitative research in sport psychology. *national Review of Sport and Exercise Psychology Volume 13, Issue 1*, 164-186.

Presenza, A., & Cipollina, M. (2010). Analysing tourism stakeholders networks. *Tourism Review*, *65*(4), 17–30. DOI: 10.1108/16605371011093845

Pritzker, S. (2020). *Doing an online ethnography: An overview of central issues and methodological strategies*. University of Alabama.

Priya, C. (2016, 09 11). *Limitations of qualitative research methods*. Retrieved from PG Project: https://www.projectguru.in/limitations-qualitative-research/

Psychology Learners. (2024, 09 07). *Discourse Analysis: Assumptions, theories and steps*. Retrieved from Psychology Learners: https://psylearners.psychotechservices.com/2015/09/solved-ignou-assignment-mpc005-q3.html

Quist-Adade, C. (2019). *Symbolic interactionism: The basics*. Vernon Press.

Rädiker, S., & Kuckartz, U. (2020). *Focused analysis of qualitative interviews with MAXQDA*. MAXQDA Press.

Rahaman, M. M., Islam, K. S., & Jahan, M. (2018). Rice farmers' knowledge of the risks of pesticide use in Bangladesh. *Journal of Health & Pollution*, *8*(20), 181203. DOI: 10.5696/2156-9614-8.20.181203 PMID: 30560002

Rahman, S. A., Tuckerman, L., Vorley, T., & Gherhes, C. (2021). Resilient research in the field: Insights and lessons from adapting qualitative research projects during the COVID-19 pandemic. *International Journal of Qualitative Methods*, *20*, 16094069211016106. DOI: 10.1177/16094069211016106

Rajasinghea, D., Garvey, B., Burt, S., Barosa-Pereira, A., & Clutterbuck, D. (2024). Innovative interpretative phenomenological analysis (IPA) approach in a coaching research project: Implications for future qualitative coaching research and beyond. *An International Journal of Theory. Coaching (Abingdon, UK)*, *17*(2), 301–318. DOI: 10.1080/17521882.2024.2358771

Rakita, M. (2017). Modernisation discourse and its contents. *Studia Ethnologica Croatica*, *29*, 103–148. DOI: 10.17234/SEC.29.4

Ramanadhan, S., Revette, A., Lee, R., & Aveling, E. (2021). Pragmatic approaches to analyzing qualitative data for implementation science: An introduction. *mplementation. Science Communication*, *2*(70), 1–10.

Ramandhan, S., Revette, A., Lee, R. M., & Aveling, E. (2021). Pragmatic approaches to analyzing qualitative data for implementation science: An introduction. *Implementation Science Communications volume 2. Article*, (70), 1–10.

Rambaree, K. (2013). Three methods of qualitative data analysis using ATLAS. ti:'A Posse Ad Esse'. In *ATLAS. ti User Conference 2013, 12-14 September 2013, Technische Universität Berlin, Germany* (Vol. 1, pp. 1-15). Universitätsverlag der TU Berlin.

Ramirez, J. (2017). *Handbook of Basel III Capital* (First edit). John Wiley & Sons Ltd.

Rashid, M., & Goldszmidt, M. (2024). Critical ethnography: Implications for medical education research and scholarship. *Medical Education*, *58*(10), 1185–1191. DOI: 10.1111/medu.15401 PMID: 38600755

Ravitch, S. M., & Riggan, M. (2017). *Reason & rigor: How conceptual frameworks guide research* (2nd ed.). SAGE Publications.

Redes, A. C. S. (2008). Networks and controversies: Tools for mapping psychosocial dynamics.

Reñosa, M. D. C., Mwamba, C., Meghani, A., West, N. S., Hariyani, S., Ddaaki, W., Sharma, A., Beres, L. K., & McMahon, S. (2021). Selfie consents, remote rapport, and Zoom debriefings: Collecting qualitative data amid a pandemic in four resource-constrained settings. *BMJ Global Health*, *6*(1), e004193. DOI: 10.1136/bmjgh-2020-004193 PMID: 33419929

Ren, Z., & Jiang, H. (2022). Risk cognition, agricultural cooperatives training, and farmers' pesticide overuse: Evidence from Shandong Province, China. *Frontiers in Public Health, 10*, 1032862. DOI: 10.3389/fpubh.2022.1032862 PMID: 36478731

Riadh, H. (2021). Intelligent tourism system using prospective techniques and the MACTOR methodology: A case study of Tunisian tourism. *Current Issues in Tourism, 25*(9), 1376–1398. DOI: 10.1080/13683500.2021.1937072

Riazi, A. M., Rezvani, R., & Ghanbar, H. (2023). Trustworthiness in L2 writing research: A review and analysis of qualitative articles in the Journal of Second Language Writing. *Research Methods in Applied Linguistics, 2*(3), 100065. DOI: 10.1016/j.rmal.2023.100065

Richards, T. J., & Richards, L. (1994). Using computers in qualitative research. *Handbook of qualitative research, 2*(1), 445-462.

Richards, L. (2002). Qualitative computing—A methods revolution? *International Journal of Social Research Methodology, 5*(3), 263–276. DOI: 10.1080/13645570210146302

Richards, T., & Richards, L. (1991). The NUDIST qualitative data analysis system. *Qualitative Sociology, 14*(4), 307–324. DOI: 10.1007/BF00989643

Roberts, J. K., Pavlakis, A. E., & Richards, M. P. (2021). It's more complicated than it seems: Virtual qualitative research in the COVID-19 era. *International Journal of Qualitative Methods, 20*, 16094069211002959. DOI: 10.1177/16094069211002959

Robertson, J. C. (1996). *Auditing*. Irwin.

Rodgers, W., Mubako, G. N., & Hall, L. (2017). Knowledge Management: The Effect of Knowledge Transfer on Professional Skepticism in Audit Engagement Planning. *Computers in Human Behavior, 70*, 564–574. DOI: 10.1016/j.chb.2016.12.069

Romanelli, G. (1998). The anthropological interview: Exchange and alterity. *Journal of the Graduate Program in Psychology of the Faculty of Philosophy, Sciences and Letters of Ribeirão Preto*, 119–133.

Ronzani, C. M., da Costa, P. R., da Silva, L. F., Pigola, A., & de Paiva, E. M. (2020). Qualitative methods of analysis: An example of Atlas. TI™ Software usage. *Revista Gestão & Tecnologia, 20*(4), 284–311. DOI: 10.20397/2177-6652/2020.v20i4.1994

Rosa, M. V. F. P. C., & Arnoldi, M. A. G. C. (2008). *The interview in qualitative research: Mechanisms for the validation of results*. Autêntica.

Rose, J., & Johnson, C. W. (2020). Contextualizing reliability and validity in qualitative research: Toward more rigorous and trustworthy qualitative social science in leisure research. *Journal of Leisure Research, 51*(4), 432–451. DOI: 10.1080/00222216.2020.1722042

Ross, G. M. (2023). Now that everyone else has panicked, well… I have to panic: Regulatory focus predicts panic buying during COVID-19. *Journal of Consumer Marketing, 40*(4), 521–534. DOI: 10.1108/JCM-06-2022-5409

Rossoni, L., Poli, I. T., de Sinay, M. C. F., & de Araújo, G. A. (2020). Materiality of sustainable practices and the institutional logics of adoption: A comparative study of chemical road transportation companies. *Journal of Cleaner Production, 246*, 119058. DOI: 10.1016/j.jclepro.2019.119058

Rotem, N. (2024). Historical ethnography: Key characteristics and the Journey before, during, and after the archival field. *Forum Qualitative Social Research, 25*(2), 1–10.

Russo, P. T., & Guerreiro, R. (2017). Perceptions about the sociomateriality of management accounting practices. *R*AE-Revista de Administração de Empresas | FGV EAESP, *57*(6), 567-584. https://doi.org/DOI: 10.1590/s0034-759020170605

Ryder, A. (2021). Critical ethnography and research relationships: Some ethical dilemmas. *Anthropology and Humanism, 46*(2), 300–314. DOI: 10.1111/anhu.12337

Saarijärvi, M., & Bratt, E. L. (2021). When face-to-face interviews are not possible: tips and tricks for video, telephone, online chat, and email interviews in qualitative research.

Saldaña, J. (2016). *The Coding Manual for Qualitative Researchers* (Seaman, J., Ed.; 3rd ed.).

Saldina, J. (2015). *The coding manual for qualitative researchers*. Sage.

Salmona, M., & Kaczynski, D. (2024). Qualitative data analysis strategies. In *How to conduct qualitative research in finance* (pp. 80–96). Edward Elgar Publishing. DOI: 10.4337/9781803927008.00012

Salmons, J. (2021). *Doing qualitative research online*. Sage Publications.

Sampieri, R. H., Collado, C. F., & Lucio, P. B. (2006). *Research Methodology*. McGraw-Hill.

Sampson, C., & Wong, L. S. (2023). Using Qualitative Data Analysis Software to Help Explore Critical Research Questions: A Tool, Not a Replacement. In Handbook of Critical Education Research (pp. 637-654). Routledge.

Sanjari, M., Bahramnezhad, F., Khoshnava Fomani, F., Sho-Ghi, M., & Cheraghi, M. A. (2014). Ethical challenges of researchers in qualitative studies: the necessity to develop a specific guideline. In *J Med Ethics Hist Med* (Vol. 7).

Sankar, J. G. (2024). AI-Driven Marketing Success Stories: A Case Note of Industry Pioneers. In *AI-Driven Marketing Research and Data Analytics* (pp. 48-66). IGI Global. DOI: 10.4018/979-8-3693-2165-2.ch003

Sankar, J. G. (2024). Travel Redefined: Future Trends in AI-Enhanced Tourism Experiences. In *AI Innovations for Travel and Tourism* (pp. 135-162). IGI Global.

Sankar, J. G. (2025). Measuring Service Quality in Generative AI Environments: A Comprehensive GAISQUAL Framework. In *Design and Validation of Research Tools and Methodologies* (pp. 317-346). IGI Global.

Sankar, J. G., & David, A. (2023). Measuring the Service Quality of Artificial Intelligence in the Tourism and Hospitality Industry. In *Handbook of Research on Innovation, Differentiation, and New Technologies in Tourism, Hotels, and Food Service* (pp. 133-155). IGI Global. DOI: 10.4018/978-1-6684-6985-9.ch007

Sankar, J. G., & David, A. (2024). A Comprehensive Examination of Mobile Augmented Reality in Tourism (MART) Adoption: Using the UTAUT2 Framework. In *Contemporary Trends in Innovative Marketing Strategies* (pp. 241-262). IGI Global.

Sankar, J. G., & David, A. (2024). Transforming the Travel Landscape: Smart Contracts in Tourism Management. In *Decentralizing the Online Experience With Web3 Technologies* (pp. 246-266). IGI Global.

Sankar, J. G., David, A., & Ramasamy, R. (2023). Assessing Water and Sanitation Service Quality at Pilgrimage Sites After COVID-19: An Empirical Investigation. In *Prospects and Challenges of Global Pilgrimage Tourism and Hospitality* (pp. 160-179). IGI Global.

Sankar, J. G., David, A., & Shrutimeena, G. (2024). From Hesitation to Integration: Understanding Digital Healthcare Adoption. In *Multi-Sector Analysis of the Digital Healthcare Industry* (pp. 1-32). IGI Global.

Sankar, G., & Naveenkumar, K. (2018). *Material Handling Management in Third Party Logistics Warehouse. International Journal for Research in Engineering Application & Management*. IJREAM.

Sankar, J. G. (2020). Customer perception about innovative safety food delivery during lockdown. *Journal of Contemporary Issues in Business and Government, 26*(2).

Sankar, J. G. (2020). Innovative celebrity endorsement and consumer purchase. *Journal of Contemporary Issues in Business and Government, 26*(2).

Sankar, J. G. (2020). Innovative consumer behaviour towards sportswear. *Journal of Contemporary Issues in Business and Government, 26*(2).

Sankar, J. G. (2020). Service Factors Influence on Customer Satisfaction in Innovative Ecommerce. *SANKAR, D.(2020). Service factors influence on customer satisfaction in innovative e-commerce.Journal of Contemporary Issues in Business and Government, 26*(2), 651–656.

Sankar, J. G. (2020). Service quality of e-Commerce logistics services. *Journal of Contemporary Issues in Business and Government, 26*(2).

Sankar, J. G., & KUMAR G, H. E. M. A. N. T. H. (2020). Customer perception towards electric two-wheeler innovation. *Journal of Contemporary Issues in Business and Government, 26*(2).

Sankar, J. G., David, A., & Ganesh Kumar, C. (2022). Adaptation of Blockchain Technology In HRM. *Available atSSRN* 4515408.

Sankar, J. G., David, A., & Valan, P. (2023). Examining User Understanding and Perceptions of E-Commerce Data Privacy, Security, and Protection. In *Confronting Security and Privacy Challenges in Digital Marketing* (pp. 159–185). IGI Global. DOI: 10.4018/978-1-6684-8958-1.ch008

Sankar, J. G., & Devanand, K. (2020). Social media innovation on consumer buying behaviour. *SANKAR, JG (2020). Social media innovation on consumer buying behaviour.Journal of Contemporary Issues in Business and Government, 26*(2).

Sankar, J. G., & Ilangovan, K. (2025). The Role of Smart Technology in Enhancing Sustainable Tourism Experiences. In *Integrating Architecture and Design Into Sustainable Tourism Development* (pp. 379–408). IGI Global Scientific Publishing. DOI: 10.4018/979-8-3693-7388-0.ch014

Sankar, J. G., & Kant, S. (2025). Assessing Consumer Trust and Perception in Online Food Marketplaces. In *Innovative Trends Shaping Food Marketing and Consumption* (pp. 375–394). IGI Global Scientific Publishing. DOI: 10.4018/979-8-3693-8542-5.ch015

Sankar, J. G., & Kiruba Sagar, M, V. (2020). Examining the Risk Factor Associated with UTAUT Model. *Solid State Technology, 63*(2), 1196–1198.

Sankar, J. G., Rajeswari, M., & Dwarak, M. V. S. (2020). To Measure the Service Quality of Film and Television Technology. *Solid State Technology, 63*(2), 1192–1195.

Sankar, J. G., & Srinivasulu, Y. (2014). A Study of Tourism Industry Impact on Macroeconomic Indicators. *GE-International Journal of Management Research, 2*(11), 2321–1709.

Sankar, J. G., & Srinivasulu, Y. (2014). Contemporary Sustainable Development Practices in Indian Hospitality Sector: Triple Bottom Line Approach. *SSRN*, , 2. DOI: 10.2139/ssrn.4204732

Sankar, J. G., Valan, P., & Siranjeevi, M. S. (2020). Various Models to Evaluate Quality in the Service Industry. In *Digital Transformation and Innovative Services for Business and Learning* (pp. 181–194). IGI Global.

Santhosh, L., Rojas, J. C., & Lyons, P. G. (2021). Zooming into Focus Groups: Strategies for Qualitative Research in the Era of Social. *ATS Scholar, 2*(2), 176–184. DOI: 10.34197/ats-scholar.2020-0127PS PMID: 34409412

Sarfo, J., Debrah, T., Gbordzoe, N., Afful, W., & Obeng, P. (2021). Qualitative research designs, sample size and saturation: Is Enough Always Enough? *Journal of Advocacy. Research in Education, 8*(3), 60–67.

Sarma, P. (2022). Farmer behavior towards pesticide use for reduction production risk: A Theory of Planned Behavior. *Cleaner Circular Bioeconomy, 1*, 100002. DOI: 10.1016/j.clcb.2021.100002

Sasikala, J., & Jeganathan, G. S. (2017). A Study on Employee Engagement in Information Technology (IT) Industry. *SSRN, 1*(1), 233–239. DOI: 10.2139/ssrn.3536354

Saunders, M., & Lewis, P. (2017). *Doing research in business and management.* Pearson.

Saunders, M., Lewis, P., & Adrian, T. (2019). *Qualitative Research in Business.* Pearson.

Saunders, M., Lewis, P., & Thornhill, A. (2019). *Research methods for business students.* Pearson.

Scapens, R. W. (2006). Understanding Management Accounting Practices: A Personal Journey. *The British Accounting Review, 38*(1), 1–30. DOI: 10.1016/j.bar.2005.10.002

Scispace. (2024, 10 23). *What is the weakness of discourse analysis in accurately identifying discourses?* Retrieved from Scispace: https://typeset.io/questions/what-is-the-weakness-of-discourse-analysis-in-accurately-3s3ehn0s4l

Scott, W. Richard (2014). Institutions and organizations: ideas, interests, and identities. In *Sage Publications* (Vol. 4).

Seatle University. (2024, 10 31). *Existential Phenomenological Psychology*. Retrieved from Seatle University: https://www.seattleu.edu/academics/all-programs/psychology-ma/epp/

Seidman, I. (2013). *Interviewing as qualitative research: A guide for researchers in education and social sciences* (4th ed.). Teachers College Press.

Semigina, T. (2022). Critical ethnography: Opportunities for social work research. *Social Work Education, 9*(3), 405–421. DOI: 10.25128/2520-6230.22.3.7

Seo, M. G., & Creed, W. D. (2002). Institutional contradictions, praxis, and institutional change: A dialectical perspective. *Academy of Management Review, 27*(2), 222–247. DOI: 10.2307/4134353

Seskir, Z. C., Umbrello, S., Coenen, C., & Vermaas, P. E. (2023). Democratization of quantum technologies. *Quantum Science and Technology, 8*(2), 024005. DOI: 10.1088/2058-9565/acb6ae

Seyedi, S., Griner, E., Corbin, L., Jiang, Z., Roberts, K., Iacobelli, L., Milloy, A., Boazak, M., Bahrami Rad, A., Abbasi, A., Cotes, R. O., & Clifford, G. D. (2023). Using HIPAA (health insurance portability and accountability act)–compliant transcription services for virtual psychiatric interviews: Pilot comparison study. *JMIR Mental Health, 10*, e48517. DOI: 10.2196/48517 PMID: 37906217

Sharifzadeh, M. S., Abdollahzadeh, G., Damalas, C. A., Rezaei, R., & Ahmadyousefi, M. (2019). Determinants of pesticide safety behavior among Iranian rice farmers. *The Science of the Total Environment, 651*, 2953–2960. DOI: 10.1016/j.scitotenv.2018.10.179 PMID: 30463146

Sharma, A., Kumar, V., Shahzad, B., Tanveer, M., Sidhu, G. P. S., Handa, N., Kohli, S. K., Yadav, P., Bali, A. S., Parihar, R. D., Dar, O. I., Singh, K., Jasrotia, S., Bakshi, P., Ramakrishnan, M., Kumar, S., Bhardwaj, R., & Thukral, A. K. (2019). Worldwide pesticide usage and its impacts on ecosystem. *SN Applied Sciences, 1*(11), 1446. DOI: 10.1007/s42452-019-1485-1

Sheehan, S. (2014). A conceptual framework for understanding trancsendental phenomenology through the lived experiences of biblical leaders. *Emerging Leadership Journeys, 7*(1), 1–6.

Shree, D., Singh, R. K., Paul, J., Hao, A., & Xu, S. (2021). Digital platforms for business-to-business markets: A systematic review and future research agenda. *Journal of Business Research, 137*, 354–365. DOI: 10.1016/j.jbusres.2021.08.031

Sikirivwa, M. (2020). Deconstruction theory and its background. *American Journal of Humanities and Social Sciences Research (AJHSSR) Volume-4, Issue-4*, 44-72.

Silva, L. S., de Oliveira, G. S., & Neves, E. H. C. (2021). Interview in research in education with a qualitative approach: Some theoretical and practical considerations. *Prisma Magazine*, *2*(1), 110–112.

Silverman, D. (2014). *Interpreting Qualitative Data*. SAGE Publications.

Simitha, S., Mugila, P., & Sankar, J. G. (2020). A Study on Customer Relationship Management in Siemens Healthineers. *Solid State Technology*, *63*(2), 1199–1203.

Sin, C. H. (2007). Using software to open up the 'Black Box'of qualitative data analysis in evaluations: The experience of a multi-site team using NUD* IST version 6. *Evaluation*, *13*(1), 110–120. DOI: 10.1177/1356389007073684

Sinfield, G., Goldspink, S., & Wilson, C. (2023). Waiting in the Wings: The Enactment of a Descriptive Phenomenology Study 22. *International Journal of Qualitative Methods*, *22*, 1–11. DOI: 10.1177/16094069231207012

Singh, S., & Estefan, A. (2018). Selecting a Grounded Theory approach for nursing research. *Global Qualitative Nursing Research*, *5*, 1–9. DOI: 10.1177/2333393618799571 PMID: 30306099

Sipes, J. B., Roberts, L. D., & Mullan, B. (2022). Voice-only Skype for use in researching sensitive topics: A research note. *Qualitative Research in Psychology*, *19*(1), 204–220. DOI: 10.1080/14780887.2019.1577518

Skinner, E. A., Edge, K., Altman, J., & Sherwood, H. (2003). Searching for the structure of coping: A review and critique of category systems for classifying ways of coping. *Psychological Bulletin*, *129*(2), 216–269. DOI: 10.1037/0033-2909.129.2.216 PMID: 12696840

Smit, B., & Scherman, V. (2021). Computer-assisted qualitative data analysis software for scoping reviews: A case of ATLAS. ti. *International Journal of Qualitative Methods*, *20*, 16094069211019140. DOI: 10.1177/16094069211019140

Smith, S. M., Cotterill, S., & Brown, H. (2020). An interpretative phenomenological analysis of performance influencing factors within the practice environment. *Journal of Physical Education and Sport ® (JPES), Vol.20 (4)*, 1646 - 1657.

Smith, A. H., Touchett, H., Chen, P., Fletcher, T., Arney, J., Hogan, J., Wassef, M., Cloitre, M., & Lindsay, J. A. (2024). Patient satisfaction with a coach-guided, technology-based mental health treatment: Qualitative interview study and theme analysis. *JMIR Mental Health*, *11*(1), e50977. DOI: 10.2196/50977 PMID: 38306167

Smith, J. A. (2004). Reflecting on the development of interpretative phenomenological. *Qualitative Research in Psychology*, *1*, 39–54.

Smith, J. A., & Osborn, M. (2015). Interpretative phenomenological analysis as a useful methodology for research on the lived experience of pain. *British Journal of Pain*, *9*(1), 41–42. DOI: 10.1177/2049463714541642 PMID: 26516556

Smith, M. (2003). *Research Methods in Accounting*. SAGE Publications. DOI: 10.4135/9781849209809

Soehardi, F., Putri, L. D., & Dinata, M. (2021). NVivo Software Training for young researchers. *Mattawang: Jurnal Pengabdian Masyarakat*, *2*(1), 8–13. DOI: 10.35877/454RI.mattawang265

Soeiro, T. M., & Wanderley, C. A. (2019). Institutional Theory in Accounting Research: A Review. *Organizações & Sociedade Journal*, *26*(89), 291–316.

Sol, K., & Heng, K. (2022). Understanding epistemology and its key approaches in research. *Cambodian Journal of Educational Research*, *2*(2), 80–99. DOI: 10.62037/cjer.2022.02.02.05

Song, L., Son, J., & Lin, N. (2011). Social support. In Scott, J., & Carrington, P. J. (Eds.), *The SAGE handbook of social network analyses* (pp. 116–128). SAGE Publications Ltd.

Soratto, J., Pires, D. E. P. D., & Friese, S. (2020). Thematic content analysis using ATLAS. ti software: Potentialities for researchs in health. *Revista Brasileira de Enfermagem*, *73*(3), e20190250. DOI: 10.1590/0034-7167-2019-0250 PMID: 32321144

Souza, S. A. F. (2006). *Understanding Discourse Analysis: Language, Society, and Ideology*. Valer Publishing.

Souza, S. A. F. (2014). *Discourse Analysis: Methodological Procedures*. Census Institute.

SP, M. S., & Sankar, J. G. An Overview of Measurement model in Marketing Research.

Spence, C., & Carter, C. (2014). An Exploration of the Professional Habitus in the Big4 Accounting Firms. *Work, Employment and Society*, *0*(0), 1–17.

Spencer, R., Pryce, J., & Walsh, J. (2014). Philosophical approach to qualitative research. In Leavy, P. (Ed.), *The Oxford of qualitative research* (pp. 81–97). OUP.

SS&C Algorithmics. (2022a). *FRTB Best Practices Guide: Lessons Learned in the Implementation of FRTB*. https://www.ssctech.com/resources/view/frtb-best-practices - guide-lessons-learned-implementation-frtb?hstc=76820298.e828e7530078f-8a2621332dbc17de5a6.1676123942561.1676150316323.1676927739039.3&hss-c=76820298.3.1676927739039&hsfp=296860102&submissionGuid=4cc153d5-46e5-4e6e-8293-dbb79c7b53cf

SS&C Algorithmics. (2022b). *FRTB implementation: key insights and learnings - Risk.net*. August 9. https://www.risk.net/market-access/investing/7953201/frtb-implementation- key-insights-and-learnings

St. Pierre, E. A., & Jackson, A. Y. (2014). Qualitative data analysis after coding. *Qualitative Inquiry*, *20*(6), 715–719. DOI: 10.1177/1077800414532435

Stahl, N., & King, J. (2020). Expanding approaches for research: Understanding and Using Trustworthiness in Qualitative Research. *Journal of Developmental Education*, *44*(1), 26–28.

Stahl, N., & King, J. (2020). Expanding approaches to research: Understanding and using trustworthiness in qualitative research. *Journal of Developmental Education*, *44*(1), 26–28.

Stanford Encyclopedia of Philosophy. (2009, 03 27). *The history of utilitarianism*. Retrieved from Stanford Encyclopedia of Philosophy: https://plato.stanford.edu/entries/utilitarianism-history/

Statistics Canada. (2024, 10 6). *Non-probability sampling*. Retrieved from Statistics Canada: https://www150.statcan.gc.ca/n1/edu/power-pouvoir/ch13/nonprob/5214898-eng.htm

Stieglitz, S., Mirbabaie, M., Ross, B., & Neuberger, C. (2018). Social media analytics–Challenges in topic discovery, data collection, and data preparation. *International Journal of Information Management*, *39*, 156–168. DOI: 10.1016/j.ijinfomgt.2017.12.002

Stolz, S. (2022). The practice of phenomenology in educational research. *Educational Philosophy and Theory*, *55*(2), 1–35.

Stone, A., Helder, L., & Schneider, M. (1988). Coping with stressful events: Coping dimensions and issues. In Cohen, L. (Ed.), *Life events and psychological functioning: Theoretical and methodological issues* (pp. 182–210). Sage.

Strauss, A., & Corbin, J. (1998). *Basics of qualitative research techniques*.

Subedi, K. (2021). Determining sample size in qualitative research. *Scholars'. Journal*, *4*, 1–13.

Sudartini, S. (2024). Postmodernism paradigm and its use on the existence of language. *Rev. Gest. Soc. Ambient.*, *18*(7), 1–26.

Suddick, K., Cross, V., Stew, G., & Stew, G. (2020). The work of hermeneutical phenomenology. *International Journal of Qualitative Methods*, •••, 1–14.

Sundaray, B. K., Sarangi, P., & Patra, S. K. (2021). Psychological vulnerability and coping among management students during COVID-19 pandemic. *The Journal of Mental Health Training, Education and Practice, 16*(6), 454–467. DOI: 10.1108/JMHTEP-05-2021-0041

Sweeney, J. T., Suh, I. S., Dalton, K. C., & Meljem, S. (2017). Are Workpaper Reviews Preparer-Specific? *The British Accounting Review, 49*(6), 560–577. DOI: 10.1016/j.bar.2017.09.003

Sy, M., O'Leary, N., Nagraj, S., El-Awaisi, A., O'Carroll, V., & Xyrichis, A. (2020). Doing interprofessional research in the COVID-19 era. *Discussion Paper (Deutsches Institut für Entwicklungspolitik)*. PMID: 32718262

Szablewska, N., & Kubacki, K. (2023). Empirical business research on modern slavery in supply chains: A systematic review. *Journal of Business Research, 164*, 113988. DOI: 10.1016/j.jbusres.2023.113988

Szymkowiak, A., Gaczek, P., Jeganathan, K., & Kulawik, P. (2021). The impact of emotions on shopping behavior during epidemic: What a business can do to protect customers. *Journal of Consumer Behaviour, 20*(1), 48–60. DOI: 10.1002/cb.1853 PMID: 38607848

Taghdisi, M. H., Besheli, B. A., Dehdari, T., & Khalili, F. (2019). Knowledge and practices of safe use of pesticides among a group of farmers in northern Iran. *The International Journal of Occupational and Environmental Medicine, 10*(2), 66–72. DOI: 10.15171/ijoem.2019.1479 PMID: 31041923

Takahashi, A., & Araujo, L. (2020). Case study research: Opening up research opportunities. *RAUSP Management Journal, 55*(1), 100–111. DOI: 10.1108/RAUSP-05-2019-0109

Takahashi, K., Muraoka, R., & Otsuka, K. (2020). Technology adoption, impact, and extension in developing countries' agriculture: A review of the recent literature. *Agricultural Economics, 51*(1), 31–45. DOI: 10.1111/agec.12539

Taylor, J., & Pagliari, C. (2018). Mining social media data: How are research sponsors and researchers addressing the ethical challenges? *Research Ethics, 14*(2), 1–39. DOI: 10.1177/1747016117738559

Teixeira, A. P. (2017). *Avaliação dos atributos implementados no processo de Candidatura para utilização de Modelos Internos no cálculo do capital regulatório dos Riscos de Mercado que contribuem para a criação de Vantagem Competitiva Sustentável : Caso Banco Bradesco* [Master's Dissertation of the Professional Master's Course in Controllership and Finance], Faculdade FIPECAFI.

Teixeira, A. P. (2023). O dilema da escolha entre os modelos internos e padronizados na busca pela eficiência na gestão de capital regulatório dos bancos no Brasil: uma análise sob a luz das lógicas conflitantes e contradições institucionais. [Doctoral Thesis. Postgraduate Program in Administration, Mackenzie Presbyterian University - UPM]. https://dspace.mackenzie.br/handle/10899/33719

Telles, R. (2001). A efetividade da "matriz de amarração" de Mazzon nas pesquisas em Administração. *Revista ADM*, *36*(4), 64–72. http://www.spell.org.br/documentos/download/16589

Teng, Y., Chen, X., Jin, Y., Yu, Z., & Guo, X. (2022). Influencing factors of and driving strategies for vegetable farmers' green pesticide application behavior. *Frontiers in Public Health*, *10*, 907788. DOI: 10.3389/fpubh.2022.907788 PMID: 36159273

Tenny, S., Brannan, J., & Brannan, G. (2022). Qualitative study. *National Library of Medicine*, 1-4.

The Editors of Encyclopaedia. (2024, 09 7). *Deconstruction*. Retrieved from Encyclopedia Britannica: https://www.britannica.com/topic/deconstruction

Thomas, C. G. (2021). *Research methodology and scientific writing*. Springer. DOI: 10.1007/978-3-030-64865-7

Thomas, P. A., & Kim, S. (2021). Lost touch? Implications of physical touch for physical health. *The Journals of Gerontology. Series B, Psychological Sciences and Social Sciences*, *76*(3), e111–e115. DOI: 10.1093/geronb/gbaa134 PMID: 32845008

Thomas, S., & Sohn, B. (2023). From uncomfortable squirm to self-discovery: A phenomenological analysis of the bracketing experience. *International Journal of Qualitative Methods*, *22*, 1–12. DOI: 10.1177/16094069231191635

Thompson, S. K. (2012). *Sampling*. John Wiley & Sons. DOI: 10.1002/9781118162934

Thornton, P. H., & Ocasio, W. (2008). Institutional Logics in The Sage Handbook of Organizational Institutionalism. In SAGE (Ed.), *International Association of Scientific Hydrology. Bulletin*. https://doi.org/DOI: 10.1080/02626667109493051

Thornton, P. H., Ocasio, W., & Lounsbury, M. (2012). *The Institutional Logics Perspective* (Issue July). Oxford University Press. https://doi.org/DOI: 10.1093/acprof:oso/9780199601936.001.0001

Thornton, P. H., & Ocasio, W. (1999). Institutional logics and the historical contingency of power in organizations: Executive succession in the higher education publishing industry, 1958-1990. *American Journal of Sociology*, *105*(3), 801–843. DOI: 10.1086/210361

Thunberg, S., & Arnell, L. (2022). Pioneering the use of technologies in qualitative research–A research review of the use of digital interviews. *International Journal of Social Research Methodology*, *25*(6), 757–768. DOI: 10.1080/13645579.2021.1935565

Tian, Q., Yu, Y., Xiang, Z., & Li, C. (2021). Agricultural technical education, interpersonal trust, and pesticide use by vegetable farmers in China. *Journal of Agricultural Education and Extension*, *27*(2), 211–227. DOI: 10.1080/1389224X.2020.1844769

Tolbert, P. S., & Zucker, L. G. (1983). Institutional Sources of Change in the Formal Structure of Organizations: The Diffusion of Civil Service Reform, 1880-1935. *Administrative Science Quarterly*, *28*(1), 22–39. DOI: 10.2307/2392383

Tomas, C. W., & Henke, E. O. (1983). *Auditing: Theory and Practice*. Kent Publishing Company.

Tonelli, D. F. (2016). Origins and epistemological affiliations of Actor-Network Theory: Implications for organizational analysis. *Cadernos EBAPE.BR*, *14*, 377–390. DOI: 10.1590/1679-395141596

Tourangeau, R., Rips, L. J., & Rasinski, K. (2000). *The Psychology of Survey Response*. Cambridge University Press. DOI: 10.1017/CBO9780511819322

Toutkoushian, R., Riffe, K., Sanford, P., & Ness, E. (2022). How do faculty select between defined benefit and defined contribution retirement plans? A qualitative investigation. *Journal of Education Finance*, *47*(3), 250–274. DOI: 10.1353/jef.2022.a852709

Tracy, S. J. (2024). *Qualitative research methods: Collecting evidence, crafting analysis, communicating impact*. John Wiley & Sons.

Trompeter, G., & Wright, A. (2010). The World Has Changed: Have Analytical Procedure Practices? *Contemporary Accounting Research*, *27*(2), 669–700. DOI: 10.1111/j.1911-3846.2010.01021.x

Tshabangu, I., Ba, S., & Madondo, S. M. (Eds.). (2020). *Approaches and processes of social science research*. IGI Global.

Tsoukas, H. (2010). Practice, strategy making, and intentionality: A Heideggerian onto-epistemology for strategy as practice. In Golsorkhi, D., Rouleau, L., Seidl, D., & Vaara, E. (Eds.), *Cambridge Handbook of Strategy as Practice* (pp. 47–62). Cambridge University Press. DOI: 10.1017/CBO9780511777882.004

Tsoukas, H., & Chia, R. (2002). On organizational becoming: Rethinking organizational change. *Organization Science*, *13*(5), 567–582. DOI: 10.1287/orsc.13.5.567.7810

Tufford, L., & Newman, P. (2010). Bracketing in qualitative research. *Qualitative Social Work: Research and Practice, 11*(1), 80–96. DOI: 10.1177/1473325010368316

Tureta, C., Américo, B. L., & Clegg, S. (2021). Controversies as a method for ANTi-history: An inquiry into public administration practices. *Organization, 28*(6), 1018–1035. DOI: 10.1177/13505084211015375

Turley, S., Humphrey, C., Samsonova-Taddei, A., Siddiqui, J., Woods, M., Baioudis, I., & Richard, C. (2016). *Skills, Competencies, and the Sustainability of Modern Audit*. The Institute of Chartered Accountants of Scotland and the Financial Reporting Council.

Turner, D. (2020, 09 29). *Analysis: Introducing discourse analysis for qualitative research*. Retrieved from Quirkos: https://www.quirkos.com/blog/post/discourse-analysis-qualitative-research/

Turner, D., Ting, H., Wong, M., Lim, T.-Y., & Tan, K.-L. (2021). Asian Journal of Business Research Volume 11, Issue 3. *Applying Qualitative Approach in Business Research*, 1-13.

Tu, Z. G., Cheng, L., & Zhang, M. (2022). Study on the regional difference and characteristic of temporal-spatial evolution of Chinese business environment. *Research on Quantitative Economics and Technological Economics, 39*, 3–25.

Tyagi, A. K., & Chahal, P. (2020). Artificial intelligence and machine learning algorithms. In *Challenges and applications for implementing machine learning in computer vision* (pp. 188–219). IGI Global. DOI: 10.4018/978-1-7998-0182-5.ch008

Übellacker, T. (2024). AcademiaOS: Automating Grounded Theory Development in Qualitative Research with Large Language Models. arXiv preprint arXiv:2403.08844.

Ugwu, C., & Eze, V. (2023). Qualitative research . *International Digital Organisation for Scientific Research vol 8 Issue (1)*, 20-35.

Ullah, F., Sepasgozar, S. M., & Wang, C. (2018). A systematic review of smart real estate technology: Drivers of, and barriers to, the use of digital disruptive technologies and online platforms. *Sustainability (Basel), 10*(9), 3142. DOI: 10.3390/su10093142

Vaismoradi, M., Jones, J., Turunen, H., & Snelgrove, S. (2016). Theme development in qualitative content analysis and thematic analysis.

Van de Ven, A. H. (2007). *Engaged scholarship: A guide for organizational and social research*. Oxford University Press. DOI: 10.1093/oso/9780199226290.001.0001

van Dijk, T. (2001). Critical discourse analysis. In Scchiffrin, T. D., & Hamilton, H. (Eds.), *Handbook of discourse analysis* (pp. 352–371). Blackwell.

Van Dijk, T. A. (1993). Principles of critical discourse analysis. *Sage (Atlanta, Ga.), 4*(2), 249–283.

Van Maanen, J. (2011). *Tales of the field: On writing ethnography*. University of Chicago Press. DOI: 10.7208/chicago/9780226849638.001.0001

van Manen, M. (1997). From meaning to method. *Qualitative Health Research, 7*(3), 345–369. DOI: 10.1177/104973239700700303

Varpio, L., Paradis, E., Uijtdehaage, S., & Young, M. (2020). The distinctions between theory, theoretical framework and conceptual framework. *Academic Medicine, 95*(7), 989–994. DOI: 10.1097/ACM.0000000000003075 PMID: 31725464

Vears, D. F., & Gillam, L. (2022). Inductive content analysis: A guide for beginning qualitative researchers. Focus on Health Professional Education: A Multi-Disciplinary Journal, 23(1),111–127.

Vellas, F., & Becherel, L. (2020). *Tourism Marketing: A Strategic Approach*. Routledge.

Velmurugan, T. A., & Sankar, J. G. (2017). A comparative study on motivation theory with Maslow's hierarchy theory and two factor theory in organization. *Indo-Iranian Journal of Scientific Research, 1*(1), 204–208.

Venturini, T. (2010). Diving in magma: How to explore controversies with actor-network theory. *Public Understanding of Science (Bristol, England), 19*(3), 258–273. DOI: 10.1177/0963662509102694

Venturini, T. (2012). Building on faults: How to represent controversies with digital methods. *Public Understanding of Science (Bristol, England), 21*(7), 796–812. DOI: 10.1177/0963662510387558 PMID: 23832559

Venturini, T., Ricci, D., Mauri, M., Kimbell, L., & Meunier, A. (2015). Designing controversies and their publics. *Design Issues, 31*(3), 74–87. DOI: 10.1162/DESI_a_00340

Vergara, S. C. (2013). *Research Projects and Reports in Administration*. Atlas Publishing.

Vespestad, M., & Clancy, A. (2021). Exploring the use of content analysis methodology in consumer research. *Journal of Retailing and Consumer Services, 59*, 1–9. DOI: 10.1016/j.jretconser.2020.102427

Vieira, M. M. F. (2004). For good qualitative research. In Vieira, M. M. F., & Zouain, D. M. (Eds.), *Qualitative research in administration* (pp. 13–28). Editora FGV.

Villar, E. G., & Roglio, K. D. D. (2019). The alignment between Actor-Network Theory and relational sociology: An onto-epistemological discussion for organizational studies. [TPA]. *Theory and Practice in Administration*, *9*(1), 77–91.

Vindrola-Padros, C., & Johnson, G. A. (2020). Rapid techniques in qualitative research: A critical review of the literature. *Qualitative Health Research*, *30*(10), 1596–1604. DOI: 10.1177/1049732320921835 PMID: 32667277

Vinuto, J. (2014). Snowball Sampling in Qualitative Research: An Open Debate. *Temáticas*, *22*(44), 203–220. DOI: 10.20396/tematicas.v22i44.10977

Vivek, R. (2022). Systematic review of modern techniques in qualitative research. *Management and Entrepreneurship: Trends of Development*, *4*(22), 34–49. DOI: 10.26661/2522-1566/2022-4/22-03

Vivek, R., Nanthagopan, Y., & Piriyatharshan, S. (2023). Beyond methodology: Theoretical foundations of triangulation in qualitative and multi-method research: A literature review. *Scientific Studios on Social and Political Psychology*, *29*(2), 53–62. DOI: 10.61727/sssppj/2.2023.53

Vivek, R., Nanthagopan, Y., & Piriyatharshan, S. (2023). Beyond methods: Theoretical underpinnings of triangulation in qualitative and multi-method studies. *SEEU Review*, *18*(2), 105–122. DOI: 10.2478/seeur-2023-0088

Volberda, H. W., Khanagha, S., Baden-Fuller, C., Mihalache, O. R., & Birkinshaw, J. (2021). Strategizing in a digital world: Overcoming cognitive barriers, reconfiguring routines and introducing new organizational forms. *Long Range Planning*, *54*(5), 102110. DOI: 10.1016/j.lrp.2021.102110

Wagenaan, H., Kiesclich, K., & Hangel, N., & Zimmerman. (2022). Collaborative comparisons: A pragmatist approach towards designing large-scale, comparative qualitative research. *SSM. Qualitative Research in Health*, •••, 1–11.

Walle, A. (2015). *Qualitative research in business: A practical overview.* Cambridge: Cambridge scholars .

Walsham, G. (1997). Actor-network theory and IS research: Current status and future prospects. *Information Systems and Qualitative Research*, 466-480.

Wangmo, T., Lipps, M., Kressig, R. W., & Ienca, M. (2019). Ethical concerns with the use of intelligent assistive technology: Findings from a qualitative study with professional stakeholders. *BMC Medical Ethics*, *20*(1), 1–11. DOI: 10.1186/s12910-019-0437-z PMID: 31856798

Wang, Y., Kung, L., & Byrd, T. A. (2018). Big data analytics: Understanding its capabilities and potential benefits for healthcare organizations. *Technological Forecasting and Social Change*, *126*, 3–13. DOI: 10.1016/j.techfore.2015.12.019

Weber, S. (2021). A step-by-step procedure to implement discrete choice experiments in Qualtrics. *Social Science Computer Review*, *39*(5), 903–921. DOI: 10.1177/0894439319885317

Wellem, A. (2022). Budgeting and cost control challenges in small and medium enterprises construction projects in Cape Town (Doctoral dissertation, Cape Peninsula University of Technology).

Wertz, F., Charmaz, K., McMullen, L., Josselson, R., Anderson, R., & McSpadden, E. (2011). *Five ways of doing qualitative analysis: phenomenological, grounded theory, discourse analysis, narrative research and intuitive inquiry*. The Guilford Press.

Westermann, K. D., Bedard, J. C., & Earley, C. E. (2015). Learning the Craft of Auditing: A Dynamic View of Auditor's On-the-Job Learning. *Contemporary Accounting Research*, *32*(3), 864–896. DOI: 10.1111/1911-3846.12107

Westland, J. C. (2017). An Empirical Investigation of Analytical Procedures Using Mixture Distributions. *Intelligent Systems in Accounting, Finance & Management*, *24*(4), 111–124. DOI: 10.1002/isaf.1405

Willig, C. (2008). *Introducing qualitative research in psychology*. Open University Press.

Wodak, R. (2008). Introduction: discourse studies - important concepts and terms. In Wodak, R., & Krzyżanowski, M. (Eds.), *Qualitative discourse analysis in the Social Sciences* (pp. 1–29). Palgrave. DOI: 10.1007/978-1-137-04798-4_1

Wood, L. A., & Kroger, R. O. (2000). *Doing Discourse Analysis: Methods for Studying Action in Talk and Text*. SAGE Publications. DOI: 10.4135/9781452233291

Woodruff, D. (2018). Phenomenology. *The Stanford Encyclopedia of Philosophy*. Retrieved 09 4, 2024, from https://plato.stanford.edu/archives/sum2018/entries/phenomenology/>

Woods, M., Paulus, T., Atkins, D. P., & Macklin, R. (2016). Advancing qualitative research using qualitative data analysis software (QDAS)? Reviewing potential versus practice in published studies using ATLAS. ti and NVivo, 1994–2013. *Social Science Computer Review*, *34*(5), 597–617. DOI: 10.1177/0894439315596311

Wu, Y. F., & Kim, E. Y. (2022). Users' perceptions of technological features in augmented reality (AR) and virtual reality (VR) in fashion retailing: A qualitative content analysis. *Mobile Information Systems*, *2022*(1), 3080280. DOI: 10.1155/2022/3080280

Xia, H., Li, C., Zhou, D., Zhang, Y., & Xu, J. (2020). Peasant households' land use decision-making analysis using social network analysis: A case of Tantou Village, China. *Journal of Rural Studies*, *80*, 452–468. DOI: 10.1016/j.jrurstud.2020.10.023

Xiang, Z., Du, Q., Ma, Y., & Fan, W. (2022). Research agenda for artificial intelligence in tourism: A case study of data-driven tourism service innovation. *Tourism Management*, *88*, 104403.

Xu, A., Baysari, M. T., Stocker, S. L., Leow, L. J., Day, R. O., & Carland, J. E. (2020). Researchers' views on, and experiences with, the requirement to obtain informed consent in research involving human participants: A qualitative study. *BMC Medical Ethics*, *21*(1), 1–11. DOI: 10.1186/s12910-020-00538-7 PMID: 33008387

Xu, Z., Zhang, K., Zhou, L., & Ying, R. (2022). Mutual proximity and heterogeneity in peer effects of farmers' technology adoption: Evidence from China's soil testing and formulated fertilization program. *China Agricultural Economic Review*, *14*(2), 395–415. DOI: 10.1108/CAER-10-2020-0250

Yadav, D. (2022). Criteria for Good Qualitative Research: A comprehensive review. *The Asia-Pacific Education Researcher*, *31*(6), 679–689. DOI: 10.1007/s40299-021-00619-0

Yang, W., Qi, J., Arif, M., Liu, M., & Lu, Y. (2021). Impact of information acquisition on farmers' willingness to recycle plastic mulch film residues in China. *Journal of Cleaner Production*, *297*, 126656. DOI: 10.1016/j.jclepro.2021.126656

Yang, Y., Li, O., Peng, X., & Wang, L. (2020). Consumption trends during the COVID-19 crisis: How awe, coping, and social norms drive utilitarian purchases. *Frontiers in Psychology*, *11*, 588580. DOI: 10.3389/fpsyg.2020.588580 PMID: 33192921

Yan-Yan, C., Shek, D., & Bu, F.-F. (2011). Applications ofinterpretive and constructionist research methods in adolescent research philosophy, principles and examples. *International Journal of Adolescent Medicine and Health*, *23*(2), 129–139. PMID: 21870675

Yap, S. F., Xu, Y., & Tan, L. (2021). Coping with crisis: The paradox of technology and consumer vulnerability. *International Journal of Consumer Studies*, *45*(6), 1239–1257. DOI: 10.1111/ijcs.12724

Yazdani, N., Murad, H., & Abbas, R. (2011). From modernity to postmodernity: A historical discourse on western civilization. *International Journal of Business and Social Science*, *2*(11), 249–256.

Yin, R. K. (2014). *Case Study Research Design and Methods* (Fifth edit). SAGE Publications Ltd.

Yin, R. K. (2016). *Case study research and applications: Design and methods* (6th ed.). SAGE Publications.

Zeng, Q., Liu, Y., & Li, L. (2023). AI-enabled smart tourism: Impacts and opportunities for destination management and marketing. *Journal of Destination Marketing & Management*, *24*, 100704.

Zhang, S., Kang, B., & Zhang, Z. (2020). Evaluation of business environment in Chinese provinces: Indicator system and quantitative analysis. *Economic Management*, *42*, 5–19.

Zhang, Y. (2023). *Use of artificial intelligence (AI) in historical records transcription: Opportunities, challenges, and future directions*. McGill University.

Zhao, L., Wang, C., Gu, H., & Yue, C. (2018). Market incentive, government regulation and the behavior of pesticide application of vegetable farmers in China. *Food Control*, *85*, 308–317. DOI: 10.1016/j.foodcont.2017.09.016

Zheng, W., Luo, B., & Hu, X. (2020). The determinants of farmers' fertilizers and pesticides use behavior in China: An explanation based on label effect. *Journal of Cleaner Production*, *272*, 123054. DOI: 10.1016/j.jclepro.2020.123054

Zhou, L., Zhang, F., Zhou, S., & Turvey, C. G. (2020). The peer effect of training on farmers' pesticides application: A spatial econometric approach. *China Agricultural Economic Review*, *12*(3), 481–505. DOI: 10.1108/CAER-01-2019-0003

Zieske, C. (2020). A brief history and overview of existential-phenomenological psychology VOL 17 Issue 2. *American Journal of Undergraduate Research*, *17*(2), 45–58. DOI: 10.33697/ajur.2020.020

Zotzmann, K., & O'Regan, J. (2023). Critical discourse analysis, critical discourse studies, and critical applied linguistics. In Zotzmann, K. (Ed.), *J. O'Regan, The Routledge of Applied Lingustics* (pp. 57–67). Routledge. DOI: 10.4324/9781003082637-7

Zou, P. X., & Xu, X. (2023). *Research methodology and strategy: theory and practice*. John Wiley & Sons. DOI: 10.1002/9781394190256

Zucker, L. G. (1987). Institutional Theories of Organization. *Annual Review of Sociology*, *13*(1), 443–464. DOI: 10.1146/annurev.so.13.080187.002303

Zulauf, K., & Wagner, R. (2022). Online shopping therapy: If you want to be happy, shop around. *Journal of International Consumer Marketing*, *34*(3), 332–345. DOI: 10.1080/08961530.2021.1955425

About the Contributors

Liliane Segura, PhD in Finance within the field of Administration from Mackenzie Presbyterian University, Master's degree in Strategy from FEA-USP. Bachelor's degree in Accounting from FEA-USP. Post-doctorate in Administration from the University of Salamanca (Spain), Specialization from CSU - Fresno in Entrepreneurship Business Plan at California State University, Fresno, United States. Certified by IACAFM (International Association of Certified Accountants and Financial Managers) in IFRS. Currently serving as the Coordinator of the Undergraduate Accounting Program at Mackenzie Presbyterian University. Works as a professor and researcher at Mackenzie Presbyterian University in undergraduate, graduate, and doctoral programs (PPG-CFE). Additionally, teaches various MBAs and in-company courses. Actively participates in national and international projects funded by CAPES and international funding agencies. Director partner at Openbook Accounting and Financial Advisory. Consultant and partner at L&L Segura in the financial and accounting area. Possesses extensive experience in the financial and accounting sectors of retail companies. Research interests include Corporate Governance, Corporate Social Responsibility, Corporate Finance, and Corporate Accounting.

Ana Lucia, Post-doctorate at the University of São Paulo (USP), Graduate Program in Controllership and Accounting at FEA/USP - PPGCC. PhD in Social Work from the Federal University of Pernambuco (UFPE). Master's degree in Accounting and Actuarial Sciences from the Pontifical Catholic University of São Paulo (PUC/SP). Bachelor's degree in Accounting from the Catholic University of Pernambuco (UNICAP). Permanent member of the Graduate Program in Accounting Sciences (PPGCC) at the Federal University of Pernambuco (UFPE); and collaborating member of the Graduate Program in Controllership and Business Finance (PPGCFE) at Mackenzie Presbyterian University (UPM).

Vilma Slomski, Post-doctorate in Controllership and Accounting from the School of Economics, Business Administration, and Accounting at EAC/FEA/USP; PhD and Master's degrees in Education from the School of Education at the University of São Paulo – FEUSP; Professor and Researcher in the Master's program in Controllership and Accounting (PPGCC) at the University Center of Fundação Escola de Comércio Álvares Penteado – Unifecap; Leader of the research group FECAP Observatory of Education and Accounting Research - CNPq.

***

M.G. Serap Atakan is a Professor of Marketing at Istanbul Bilgi University, Istanbul, Türkiye. Prof. Atakan holds undergraduate, master's, and doctorate degrees in Business Administration with a concentration in Marketing from Boğaziçi University. She teaches undergraduate and graduate-level courses in Global Marketing, Services Marketing, Retail Management, Marketing Communications, New Product Development, Special Topics in Marketing, and Business Ethics. Her research interests include consumer behavior, the internationalization of retailing, marketing ethics, business ethics, and sustainability practices in higher education institutions.

Walter Bataglia is a full professor at Mackenzie Presbyterian University in São Paulo, Brazil, where he teaches strategy, innovation, and organizational theory in undergraduate, master's, and doctoral business courses. As a visiting professor at Université de Lille and Copenhagen Business School, he has conducted collaborative research on strategic alliances, service ecosystems, and their interactions with regulatory frameworks. He currently serves as Research Coordinator at Mackenzie, focusing on Business Administration, Accounting, and Economics, and is a member of several international business journal editorial boards. He holds a Master's and a PhD in Business Administration from the University of São Paulo. With extensive experience in technology, pharmaceutical, and healthcare sectors, his research and teaching interests include strategic alliance management, dynamic capabilities, ecosystem innovation, and regulatory decision-making.

Chakabwata William holds a PhD in Curriculum and Instruction obtained from University of South Africa on 27 September 2023. He also holds a master's degree in Curriculum and arts education which he earned from University of Zimbabwe. He has taught at Midlands State University in Namibia and also at Namibia University of Science and Technology (NUST). He may be contacted at williechaka@gmail.com

Dina Darwish, Vice dean, faculty of computer science and information technology, Ahram Canadian university, Egypt. I obtained my Ph.D from Cairo university at 2009, engineering faculty. I am professor since 2020, also my special domain is artificial intelligence and I have many publications in the following topics, including: Artificial intelligence, wireless ad hoc networks, Internet of things, Big data analytics, Blockchain applications, Cloud computing applications, Web 3 technologies, Chatbots development and others.

Ricardo de Abreu Barbosa is a professor at Mackenzie Presbyterian University

Cristiane de Almeida is a Research Associate from the Graduate School of Management at University of Sao Paulo, Brazil

Tutku Eker İşcioğlu is an Associate Professor of Marketing at Piri Reis University, Istanbul, Türkiye. Dr. İşcioğlu holds an undergraduate degree in Business Administration from Istanbul Bilgi University, a master's degree in Management and Organization from Marmara University, and a Ph.D. in Business Administration with a concentration in Marketing from Boğaziçi University. She teaches courses in Marketing Management, International Marketing, Consumer Behavior, Brand Management, Corporate Social Responsibility, and Sustainability at both undergraduate and graduate levels. Her research interests span marketing ethics, institutional sustainability practices, consumer behavior, and more.

Joshua Imoniana, Professor of Auditing, School of Economics, Managment & Accounting, Department of Accounting and Actuarial Sciences. Professor of Qualitative Research Methods and Supervisor on PhD Programs.

Mustafa Kayyali, I am an ardent advocate for excellence in higher education, driven by a relentless pursuit of quality, recognition, and innovation. My journey in academia started with a Master's in Quality Management and Evaluation in Higher Education from Universitat Oberta de Catalunya, followed by a Ph.D. in Quality Management from Azteca University. These academic pursuits have laid a strong foundation for my expertise in Accreditation, Quality Assurance, and Higher Education Rankings. As an entrepreneur, researcher, translator, and publisher, I am deeply involved in various facets of the academic world. My diverse interests encompass Management, Translation, Interpretation, and Academic consulting, contributing to a well-rounded understanding of the industry. Throughout my career, I have made significant contributions to academic literature, with more than 20 published papers and 7 book chapters to my name. Additionally, I take pride in having translated 5 books, bridging language gaps, and promoting knowledge exchange on a global scale. I am committed to fostering positive change in higher education and contributing to its continuous improvement.

Ozge Kirezli received her M.A. degree in Production Management and Marketing from Marmara University in 2010. In 2015, she earned her Ph.D. in Production Management and Marketing from the same institution. She began her academic career as a Teaching Assistant in the Business Administration Department at Istanbul Bilgi University in 2010 and worked until 2018. Since 2019, she is working as an Assistant Professor at Yeditepe University, teaching courses in Marketing Principles, Marketing Management, Marketing Research, and Consumer Behavior. Her research interests include consumer decision-making and paradigm shifts in contemporary marketing.

Boukil Mehdi has over a decade of experience as a travel planner and tourism specialist, Mehdi has successfully organized, designed, and marketed diverse travel packages, tours, and events. Holding a PhD in Geography with a focus on international tourism, he has conducted extensive research and published works on sustainable tourism and cultural revitalization. A skilled educator and consultant, Mehdi specializes in crafting innovative travel solutions, establishing strategic partnerships, and promoting responsible tourism. His expertise bridges academic knowledge with practical industry experience. Here are two of his relevant publications indexed in Scopus: 1. _Cultural Revitalization and Tourist Conversion of Historical Monuments: The Cases of the Ancient Medinas of Fez and Tunis_ Published in _Vernacular Architecture: Support for Territorial Development_ (Materials Research Proceedings, 2025). 2. _Digital Journey Through Fez's Historic Medina: Embracing the Core of Responsible Tourism_ Published in _Promoting Responsible Tourism With Digital Platforms_ (IGI Global).

Octavio Mendonca Neto, Graduated in Mechanical Engineering from Instituto Mauá de Tecnologia (1972), with a Master's in Accounting and Actuarial Sciences from PUC-SP (2002) and a Ph.D. in Accounting and Actuarial Sciences from FEA/USP (2007). Currently an Assistant Professor at Universidade Presbiteriana Mackenzie and a faculty member of its Graduate Program in Management, Finance, and Technology (Master's and Professional Doctorate). Research interests include Risk Analysis and Management, Business Performance Evaluation, Pragmatism in Organizational Studies, and the Sociology of the Accounting Profession. CNPq PQ2 Research Productivity Fellow (2017–2020).

Claudio Parisi, PhD in Controllership and Accounting from the University of São Paulo (USP), a master's degree in Controllership and Accounting from USP and a bachelor's degree in Accounting from USP. He is currently Director of the

Center for Social and Applied Sciences (CCSA), Research Professor in the Doctorate and Professional Master's Program in Controllership and Corporate Finance at Mackenzie Presbyterian University (UPM) and a member of the Management Committee of the Mackenzie Center for Economic Freedom (CMLE). He was a research professor at Fundação Escola de Comércio Álvares Penteado (FECAP) and coordinator of the Master's Program in Accounting Sciences at FECAP. He was administrative-financial director and fiscal advisor to ANPCONT. He was a professor in the MBA Controller program and a researcher at Fipecafi's Gecon center. Author and co-author of scientific articles, books and book chapters. Member of editorial boards and ad hoc consultant for scientific journals. Lecturer and business consultant in the area o

Jeganathan GomathiSankar, a dedicated educator, specializes in marketing and currently teaches at the BSSS Institute of Advanced Studies. With a rich academic background, he served as a research scholar at Pondicherry University, focusing on information diffusion, ICT, Marketing technology, AI, and Data Science. Dr. Sankar's impactful research is published in reputable journals indexed in Scopus and ABDC. Formerly a faculty member at Saveetha School of Management, he contributed significantly to marketing education. Dr. Sankar's pedagogy integrates theoretical insights with practical applications, offering students a holistic understanding of contemporary marketing. Beyond teaching, he is proficient in analytical tools such as SPSS, AMOS, STATA, PowerBI, and Python, enhancing the sophistication of his research. Known for his engaging teaching style, Dr. Sankar fosters an environment encouraging critical thinking. As a member of BSSS Institute of Advanced Studies, he continues shaping future marketing professionals. His commitment to academic excellence, research, and proficiency in modern tools exemplifies his dedication to advancing knowledge in marketing and technology.

DéboraSeveriano is a professional in the field of Finance, bringing over 18 years of experience in accounts payable and receivable, with a specialization in Treasury and Financial Planning. She holds a degree in Accounting, a Master's in Management and Finance, and is pursuing a doctorate in the Corporate Management and Finance program at Universidade Presbiteriana Mackenzie. She is a guest lecturer in the Postgraduate Program in Management and Finance at Universidade Presbiteriana Mackenzie and a content professor at Universidade Metodista de São Paulo.

Satya Subrahmanyam, As a postdoctoral researcher with a double PhD, I specialize in Leadership, HR Practices, Business Analytics, Startups, and Soft Skills, contributing to academic and industry innovation. Currently, I serve as Professor and Dean (Research) at the Holy Spirit University of Kaslik (USEK), Lebanon, where I lead research and innovation activities to enhance academic quality and standards. With extensive academic and managerial experience, I have taught at institutions such as Ambedkar Open University, Cambridge University, and the University of London. As an academic entrepreneur, I am the Managing Partner and Trust Member at Vignan Institute of Technology and Management, Odisha, fostering its growth into a leading institution. At the Catholic University in Erbil, I advanced education and employment as Professor, Research Centre Director, and Lifelong Learning Centre Director, launching impactful programs. My academic qualifications include a Master's, M.Phil., an MBA in HR Management, two Ph.D.s, and postdoctoral research in Global Leadership at the Polytechnic Institute of Leiria, Portugal.

Adriano Teixeira, PhD in Controllership and Corporate Finance at Presbiterian University Mackenzie. MPhil in Controllership and Finance from Faculdade FIPECAFI (Research Line in Finance). MBA in Internal Auditing with Emphasis on Financial Institutions (FIPECAFI). Postgraduate in Accounting Auditing - Tax from the Federal University of Paraíba (UFPB). Graduated in Accounting from the Federal University of Paraíba (UFPB). International Certifications: Anti-Money Laundering Specialist (CAMS) and Certifield Risk Management Assurance (CRMA). With more than 25 years of experience in a large financial institution, the last 16 years being in the Internal Audit Department, working in process evaluation works in the Products and Services Support Areas (Back Office), Treasury Operations (Market) Financial and Capital), Controllership and Custody of Investment Funds, Bookkeeping of Shares, Management of Guarantees and Registration, Integrated Control of Corporate Risks, Administration and Management of Third Party Resources, Evaluation of Internal Models for Capital Allocation of Market Risks, Credit and Operational (Basel II), ICAAP - The Internal Capital Adequacy Assessment Process, Liquidity Risk Management (Basel III) and Preventing Money Laundering and Fighting Bribery and Corruption. Currently, leads a team of 42 professionals responsible for Internal Audit work, focusing on aspects of technology and business, in the Related Companies of the Banking Financial Institution in the branches of Insurance, Pension Plans, Capitalization, Health, Consortia, Credit Cards and Foundations.

Peter Valan, I believe Teaching, is in my DNA. The classrooms require a teacher whole blends traditional and new age learning. He/She should ensure collective growth of personal & society. I strive to explore more human development and know more insights about societal issues. Digital learning is another interesting domain I am exploring. Wanna give more digital experience to students and prepare them for future digital world.

Index

A

B

C

D

E

F

I

N

P

Q

R

S

T

www.ingramcontent.com/pod-product-compliance
Lightning Source LLC
LaVergne TN
LVHW080612300926
849162LV00005B/1128
9798337344324